OFFSHORE TIDAL SANDS

Processes and deposits

OFFSHORE TIDAL SANDS

Processes and deposits

Edited by A. H. Stride
Institute of Oceanographic Sciences
Surrey, UK

LONDON   NEW YORK
CHAPMAN AND HALL

First published 1982 by
Chapman and Hall Ltd
11 New Fetter Lane, London EC4P 4EE
Published in the USA by
Chapman and Hall
733 Third Avenue, New York NY 10017
© 1982 Chapman and Hall Ltd

Typeset in Great Britain by
Scarborough Typesetting Services
and printed by J. W. Arrowsmith Ltd, Bristol

ISBN 0 412 12970 1

All rights reserved. No part of this book
may be reprinted, or reproduced or utilized in
any form or by any electronic, mechanical or
other means, now known or hereafter invented,
including photocopying and recording, or in
any information storage and retrieval system,
without permission in writing from the
Publisher.

British Library Cataloguing in Publication Data

Offshore tidal sands
  1. Marine sediments   2. Tides
I. Stride, A. H.
551.3'04   GC380.15

ISBN 0-412-129701

Contents

Plates section	between pages 48 and 49
LIST OF CONTRIBUTORS	xiii
PREFACE	xv

1.	**Background and outline**	1
	A. H. STRIDE	
	1.1 Introduction	1
	1.2 History of research on modern offshore tidal current sedimentation	1
	1.2.1 Early work	1
	1.2.2 Post-1950 advances	2
	1.3 A depositional surface for late Holocene deposits	7
	1.4 Limits and outline of the book	7
2.	**Tidal currents of the continental shelf**	10
	M. J. HOWARTH	
	2.1 Introduction	10
	2.2 Tide generating forces and the ocean's response	10
	2.2.1 Tide generating forces	10
	2.2.2 Spring–neap cycles	11
	2.2.3 Peak astronomical tides	12
	2.2.4 Relative amplitudes of daily and twice-daily tides	12
	2.2.5 A computation of ocean tides	12

2.3	Tidal currents in shelf seas	12
	2.3.1	Amplification due to decreasing depth and width	12
	2.3.2	Resonance	13
	2.3.3	Progressive and standing waves	14
	2.3.4	Effects of the Earth's rotation	14
	2.3.5	Some effects of continental shelf width	14
	2.3.6	Tidal range at the coast	16
	2.3.7	Tidal current speeds on the continental shelf around the British Isles	16
	2.3.8	Tidal ellipse	19
2.4	Net sand transport caused by tidal current asymmetries	20
	2.4.1	Distortions to the tide	20
	2.4.2	Combination of the principal and its first harmonic	20
	2.4.3	Net sand transport by tidal currents	21
	2.4.4	Tidal current patterns in the vicinity of sand banks	21
2.5	Flow near the sea floor	22
	2.5.1	Constant stress layer	22
	2.5.2	Ekman layers	23
	2.5.3	Effects of the oscillatory nature of tidal currents	23
	2.5.4	Drag coefficient and bottom stress	23
	2.5.5	Current profiles above the logarithmic layer	24
2.6	Internal tides	24
	2.6.1	Nature	24
	2.6.2	Causes	24
	2.6.3	Measured currents of internal tides	25
2.7	Tides past	25
	2.7.1	Effects of tidal friction	25
	2.7.2	Effects of changes of bathymetry	25
2.8	Main conclusions	26

3. Bedforms 27

R. H. BELDERSON, M. A. JOHNSON AND N. H. KENYON

3.1	Introduction	27
3.2	Relevant flume bedforms	29
	3.2.1	Lower flow regime flume bedforms (sand ripples and sand waves)	30
	3.2.2	Transition bed conditions	33
	3.2.3	Upper flow regime flume bed states	33
	3.2.4	Paucity of longitudinal bedforms in flumes	34
	3.2.5	Note on theory of transverse bedforms in flumes	34
3.3	Transverse bedforms of the continental shelf	34
	3.3.1	Unlikelihood of antidunes occurring on the continental shelf	35

	3.3.2	Sand ripples	35
	3.3.3	Sand waves	36
	3.3.4	Transverse sand patches	43
3.4	Longitudinal bedforms of the continental shelf		44
	3.4.1	Scour hollows	44
	3.4.2	Longitudinal furrows	45
	3.4.3	Obstacle marks	46
	3.4.4	Sand ribbons and longitudinal sand patches	47
	3.4.5	Tidal sand banks	49
3.5	Relationship between bedforms		54
3.6	Aeolian equivalents		55
3.7	Main conclusions		55

4. Sand transport — 58
M. A. JOHNSON, N. H. KENYON, R. H. BELDERSON AND A. H. STRIDE

4.1	Introduction		58
	4.1.1	Availability of sand for offshore transport	59
4.2	Relation of sand transport rate to tidal current speed		59
	4.2.1	Sand transport rate in flumes and rivers	59
	4.2.2	Relative sand transport rate over the sea bed	60
	4.2.3	Lag effects in tidal current sand transport	62
	4.2.4	Transport of sediments with two or more modes	64
4.3	Geographical variation in sand transport rate		66
	4.3.1	Relative sand transport rate shown by mean spring peak tidal current speed	66
	4.3.2	Relative sand transport rates shown by bedforms	67
4.4	Net sand transport by tidal currents		67
	4.4.1	Net sand transport direction predicted from mean spring peak tidal currents	67
	4.4.2	Field evidence of net sand transport directions	70
	4.4.3	Regional net sand transport directions around the British Isles	75
	4.4.4	Net sand transport paths on other continental shelves	80
	4.4.5	Bed-load partings and bed-load convergences	80
	4.4.6	Origin of bed-load partings and convergences	81
	4.4.7	Bed-load partings and convergences with non-tidal currents and in deserts	83
4.5	Temporal variations of sand transport rate and direction in a tidal sea		83
	4.5.1	Variations due to the tidal cycles	83
	4.5.2	Variations due to sea surface waves	84
	4.5.3	Variations due to non-tidal currents	86

4.6	Growth, migration and decay of sand waves in the Southern Bight of the North Sea by total water movements	89
4.7	Local sand transport on modern sand banks	93
4.8	Main conclusions	94

5. Offshore tidal deposits: sand sheet and sand bank facies   95
A. H. STRIDE, R. H. BELDERSON, N. H. KENYON AND M. A. JOHNSON

5.1	Introduction	95
5.2	Late Holocene sand and gravel sheet facies	98
	5.2.1 Grain size and current speed	99
	5.2.2 Gravel sheet form, composition and structure	101
	5.2.3 Sand sheet form and texture	102
	5.2.4 Structure of a sand sheet in the Southern North Sea	102
	5.2.5 German Bight sand to mud sheet	106
	5.2.6 Irish Sea sand to mud sheet	108
	5.2.7 Regional cross-bedding dip directions within the sand sheet facies	109
	5.2.8 Sand patches	109
	5.2.9 Sand waves formed by tidal lee waves	110
	5.2.10 Facies model of an offshore tidal current sand sheet	110
5.3	Sand bank facies	113
	5.3.1 Early Holocene low sea level sand bank facies	114
	5.3.2 Late Holocene sand bank facies	115
	5.3.3 Internal structure of offshore and estuarine sand banks	117
	5.3.4 Facies models of offshore and estuarine tidal sand banks	119
5.4	Sediment and faunal indicators of shape, depth and exposure of continental shelves	121
5.5	Longer term evolution of the deposits	122
5.6	Sand and gravel deposits of non-tidal marine currents	123
5.7	Main conclusions	124

6. Shelly faunas associated with temperate offshore tidal deposits   126
J. B. WILSON

6.1	Introduction	126
6.2	Faunal associations	127
6.3	Bioturbation	130
	6.3.1 Depth of disturbance by bioturbation	130
	6.3.2 Types of bioturbation	132
6.4	Topics and areas excluded	134

6.5	Temperate water regions studied and their geological importance	135
	6.5.1 Carbonate content of sediments on the continental shelf around the British Isles	135
6.6	Faunas in shallow nearshore waters	135
	6.6.1 Temperate water calcareous algal gravels	137
6.7	Faunas of the middle and outer continental shelf	137
6.8	Faunas of a bed-load parting	138
6.9	Faunas associated with bedform zones in the Western English Channel	139
	6.9.1 Faunas from the gravel sheet	139
	6.9.2 Faunas from the sand ribbon zone	139
	6.9.3 Faunas from the zone of large sand waves	141
	6.9.4 Faunas from the zone of rippled sand	141
6.10	Faunas associated with bedform zones in the Bristol Channel	141
	6.10.1 Benthic faunas in relation to tidal bottom stress	142
	6.10.2 Faunas from the rock floor	143
	6.10.3 Faunas from the sand ribbon zone	144
	6.10.4 Faunas from the zone of large sand waves	145
	6.10.5 Faunas from the rippled muddy sands in bays	145
6.11	Faunas associated with bedform zones in the Southern North Sea	145
	6.11.1 Faunas from the zone of large sand waves	146
	6.11.2 Faunas from the zone of small sand waves	146
	6.11.3 Faunas from the zone of rippled sand	148
	6.11.4 Faunal differences from the sand wave zone to the zone of rippled sand	148
6.12	Faunas associated with bedform zones on the Atlantic continental shelf between Brittany and Scotland	149
	6.12.1 Faunas from the gravel sheet zone, Fair Isle Channel	149
	6.12.2 Faunas from the rippled sand zone	149
	6.12.3 Faunas associated with gravels in weak current areas west of Scotland	151
6.13	Faunas of active sand banks	153
6.14	Faunal evidence for stability of sand waves	154
6.15	Faunas as environmental indicators	155
	6.15.1 Faunal differences between adjacent sand transport paths	155
	6.15.2 The proximity of the open ocean	156
	6.15.3 The edge of the continental shelf	156
6.16	Factors determining the faunal composition of death assemblages in shell gravels	157
	6.16.1 Predation on shell bearing invertebrate faunas	157

x   Contents

      6.16.2 The role of borers in the breakdown of shells   159
      6.16.3 Mechanical breakage and dissolution of shells   159
      6.16.4 Differences in faunal composition between living and dead faunas   160
  6.17 Age of temperate water carbonates   161
      6.17.1 Age of shell gravels on the continental shelf around the British Isles   161
      6.17.2 Rates of deposition   161
  6.18 Relative proportions of the major carbonate producers in death assemblages of continental shelf carbonates   162
      6.18.1 Faunal composition of death assemblages in shell gravels in the strong current areas, Western English Channel and Celtic Sea   162
      6.18.2 Faunal composition of death assemblages in shell gravels on the continental shelf west of Scotland   163
  6.19 Temporal changes in the faunal composition of shell gravels   165
      6.19.1 Faunal evidence of lowered sea level   165
  6.20 Long term evolution of temperate shelf carbonates   166
  6.21 Applications to the fossil record   167
  6.22 Main conclusions   167
  Appendix 6.1 List of species mentioned in Chapter 6   168

7. Ancient offshore tidal deposits   172
P. H. BRIDGES

7.1 Introduction   172
7.2 Recognition of ancient offshore tidal current activity   172
7.3 Structures preserved in ancient offshore tidal current deposits   173
      7.3.1 Sand waves   173
      7.3.2 Sand banks   176
      7.3.3 Sand and mud sheets   178
      7.3.4 Scoured horizons and bed-load partings   180
7.4 Tidal currents aided by storm processes   180
7.5 Factors controlling the structure and composition of offshore tidal sediments through geological time   181
7.6 Some possible palaeotidal regimes   181
      7.6.1 Upper Jurassic gulf of western North America   181
      7.6.2 Upper Cretaceous epicontinental seaway of western North America   183
7.7 Sedimentology of a tidal sea: the Lower Greensand of southern England   183
      7.7.1 Lower Aptian phase   184
      7.7.2 Upper Aptian and Lower Albian phases   186

7.8 Tidal currents through geological time: implications for future studies 187
7.9 Main conclusions 189
Appendix 7.1 Possible ancient offshore tidal current deposits 189
Appendix 7.2 Estimate of the amplification of the twice-daily tidal wave in the Lower Aptian gulf of south-east England 192

REFERENCES 193
INDEX 214

List of contributors

R. H. Belderson – Institute of Oceanographic Sciences, Wormley, Godalming, Surrey, UK.

P. H. Bridges – Derby Lonsdale College of Higher Education, Kedleston Road, Derby, UK.

M. J. Howarth – Institute of Oceanographic Sciences, Bidston, Birkenhead, Wirral, Merseyside, UK.

M. A. Johnson – Institute of Oceanographic Sciences, Wormley, Godalming, Surrey, UK.

N. H. Kenyon – Institute of Oceanographic Sciences, Wormley, Godalming, Surrey, UK.

A. H. Stride – Institute of Oceanographic Sciences, Wormley, Godalming, Surrey, UK.

J. B. Wilson – Institute of Oceanographic Sciences, Wormley, Godalming, Surrey, UK.

Preface

In the early 1970s a start was made on a broad review of what was known or could be surmised about sedimentation by strong tidal currents on modern continental shelves. This task was initiated because of the need to define the next phase of research in this field by the Marine Geology Group of the Institute of Oceanographic Sciences. Related indications of the longer term evolution of the deposits were sought by close reference to the nature of modern tidal currents and the supposedly offshore tidal deposits of ancient seas.

As the review grew in completeness it became of increasing relevance to a wider audience so it was amalgamated with the new results and shaped as a book.

The fruits of the long-continued discussions within and outside the Geology Group have served to improve understanding of the processes and products of offshore tidal current sedimentation. On the other hand, the discussions have blurred the parts played by the people concerned. This applies to all chapters in varying degrees, but is especially true for Chapters 3, 4 and 5. The authorship attributed to each chapter therefore seeks to reflect those who were most concerned with it.

The book is intended for the final year geology undergraduate, the post-graduate and the professional geologist. It should have especial relevance to workers wishing to clarify their interpretation of marine sedimentary rocks or searching for stratigraphic traps of potential economic significance, including the products of unidirectional marine currents. It is also of relevance to biological and engineering workers concerned with marine sedimentation processes at the gross scale and with astronomers wanting a record of the Earth's tidal history, and hence of the evolution of the Earth–Moon system. However, the book is not concerned with the physics of grain movement in its conventional form.

The authors are grateful to many people. Their colleagues provided help in numerous ways, both at sea and in the laboratory. Particular thanks must be offered to A. R. Stubbs for his unfailing help with the short range side-scan sonar during many years of data gathering at sea, to D. J. Webb for his advice concerning the tidal currents of particularly broad continental shelves, to G. F. Caston for valuable comments on Chapters 3 and 4 and to R. Anderton for commenting on Chapter 7. C. D. Pelton is warmly thanked for the care taken in drafting the final diagrams of Chapters 1 to 5, and C. E. Darter for the three associated block diagrams. J. M. Weller drew the final diagrams of Chapter 6 and prepared some of the associated plates, while P. E. Williamson drew the associated block diagram. M. J. Conquer and A. Gray are thanked for their care in producing the photographs, especially for the more demanding material.

Numerous other workers have generously answered enquiries or supplied data. These include R. S. Aitken, A. Bastin, P. Binns, B. D'Olier, D. Eisma, D. Hamilton, H. W. Hill, J. J. H. C. Houbolt, J. W. Jardine, N. S. Jones, J. M. Kain, N. Kelland, K. Krank, N. Langhorne, J. C. Ludwick, T. K. Mallik, E. Oele, H. M. Pantin, W. R. Parker, J. W. Ramster, W. A. Read, H.-E. Reineck, R. T. E. Schüttenhelm, J. Sündermann, J. H. J. Terwindt, M. J. Visser, R. M. Warwick and F. Werner.

Finally, thanks must go to families of the authors who have inevitably taken a second place when there was the need to work on the book.

Petersfield   A. H. STRIDE

Chapter 1

Background and outline

1.1 Introduction

The compilation of a more complete history of the Earth calls for recognition of ancient seas that were swept by strong tidal currents. Few examples of these seas are known, though ancient tidal flat deposits are distributed widely in the stratigraphic record and are found in many parts of the world. Modern offshore tidal current deposits can cover larger areas of sea floor than the adjacent modern tidal flats and a similar ratio must be expected for ancient seas. The deposits of some ancient seas may hold a record of periods of relatively higher tidal energy than at present, which could indicate such events as the possible capture of our Moon (Olson, 1970) and of any near-misses of the Earth by other bodies.

There are many practical reasons for wanting information about the sands of sea floors swept by strong tidal currents, as well as of the deposits being formed by them. The safety of shipping can be dependent on knowing about the changing position of sand banks and the changing depth of water in inshore channels, especially in port-approaches where supertankers may have little clearance beneath them. The safety of submarine power and telephone cables, as well as oil or gas pipelines, can be dependent on avoiding their exposure or undermining by the erosion of sand. Such exposures can lead to their collapse or to damage by bottom trawls or anchors. The detection of stratigraphic traps can call for knowledge of their probable position in a sedimentary sequence. Information about modern analogues of these porous sands is likely to become increasingly essential as the more readily detectable structural traps for oil become exhausted.

1.2 History of research on modern offshore tidal current sedimentation

The aim of this brief historical review is to give a general impression of the development of the subject. Some of the earliest references must be given, however slight their relevant content, whereas later ones are treated much more selectively. Indeed, from the 1960's onwards reference to sources is largely omitted in this chapter so as to be able to present some of the main advances as briefly as possible. This obvious deficiency is made good by the many citations to original work that will be found in subsequent chapters, where there is space for their vital data and ideas to be developed more fully. Some of the more significant misinterpretations are discussed separately (Section 1.2.2(d)).

1.2.1 EARLY WORK

The mariner has long been aware of the menace

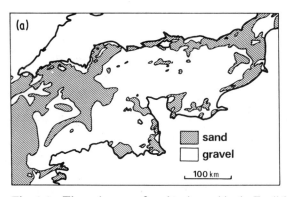 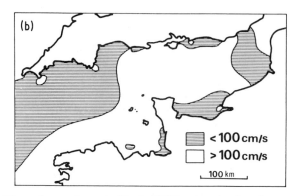

Fig. 1.1 The main areas of sand and gravel in the English Channel were shown by Pratje (1950) to be related to the speed of the mean peak springs, near-surface tidal currents.

created by the shifting sand banks of tidal seas. Early geologists such as de la Beche (1851) and Lyell (1853) drew attention to the importance of modern tidal currents as an agent of sediment transport, while Reade (1888) suggested that they should be able to move sediment far from its coastal sites of origin and that the passage of the sands should abrade the underlying floor. Dangeard (1925) seems to have been the first worker to point out, for the English Channel, that there was a rough geographical correspondence between the grain size of the sediment on the surface of the sea floor and the strength of the tidal currents sweeping over it. By about two decades later Pratje (1950) had used improved data to confirm this empirical correlation for the English Channel (Fig. 1.1) and suggested that it had widespread applicability to the seas of north-western Europe.

Rather earlier, van Veen (1935, 1936) had drawn attention, for the Southern Bight of the North Sea, to the numerous sand banks lying approximately parallel with the strongest tidal flow and to the associated transverse sand waves; he made the fruitful suggestion that the latter were being moved northwards past Holland by the stronger northward-flowing ebb tidal current and he also drew useful analogies between the North Sea sand bodies and aeolian dunes. The deepening of narrow seaways by tidal scouring had been referred to by a number of workers (e.g. Kuenen, 1950).

1.2.2 POST-1950 ADVANCES

The period after the Second World War until the early 1950's was notable for the increased numbers of samples of the sea floor taken by scientists. Despite this there was increasing dissatisfaction with the results, because they largely failed to reveal deposits that resembled the supposedly shallow marine sediments of the stratigraphic record. Not only were these supposedly modern sediments patchy but also they were unexpectedly variable in grain size. This dissatisfaction was accompanied by increasing numbers of attempts to find explanations for these failures. Of particular importance in these studies was the growing appreciation that the purely descriptive approach to sediments would have to be replaced by one that was based on an understanding of the processes that were affecting the sea floor.

Two discoveries of the early 1950's did much to encourage the effective study of modern offshore tidal current sedimentation processes. Both of them were concerned with bedforms. The first finding was that side-scan sonar could be used as a means of revealing the form and composition of the sea floor in plan view, for any given area (Chesterman, Clynick and Stride, 1958; Stride, 1963a). Once the considerable initial difficulties of interpreting the records had been overcome this method was quick and easy to use. It provided a wealth of information about the ground between the isolated echo-sounder profiles and widely scattered samples which were the best data that were previously available. The new approach enabled interpretation of large areas of the sea floor surface to be made in geological terms (such as are summarized in Belderson, Kenyon, Stride and Stubbs, 1972), in much the same way as was already possible for aerial pictures of the surface of the land. The seemingly random variation in the nature of the sediments on the continental shelf, which had so puzzled earlier workers (e.g. Shepard, 1932), was now seen as the norm for large areas of that ground. It was recognized as evidence of a highly organized pattern of sand in transit. The second finding was that the shape of a suite of bedforms indicated the sand transport directions. These discoveries, in turn, allowed the mechanism of net sand transport by tidal currents to be recognized (Stride, 1963a). They also stimulated an increasingly quantitative approach to the sedimentation studies. These include numerical simulation of bed shear stress and sand transport by using observed or numerically-modelled tidal currents. By understanding the processes of offshore tidal current sedimentation in regional terms it proved possible to locate the resulting modern deposits. Thus, a general study of processes had to precede the rather more local study of deposits.

Many of the general conclusions derived from the sedimentation studies of the offshore tidal realm also have relevance to seas with strong non-tidal currents. Indeed, these conclusions provided a much needed stimulus to reappraise the value of some of the geological dogmas current in the 1950's.

(a) *Recognition of net sand transport directions*

Two main controversies were associated with the recognition that net sand transport directions existed in seas dominated by tidal currents. The first was concerned with the significance of the asymmetrical profile of the majority of sand waves in these seas. Some workers considered, by analogy with the observed migration direction of sand waves in rivers and tidal estuaries, that the asymmetry was an indicator of their direction of advance and thus of the net sand transport direction. In contrast a few other workers took the unpractical view that the asymmetry was of no use unless it could be shown that each sand wave of interest actually moved in the direction supposed.

The second main controversy concerned the best way of determining the net sand transport direction in a tidal sea. The geologists showed empirically that the polarity of bedform morphology was generally in keeping with the ebb or flood direction in which the mean springs tidal current reached the higher peak speed (Section 4.4). Other workers scorned such an empirical approach and demanded one that was firmly based on what was known of the mechanics of grain movement. This would require analysis of flume and river data to fix the value of coefficients. Then sand transport on continental shelves would be predicted from knowledge of grain size and total water movements. The first approach provided a practical solution of immediate value to the geologists. The latter approach will one day provide a sound theoretical basis for their general conclusions.

The widespread occurrence around the British Isles of sand waves with asymmetrical profiles and the finding of a suite of new bedforms (Chapter 3) provided empirical proof of sand (and some gravel) transport over large areas of continental shelf (Chapter 4). These new findings made it possible to deduce sand transport paths for much of a sea dominated by tidal currents, and even to deduce tidal current speeds when unknown. Furthermore, by comparing these results with those from neighbouring tidal seas, it proved possible to distinguish between the regional and the more local sand transport paths and between the effects of tidal and non-tidal currents. Novel aspects became clear from the pattern of the net sand transport by tidal currents around the British Isles. These were, first, the relative shortness of most of the paths (compared to some expected sand transport paths due to ocean currents) and secondly their arrangement such that in some areas the directions of transport diverge from bed-load partings and in other areas the transport paths meet at bed-load convergences. On one side of such a parting or convergence the ebb current was shown to move most sand, while on the other side it is the flood tidal current that is the more effective (Chapter 4). The bed-load partings were seen as regions of net erosion, while the convergences were regions of net deposition. Later work has further clarified the nature of these zones.

The success of the empirical correlation of bedform and tidal current data around the British Isles encouraged workers to attempt to make similar field studies for other widely different regions. It also encouraged workers to observe bedform movement and to try to estimate sand transport rates, despite the obvious difficulties of short term observational work and limited theoretical understanding of the processes involved. As a result of these various approaches it is now certain that the main offshore effect of storm wave incidence, in a sea dominated by tidal currents, is to increase sand transport rates in the direction of the tidal currents. Independently of waves, the well established, marked increase of sand transport rate (as bed-load or suspension) with a relatively small increase in current strength ensures that the stronger of the peak ebb or flood tidal currents generally transports the greater amount of material and so determines the net sand transport direction. Only very locally will the slightly weaker of the peak ebb or flood tidal current, flowing for a longer period, be able to move more sand (Chapter 4). The effect of a net flow of water on sediment transport is discussed in Sections 2.4 and 4.5.3. It is emphasized that the net sand transport direction caused by the peak tidal current can be quite different from that caused by the net (residual) flow of water which transports the silt and clay.

(b) *Evolution of modern deposits in offshore tidal seas*

A period such as the present, when the Holocene marine transgression has only recently been completed, is a particularly good one for making observations of sand transport and deposition on the sea floor. This is because there is so much sand still being moved along the transport paths that some aspects of the processes giving rise to the associated deposits can be readily and unambiguously discerned. The same processes will be at work long into the future, using new material won from coasts, brought down by rivers, eroded from the sea floor, or derived from the biota.

It is now common knowledge that the existing continental shelves do not everywhere offer a finished pattern of modern offshore tidal current (and other) deposits, that have merely to be mapped before they can be used as a guide to aid recognition of similar deposits in the stratigraphic record. Increasingly it is becoming appreciated that each part of the existing continental shelf has reached its own particular stage of development which has to be interpreted correctly before the depositional products of the present sea can be located and studied. The material involved in this process was not simply eroded yesterday, for transport today and deposition tomorrow. Instead, deposition is achieved after a complex but progressive decrease in overall activity, during the latter part of which the grains move for progressively shorter periods and are static for ever longer periods until they are finally buried deep enough to be no longer affected, even by the rare but most powerful water movements. Thus the signs of transport and deposition are available at the same sites (Chapter 5). Failure by some workers to make allowances for the continuing evolution of continental shelves and their modern deposits has been responsible for a lot of disappointment and misinterpretation that, even now, spoils many otherwise excellent accounts of modern marine sediments.

(c) *Description of offshore tidal current deposits*

The modern sand or gravel sheet facies, with grain size increasing with current strength, was at first demonstrable only in rather general terms (Jarke, 1956; Stride, 1963a; Houbolt, 1968). It was missed by some workers (e.g. Klein, 1977b). Later work began to show that some of the associated sand waves, although serving as valuable indicators of net sand transport direction, were at the same time being partly incorporated into the deposits (Johnson, Stride, Belderson and Kenyon, 1981). Information about the living shelly faunas from some of these deposits around the British Isles had been available for many years (e.g. Davis, 1925). However, the faunas of the Bristol Channel were not integrated with the deposits until recently (Warwick and Davies, 1977) and for the remainder of these seas until the present book was being written (Chapter 6). The associated fragmental calcareous material in sands and gravels had not been described previously.

The modern sand bank facies was also recognized (Off, 1963) when it was shown, from inspection of navigational charts, that sand banks cluttered the tidal estuaries and embayments in many parts of the world. It was argued that they had a high preservation potential. The size and porosity of the modern sand banks in the North Sea led to the suggestion that the analogous deposits in the stratigraphic record could have considerable economic significance: their internal structure and fauna provided diagnostic criteria for their recognition (Houbolt, 1968; Reineck and Singh, 1973). Fuller details are given in Chapters 5 and 6 and misinterpretations of some of the data on internal structure (Klein, 1977b) are corrected in Section 5.3.4.

Recent interpretations of some ancient shallow marine deposits, showing that they resulted from offshore tidal current activity (Chapter 7), have provided a welcome stimulus to broaden and sharpen understanding of the nature and longer term evolution of modern offshore tidal current deposits. However, the interesting suggestion that the numerous examples of cross-bedded marine sands of the geological past could be correlated (for genetic purposes) only with broad fields of sand waves like those of modern seas (Pettijohn, Potter and Siever, 1972) has to be reconsidered, as will be shown in Chapter 5, as cross-bedding is also associated with the growth of sand banks, tidal deltas, tidal flats and beaches.

(d) *Some misinterpretations revealed*

The widespread indications of erosion, transport and deposition of sand by strong tidal currents in the seas of north-west Europe provided good reasons to reassess some commonly held dogmas which (though valuable soon after their introduction) did much during the 1960's to early 1970's to hold back attempts to understand the processes that control sedimentation on modern continental shelves. Each dogma has some truth in it but has been used too widely and out of context. They can only be stated and discussed briefly. One of these misinterpretations implied that the modern continental shelf was largely an old land surface, drowned but almost unaffected by the present sea (e.g. Baak, 1936). In regions of strong currents this is manifestly untrue as will be evident in later chapters. Another misinterpretation was that waves had an effective 'wave base' of a few metres water depth, whereas it is now known that storm waves can do significant work even on the deeper-lying parts of the continental shelf. Thirdly, some workers considered that during a marine transgression the land deposits were reworked in the surf zone as it swept inland, and that much of the sand so liberated would be carried forward in a migrating near-shore sand prism. Doubtless some reworking did take place, but much material remained behind and is now still being reworked by the modern tidal currents. Fourthly, it was argued that during periods with stable sea levels there was little sand lost from the coast to the open reaches of the continental shelf. In practice some rivers are still supplying sand to the sea and there is much sand being supplied as a result of coast erosion (Section 4.1.1). Fifthly, there was an implicit misinterpretation that (all) the present day continental shelf deposits should show a seaward decrease in grain size with an increase in water depth (e.g. D. W. Johnson, 1919) as had been assumed in making palaeogeographical reconstructions of past seas. This is by no means necessarily true in a sea with strong tidal currents. Some of these misinterpretations arose because the geologists concerned took little account of what was known of modern water movements or because there were so few available observations of sand mobility to guide them. Indeed, their task was not made easy as there was a substantial data gap concerning continental shelf water movements. This was because physical oceanographers were showing preferential interest in the currents of deeper water, and hydraulics workers were interested mainly in the physics of grain movement in flumes and in local studies of sediment transport in rivers, estuaries and close to the coasts.

Attempts by workers to provide a system of terms to describe major sediment types on the continental shelf stultified some of the thinking about modern sedimentation processes. For example, doubt was cast on the relevance of material that was being derived by the present sea from floor that had become submerged during the post-glacial transgression. Thus, Emery (1968) and some later workers sought to restrict the term 'modern sediment' to deposits made of material won from the present coasts or carried to the sea by rivers (during the past 5000 years, while sea level has been approximately at its present height). Although this idea is useful in some ways it included a pointless restriction that allows some ludicrous consequences. For example, some cliffs and the adjacent floor of the North Sea

6 Offshore Tidal Sands

Fig. 1.2 Bathymetry of the sea floor around the British Isles and adjacent parts of mainland Europe, with depths in metres. Note the varied contour interval. The relative depths of small patches of shallower or deeper ground are shown by + or − signs, respectively.

are made of glacial material of Quaternary age. Yet use of the proposed term would mean that new marine deposits made from those cliffs would be called 'modern' whereas other new marine deposits made from adjacent submarine glacial material would not be modern but would be called 'relict'.

Moreover, in practical terms one cannot assess whether the new marine deposits would necessarily be derived from one or the other locality, because of the similar composition of the two sources and because there will be sediment exchange between the two localities in one direction or the other depending on whether the sea is rough or calm. Furthermore, such an artificial distinction, if used logically, would have to be applied to all sea floors being eroded after a rise of sea level, and not just to those made of Quaternary deposits. For example, Jurassic and Cretaceous rocks are still being eroded at the coast or beneath the sea in the English Channel and the Bristol Channel 5000 years after the latest marine transgression has been completed. Indeed, it is argued in this book that recycling of older material by a sea is the norm, whether that material is provided by modern rivers, volcanoes, wind or ice, plants or animals or whether it is obtained by the sea's attack on coasts or any other part of the sea floor. Thus, in this book there will be no use of such misleading and therefore outmoded terms as 'relict' or even of the subsequent and somewhat more realistic ones, 'palimpsest', 'allochthonous' or 'autochthonous'. Nevertheless, there is interest in where material has been derived from and what it is made of, because of the temporal changes of facies that can result.

1.3 A depositional surface for late Holocene deposits

The continental shelf around the British Isles (Fig. 1.2) can serve as a good example of the type of depositional surface to expect in cool temperate middle latitudes of the present time. A few generalizations and examples will set the scene. This continental shelf has already had a complex history, with its nature, origin and age varying locally and from region to region. Some of its oldest parts are typified by the flat, extremely low gradient (about 1:1000) rock floor of the western half of the English Channel and adjacent Celtic Sea, which was evolving during the Upper Tertiary and was repeatedly shaved during Pleistocene low sea levels (e.g. Donovan and Stride, 1975; Wood, 1974). In contrast fault troughs are obvious in the continental shelf lying west of Scotland and there are also areas of ragged rock separated by sediment ponds (Plate 1.1). Much of the floor around the British Isles has been glaciated and still shows clear signs of moraines and tunnel valleys (e.g. Eisma, Jansen and van Weering, 1979), although these are small in height compared with the 1 km maximum thickness of Quaternary deposits (mostly glacial) in the North Sea (V. N. D. Caston, 1972). On the outer part of the continental shelf around Scotland and Norway occur the associated Pleistocene iceberg plough marks (Plate 1.2), up to a few metres deep (Belderson, Kenyon and Wilson, 1973; Belderson and Wilson, 1973). In addition, there are many examples of drowned river valleys (Fig. 1.3) and of hollows attributable to the tidal scour associated with periods of low sea level (Section 3.4.1). Massive sand banks were constructed by tidal currents at the beginning and during the period of rising sea levels of the Holocene (Flandrian) transgression (Sections 3.4.5, 5.3.1). The maximum known dimensions of some of the bed features of the depositional surface below modern deposits are shown in Table 1.1.

Table 1.1 The maximum known dimensions of relief features predating modern marine deposits on the continental shelf of north-western Europe.

Relief Type	*Length (km)*	*Breadth (km)*	*Vertical dimension (m)*
Moraines	330	110	60
Iceberg plough marks	5	0.3	10
Low sea level sand banks	120	15	56
Tunnel valleys	60	3	100
Erosional trenches	145	6	90

Some of these old features are being buried now (Chapter 5). Others have hardly been affected by the sea while elsewhere there are parts of the continental shelf still being eroded by the tidal currents (Chapter 4).

1.4 Limits and outline of the book

The scope of the present book was partly determined by the nature of recent publications. Thus, there was no need to review modern or ancient tidal flat

8 Offshore Tidal Sands

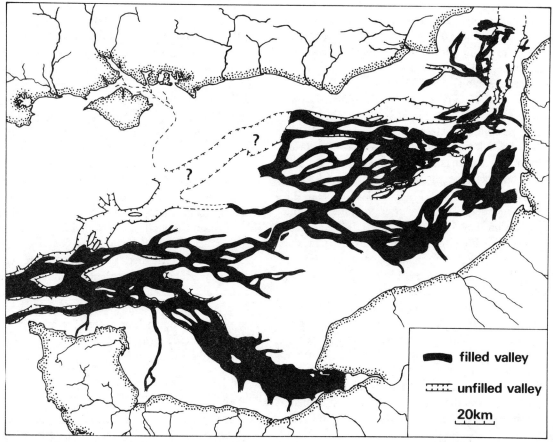

Fig. 1.3 Drowned river valleys in the eastern half of the English Channel. Unfilled valleys are shown in outline and filled ones in black (after Auffret, Alduc, Larsonneur and Smith, 1980).

deposits as there are already good recent accounts of these (e.g. van Straaten, 1956; Reineck and Singh, 1973; Ginsburg, 1975; Elliott, 1978). Similarly, there are good general accounts of sedimentation that set the scene for continental shelves where the processes are dominated either by tidal currents or by storms and waves (Reineck and Singh, 1973; Swift, 1976; Klein, 1977b; Vanney, 1977; Johnson, 1978; Walker, 1979). When taken together these reviews provide a useful amalgam of ancient and modern examples.

The present book is concerned largely with sedimentation by strong tidal currents on continental shelves but it also makes allowance for significant occasional non-tidal water movements. It uses published data as well as drawing heavily on the authors' unpublished material. Most of the examples refer to floors of sand and gravel, including carbonates, in temperate latitudes, especially those occurring around the British Isles. It is this portion of the offshore tidal realm that is best known and it is here that there is a high rate of loss of tidal energy (Flather, 1976) much of which goes into moving bed sediment. The emphasis is on understanding sedimentation processes at a general level, for without such an approach the sediments are largely unintelligible. The sands are seen as the most significant product of the offshore tidal current environment. Muds have been largely excluded as they can travel continuously in suspension for many tidal cycles, so that they indicate the net flow of water whether tidal or otherwise.

Chapter 2 gives a brief worldwide outline of the principles of tides and tidal currents, that is essential background reading for a review of the associated offshore tidal current sedimentation. Chapter 3 provides a general description of the bedforms of those seas and of the known conditions required for their formation. This enables bedform descriptions to be reasonably full yet kept separate from their usage in later parts of the book. Chapter 4 makes use of data on tidal currents, bedforms and other lines of evidence to demonstrate the net sand transport direction in some modern tidal seas and to discuss the variability of transport. The effects of the occasional non-tidal currents are also included, but it would be inappropriate in this book to deal with seas where tidal currents are not at present dominant. Chapter 5 describes what is known or can be surmised about the late Holocene offshore deposits of tidal seas and draws particular attention to what is known about the processes controlling deposition at present sea level. It also summarizes what is known of the low-sea-level deposits of the earliest part of the Holocene. Chapter 6 provides an account of the shelly faunas that are associated with the deposits and bedforms described above and mentions some of the publications of historical importance. Although the faunas are not uniquely tidal in origin they are an essential part of the deposit and so merit largely new description because their variety and abundance are an expression of significant variations of that depositional environment. The final chapter makes use of the modern analogues to make a literature search for offshore tidal deposits of the geological past. The list is in no way exhaustive in coverage but should serve to attract attention to numerous other possible offshore tidal current deposits whose depositional environment merits re-examination. The ancient deposits are not only of interest for what they tell us about geological history but are also a help because of their indications of the future, long term development of deposits of sand now accumulating in modern tidal seas.

Chapter 2

Tidal currents of the continental shelf

2.1 Introduction

Where tidal currents are strong their importance in sedimentation has long been acknowledged, leading to erosion and deposition, controlling the shape of sandy sea floors and inducing net, as well as oscillatory, sediment movement. This net transport of sand arises from distortions to the tide caused by bathymetry (a term which will be used to include the outline of the coast as well as the depth distribution) since a symmetrical tidal wave, like the tide in the deep ocean, does not generate net sediment transport. The dynamics of many continental shelf seas, for example most areas around the coast of the British Isles, are dominated by tides. Even in areas of small tidal range strong local tidal currents can occur, either near to a 'node' in a stationary tidal oscillation (Section 2.3) or because of bathymetry, particularly in channels or straits connecting two regions − for instance the Strait of Messina in the Mediterranean.

The theoretical basis for tidal motion has been known for several centuries but tidal observations have been limited to coastal gauges at ports, with very few offshore elevation or current measurements. Seas with better studied tidal currents than most are those on the continental shelf around the British Isles and from these will come some of the illustrations for this chapter. The chapter contains an introductory account, with a minimum of mathematics, of the generation of tides in the world's oceans and their propagation into shelf seas, the origin of net sediment transport by tidal currents, the variations of tidal currents with depth near the sea floor, internal tides and the changes in tides that are likely to have occurred on a geological time scale. A general introduction to the subject of tides is that of Macmillan (1966), with more detail given by Doodson and Warburg (1941), Defant (1961) and Cartwright (1978). The effects of non-tidal currents and waves are referred to in later chapters.

2.2 Tide generating forces and the ocean's response

2.2.1 TIDE GENERATING FORCES

The tide is a periodic movement of the sea or ocean due to periodic forces. Since the time of Isaac Newton it has been known that tides are generated by the gravitational forces of the Sun and Moon on the Earth. Considering first the Earth and the Moon (Fig. 2.1) take a point A on the Earth (mass e, radius a) with latitude θ and suppose the Moon (mass m) is a distance d from the centre of the Earth and above the Earth's equator. To produce movement in the ocean

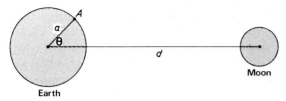

Fig. 2.1 Symbols used for the Earth–Moon system.

the tide generating force must vary over the Earth's surface. At point A the force arises from the difference between the gravitational attractive force there due to the Moon and that experienced at the centre of the Earth. (The latter is just the force required to keep the Moon in its orbit around the Earth.) The tide generating force at A can be resolved into local vertical and horizontal components. The vertical force is negligible in comparison with the Earth's gravitational attraction at the surface, g (a maximum of $10^{-7}\,g$). The dominant term in the formula for the horizontal force is

$$1.5\,g\,\frac{m}{e}\left(\frac{a}{d}\right)^3 \sin 2\theta.$$

(The other terms are small and have been ignored in this chapter.) Fig. 2.2 shows schematically how this force varies over the Earth's surface. It generates two bulges in the oceans, one directly underneath the Moon and the other on the opposite side of the Earth, and so, because the Earth rotates, there will be two high and two low tides a day. However, Figs 2.1 and 2.2 represent a simplification since the Moon's orbit does not lie in the plane of the Earth's equator but oscillates about it. The formula for the horizontal force is then split into three terms, respectively of several long periods, a period of one lunar day and a period of half a lunar day. A lunar day is 50 minutes longer than a solar day of 24 hours since the Moon orbits the Earth every 29 (solar) days. The long period term has a smaller amplitude than the other two terms. The amplitude of the daily, 'diurnal', term is approximately proportional to the Moon's declination (the angle between the plane of the Moon's orbit and the plane of the Earth's equator) whereas the amplitude of the twice daily, 'semi-diurnal', tide is at a maximum when the Moon is above the Earth's equator (zero declination). Both the daily and twice daily terms are affected by the

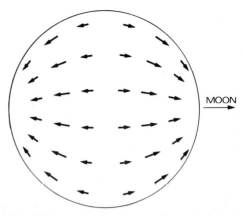

Fig. 2.2 Schematic representation of the variation of the horizontal tide raising force over the Earth's surface, for positions at tips of arrows.

Moon's distance from the Earth, because of the $(a/d)^3$ factor, which varies since the Moon's orbit is elliptic.

Similar forces are generated by the Sun with periods corresponding to solar days. Although the Sun is more massive than the Moon (by a factor of 27×10^6) it is further away (by a factor of 389) and so the ratio of the amplitudes of their tide generating forces is

$$\text{Sun/Moon} = 27 \times 10^6 \times (389)^{-3} = 0.46.$$

Hence the Sun generates a tide raising force less than half as strong as the Moon's. No other heavenly bodies are significant.

2.2.2 SPRING–NEAP CYCLES

The combination of the Sun's and Moon's tide raising forces leads to the 'spring-neap' cycle with a period of 14.77 days for the twice-daily tides. Springs occur when the forces add, i.e. when the Sun, Moon and Earth are in line (at new Moon and full Moon), or up to several days after, due to a lag in the ocean's response which varies over the Earth's surface. Neaps occur at, or just after, half moon when the Moon's and Sun's forces subtract. Theoretically the tidal range at mean neaps is approximately $(1 - 0.46)/(1 + 0.46)$ or 37% of that at mean springs, but there are local variations in the ocean's response. Variations in the Earth's and Moon's orbits generate small perturbations to this spring–neap cycle and, for the twice daily tides, maximum spring range occurs near the spring and autumn equinoxes when the Moon is above the Earth's equator.

For the daily tides there is also a fortnightly variation, this time with a period of 13.66 days, and a yearly variation such that their maximum spring range occurs near the mid-summer and mid-winter solstices when the Moon's declination is greatest.

2.2.3 PEAK ASTRONOMICAL TIDES

The long-period term has contributions from variations in the Moon's orbit around the Earth and the Earth's orbit around the Sun. Only five separate periods occur – a lunar month, a year, 8.85 years, 18.61 years and 21 000 years. Cartwright (1974) computed the times of near-maximum twice-daily tide-raising forces for the period 1–4000 AD. These will be at the equinoxes, as mentioned above, and the most recent was in September 1922 when the observed range at Newlyn, Cornwall, was 1.22 and at Brest, France, 1.28 times the mean spring range, respectively 1.64 and 1.72 times the dominant lunar twice-daily (M_2) range. Cartwright also calculated that there were or will be peak tides in the years 135, 1020, 1113, 1745, 2192, 2732, 2825 and 3002 AD. However, ranges within about 3% of these are predicted several times per century and also variations in total current speeds due to non-tidal contributions are much larger than this.

2.2.4 RELATIVE AMPLITUDES OF DAILY AND TWICE-DAILY TIDES

The theory predicts that the daily tides should have approximately the same amplitude as the twice-daily tides. However, observations show that, for the present ocean configuration, the twice-daily tides have a magnified response and dominate most areas. Predominantly daily tides are found in the Gulf of Mexico and much of the Gulf of Thailand and the South China and Java Seas, and locally elsewhere where twice-daily tides are reduced by tidal wave interference. In many parts of the Pacific and Indian Oceans both daily and twice-daily tides are important. In such regions a few days of twice-daily tides can alternate with a few days of daily tides, especially in spring and autumn (Webb, 1976a). Even in areas of twice-daily tides there will be a daily modulation of the tide and tidal current, which will be most marked when the Moon's declination is large.

2.2.5 A COMPUTATION OF OCEAN TIDES

The variation over the deep oceans of the amplitude (half range) of the principal lunar twice-daily (M_2) tide is approximately as shown in Fig. 2.3. It has been taken from Accad and Pekeris (1978), and was obtained from a computer prediction based on the governing equations. The results agree fairly well with the few deep ocean observations that exist, mainly taken from coastal gauges on islands. The ocean water depth, h, roughly 5 km, is small compared with the corresponding tidal wavelength which is given by $T(gh)^{\frac{1}{2}} = 10\,000$ km, where T is the tidal period.

The amplitude of the current, u, of a deep ocean tide with elevation amplitude ζ is, making several simplifications, approximately

$$u = (g/h)^{\frac{1}{2}}\zeta \qquad (1)$$

From Fig. 2.3, ζ is less than 0.75 m, giving depth-averaged M_2 currents of less than 4 cm/s in the deep ocean.

2.3 Tidal currents in shelf seas

The continental shelf seas are too small for the tide raising forces to generate a significant response in them. Thus the tides observed in shelf seas result from the tidal wave propagating onto the shelf from the adjacent ocean. Hence, if the sea has only a narrow connection with the ocean, such as the Mediterranean, it cannot acquire a significant tidal range.

2.3.1 AMPLIFICATION DUE TO DECREASING DEPTH AND WIDTH

A first approximation to the effect on the wave amplitude of reducing the water depth can be obtained by assuming that no energy is lost. The

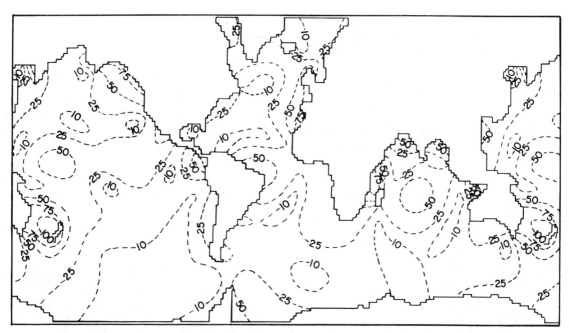

Fig. 2.3 Map of computed amplitude (half range) in cm of the principal lunar twice-daily (M_2) tide for the World's oceans, without continental shelves (after Accad and Pekeris, 1978).

calculation shows that wave amplitude is inversely proportional to the fourth root of the water depth:

$$\zeta^4 \propto \frac{1}{h}$$

For example, if the tide has an amplitude of 0.5 m in 5 km water depth and propagates onto a shelf with depth 50 m it will have an amplitude of 1.6 m and a maximum current speed of 70 cm/s. A similar argument for a channel or estuary decreasing in width, b, with uniform water depth, leads to the wave amplitude being inversely proportional to the square root of the width:

$$\zeta^2 \propto \frac{1}{b}$$

These two amplifications, due to decreasing water depth and decreasing width, are common in continental shelf seas. For instance they contribute to the high tidal ranges in the Bristol Channel and central English Channel. Since the amplification arises from variations in geometry it does not depend on the frequency of the wave and will affect each of the constituents of the tide identically.

2.3.2 RESONANCE

Another type of amplification, which is frequency dependent, is resonance. Consider a rectangular tank of water. If it is rocked repeatedly waves will appear on the surface. For several distinct frequencies of rocking the waves will grow in amplitude and appear not to move across the tank. For the lowest growth frequency the water rises at one end when it falls at the other (the two ends are out of phase) and there is no movement of the water surface in the middle, a nodal line. Since a wavelength is defined as being the length from one wave peak to the next, in this case the tank is half a wavelength long. When considering tides that are directly approaching the shore across a shelf or bay there is only one reflecting boundary, not two, and maximum amplification will occur for a shelf width or bay length of a quarter wavelength, so that whilst there is little vertical movement of water surface at the shelf edge (corresponding to the middle of the tank) there is much greater vertical movement at the shore line. This sort of amplification occurs in the Bay of Fundy (in which occur the world's largest tides with a range of up to 15 m) and also in the Bristol Channel.

A rectangular basin with uniform depth 75 m (the average for the Bay of Fundy) would be in resonance if its length were a quarter wavelength of the twice-daily tide:

basin length = quarter wavelength =
$$0.25 \times T(gh)^{\frac{1}{2}} = 300 \text{ km.}$$

Allowing for the deviations of the Bay from the idealized rectangular shape this value is close to its actual length of 270 km. In reality, the resonant system is a combination of the Bay of Fundy with the adjacent, deeper Gulf of Maine and has a resonant period of between 13 and 14 hours (Garrett, 1972). There is evidence from sand deposits (Amos, 1978) that current speeds in the Bay of Fundy have been gradually increasing, i.e. its dimensions have been changing (length decreasing and/or depth increasing), perhaps due to changing sea level, so that its longest free oscillation period has been approaching the resonance value of half a day.

2.3.3 PROGRESSIVE AND STANDING WAVES

Two types of wave have been mentioned: the first was a tidal wave in the ocean and the second was set up in a tank, by reflection. The first type is called a progressive wave (it can have any direction of propagation). In it the strongest currents at a particular place occur at high and low water (the current and elevation are in phase) and there is a transport of energy. The formula (1) in Section 2.2 relating current and elevation amplitudes is for a simple version of this kind of wave.

The second type is called a standing wave. In it the strongest currents are at mean water level (the current and elevation are in quadrature) and there is no transport of energy. A standing wave is the combination of two progressive waves with equal amplitudes and frequencies but travelling in opposite directions. For a standing wave the amplitudes of current and elevation vary in space with maximum current amplitude where the elevation amplitude is zero (a nodal line). (Referring to the rocking tank at its lowest resonance frequency, the fastest currents occur at the middle of the tank where the water surface does not move, whilst the current is zero at the ends where the surface moves up and down the most.) In most shelf seas the tide is a combination of standing and progressive waves.

2.3.4 EFFECTS OF THE EARTH'S ROTATION

For a narrow sea, like the Bay of Fundy, the Earth's rotation does not significantly affect tidal currents or elevations. However, for larger seas, like the North Sea, and the open ocean the effects of the Earth's rotation are important.

Where the tidal wave in the adjacent part of the ocean is progressive and travels parallel to the coast, the Earth's rotation modifies the wave in such a way that its amplitude increases towards the shore and so that it can only propagate in one direction – with the shore on its right in the Northern Hemisphere and on its left in the Southern Hemisphere. This type of wave is called a Kelvin wave, after its proposer. (His original assumption of uniform water depth can be relaxed since Kelvin wave properties are similar for gradually sloping continental shelves.) The twice-daily and daily tides along many open shelves are of this form. With such waves another kind of resonance can occur – along the continental shelf instead of across it. This may contribute to the high tidal ranges on the Patagonian continental shelf (Webb, 1976b).

Where the tide propagates into a wide shelf sea its reflection pattern will also be altered by the Earth's rotation. Instead of lines of no vertical motion (cf. the centre of the tank) there are points of no vertical motion, called 'amphidromic' points, about which the tide appears to rotate. If the sea is long enough there will be several amphidromic points separated by distances of about half the wavelength corresponding to the tidal period. At amphidromic points the tidal currents will tend to be larger than in the surrounding area in the same way as the currents were a maximum at the nodal line in the middle of the rocking tank.

2.3.5 SOME EFFECTS OF CONTINENTAL SHELF WIDTH

Several authors (e.g. Silvester, 1974; Klein and Ryer, 1978) have suggested that tidal elevation and current amplitudes are proportional to continental

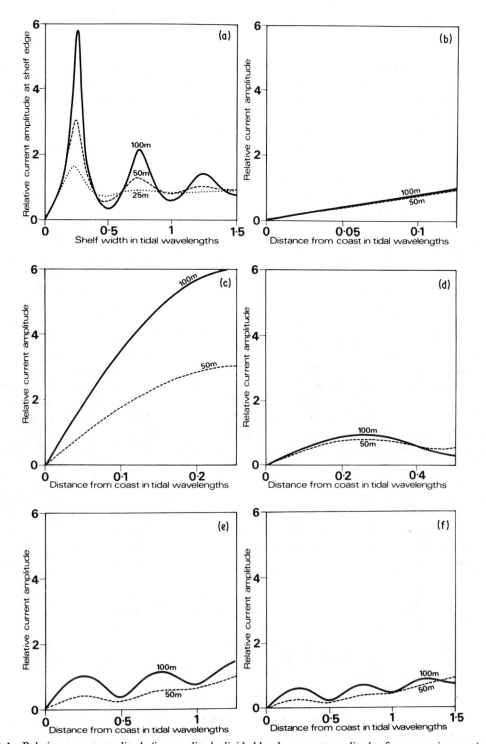

Fig. 2.4 Relative current amplitude (i.e. amplitude divided by the current amplitude of a progressive wave) which would have applied for a constant depth continental shelf and linear friction.
(a) Relative current amplitude at the shelf edge vs shelf width in tidal wavelengths for depths 100, 50 and 25 m.
(b)–(f) Current amplitude versus distance (in tidal wavelengths) from the coast for shelf widths (b) 1/8, (c) 1/4, (d) 1/2, (e) 5/4, (f) 3/2 wavelengths and for depths 50 and 100 m.

shelf width. This relationship would be particularly relevant to the very wide shelves that have occurred during certain periods of the Earth's existence. However, for a standing wave type of tide, which will often be the case for a wide continental shelf, maximum amplitudes will occur when the shelf width allows the shelf sea to resonate with the open ocean tide. Resonance occurs for a shelf width of a quarter of the tidal wavelength and also for widths 3/4, 5/4, etc. of the tidal wavelength, but tidal friction will tend to cause lower elevation and current amplitudes for very wide shelves (Fig. 2.4). This figure is based on theory (Proudman, 1953, article 158) and applies to an idealized, long, straight ancient continental shelf assumed to have a constant depth and for friction inversely proportional to the shelf depth. The effects of the Earth's rotation have been neglected. The resonances would have similar properties if the Earth's rotation and the actual depth variation across the continental shelf were included. Fig. 2.4a plots the current amplitude at the shelf edge against shelf width in tidal wavelengths for shelf depths of 100, 50 and 25 m and shows the peaks at the resonant widths. These peaks become smaller for wider continental shelves with greater frictional losses. Tidal wavelength for the M_2 constituent is approximately 1400, 990 and 700 km for respective depths of 100, 50 and 25 m. Fig. 2.4b–f shows how the current amplitude varies across the shelf for shelf widths of 1/8, 1/4, 1/2, 5/4 and 3/2 wavelengths. If the shelf width is much less than 1/4 wavelength (as in b) the current amplitude varies linearly with distance from the shore. Measurements which illustrate this from the continental shelf off eastern USA are discussed by Redfield (1958) and for the Gulf of Panama and adjoining continental shelf by Fleming (1938). If the shelf is much wider than 1/4 wavelength, there are local maxima at distances 1/4, 3/4, 5/4 . . . wavelength from the coast and local minima at 1/2, 1, 3/2 . . . wavelengths from the coast (Fig. 2.4d–f). For wider shelves friction is also important, tending to reduce the differences between maxima and minima and to cause maximum currents to be at the shelf edge near the energy input.

For present day continental shelf configurations there is, indeed, an approximately linear relationship between shelf width and tidal range at the shore for many individual shelves, as summarized by Cram (1979), but the factor of proportionality varies from shelf to shelf. Present continental shelf widths are, in the main, less than the smallest (1/4 wavelength) necessary for resonance and so, for each area, the wider the shelf the nearer that part of the sea is to resonance and hence the larger the tidal range at the shore.

2.3.6 TIDAL RANGE AT THE COAST

The importance of each of the effects listed above varies as the bathymetry varies and causes large variations in the tidal range around the world's coasts. Tidal range is important in determining coastal morphology and stability and so Fig. 2.5 shows where the mean spring tidal range at the shore is greater than 3 m and where it is greater than 5 m. As discussed already, the tidal current amplitude depends on other factors as well as tidal range, so that the maps do not necessarily also show areas of strong tidal currents, either at the shelf edge or nearer shore. Further, daily tides (but not twice-daily tides) at latitudes above about 30° can generate not only Kelvin waves but also continental shelf waves (Huthnance, 1975; Cartwright, Huthnance, Spencer and Vassie, 1980) and double Kelvin waves centred on the shelf break (Longuet-Higgins, 1968), all of which progress parallel with the shore and hence affect current amplitude.

In summary, all the factors discussed above mean that there can be no simple general relationship between tidal current speeds, at the shelf edge or elsewhere, and tidal ranges at the coast, which would involve the width of the continental shelf.

2.3.7 TIDAL CURRENT SPEEDS ON THE CONTINENTAL SHELF AROUND THE BRITISH ISLES

Contours of the mean spring near-surface peak tidal current speeds for off-shore areas of the seas around the British Isles are shown in Fig. 2.6. This illustrates many of the features already discussed, such as the faster currents in shallower water, amplification in the Bristol Channel and the effect of the Earth's rotation. It is also apparent that there are local

Tidal currents 17

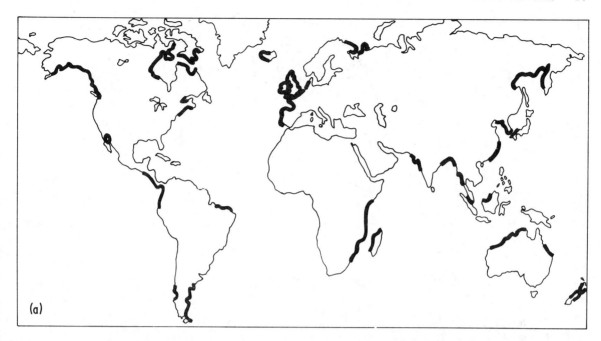

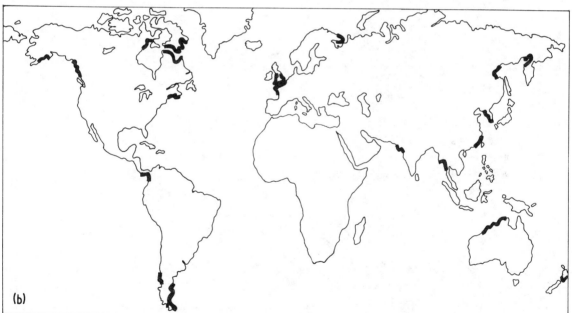

Fig. 2.5 Portions of the World's coastline where mean spring tidal range is greater than 3 m (a) or 5 m (b).

bathymetric effects and these are pronounced close to the coast. These are shown in more detail in Sager and Sammler (1975, Chart 1).

Friction has not been mentioned but it will gradually remove energy from the tides at the bottom of the water column. In large seas the total attenuation can be severe. For instance in the North Sea, the tidal wave propagates in an anticlockwise sense

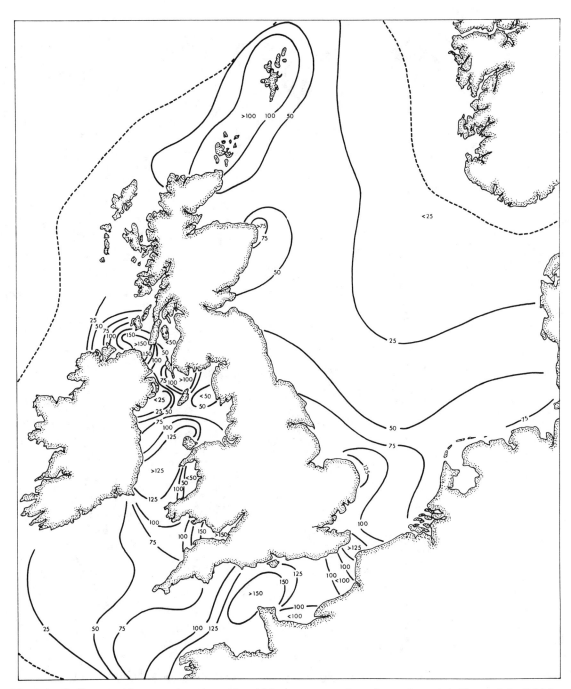

Fig. 2.6 Outline map of mean spring near-surface tidal current strength on the continental shelf around the British Isles (after Sager and Sammler, 1975), with additional data from analysis of moored current meter recordings in the Irish Sea, Celtic Sea and Southern North Sea. The edge of the continental shelf is shown by a pecked line. Current speed is in cm/s.

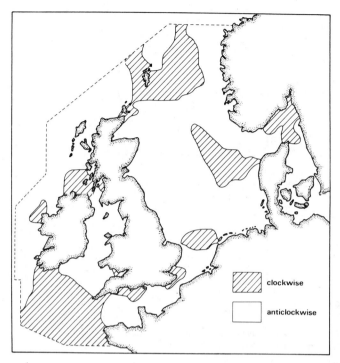

Fig. 2.7 Sense of rotation of near-bottom tidal ellipses of (the principal lunar twice-daily tide) M_2 on the European continental shelf from a numerical model (after Davies and Furnes, 1980).

(as a Kelvin wave should), entering by travelling southward along the east coast of Scotland, and the tidal currents and elevations are much greater near Scotland than near Denmark and Norway, at the end of the tide's travels. The transfer of energy through Dover Strait is not significant in this respect.

2.3.8 TIDAL ELLIPSE

The current vector at a particular place and time is specified by its speed and direction. Since tidal currents are generated by periodic forces both the speed and direction will oscillate. The speed oscillates like a sine wave but the direction oscillation can vary from rectilinear flow (first one way, then the reverse), to circular motion, with a constant rate of direction change. These are extreme forms of motion round an ellipse and for each tidal constituent at a point both the tide generating force vector and the current vector trace out ellipses. The axial ratio of the current ellipse and its sense of rotation are determined by three factors – the tide generating forces, the Earth's rotation and the topography. For the twice-daily constituents both the tide generating forces and the Earth's rotation drive the currents in the same sense, clockwise in the Northern Hemisphere and anticlockwise in the Southern Hemisphere. The pattern is more complex for the daily constituents and will not be described here, as in most of the oceans the twice-daily terms dominate. The effect of topography can be to drive the ellipses either clockwise or anticlockwise. So, even for the twice-daily constituents, a complicated picture emerges for a continental shelf sea, especially as the ellipses from several twice-daily constituents will be combined to obtain the mean spring tidal current envelope, which is commonly distorted from an elliptic shape. As well as varying horizontally the sense of rotation of the ellipse can change through the water column, as found from analysis of current meter recordings for some Irish and Celtic Sea positions (Robinson, 1979) and in the North Sea near Aberdeen (Davies and Furnes, 1980). Fig. 2.7 shows the sense of rotation for the near-bottom currents in European shelf seas as computed in a three-dimensional mathematical model of the M_2 tide (Davies and Furnes, 1980). Discrepancies between this figure and Robinson's figure 2c, and

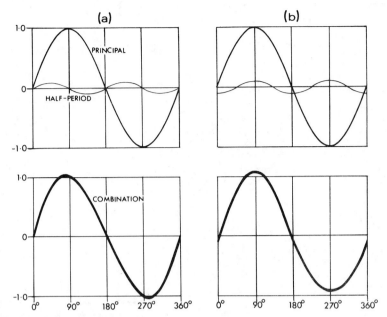

Fig. 2.8 Examples of combinations of principal and half period waves (e.g. tidal constituents M_2 and M_4).
(a) Peak current of half period wave 90° ahead of principal wave.
(b) Peak current of half period wave in phase with principal wave.

map 5 of Sager and Sammler (1975), result partly from the use in the latter of near-surface mean spring, rather than near bottom M_2 currents.

2.4 Net sand transport caused by tidal current asymmetries

2.4.1 DISTORTIONS TO THE TIDE

The generation and progression onto the shelf of the long period, daily and twice-daily tides so far outlined does not contain any mechanism for net transport of sand, since each tidal constituent is oscillatory and symmetrical. However, distortions of the tidal curve do arise, especially if current speeds are high, when the tidal wave enters shallow or constricted water or when the direction of the tidal current varies with geographical position. When the total water depth varies significantly over a tidal cycle the interval from low to high water becomes shorter than that from high to low water. The distortion from symmetrical oscillatory motion can be expressed mathematically as a mean (tidal residual) and higher harmonics of the principal period. That is, there are terms with periods of a half, a third, a quarter ... of the principal (e.g. M_4, M_6, M_8 etc. for the M_2 constituent). Tide gauge and current meter records show that only the first two higher harmonics are important, with the one with a half period larger than the one with a third period.

2.4.2 COMBINATION OF THE PRINCIPAL AND ITS FIRST HARMONIC

Consider a principal wave M_2 and one with half its period M_4, for a tidal current alternating in direction. If the two terms have a peak flow in phase, in one direction or the opposite, the effect is for that flow to be faster and of shorter duration, while the opposing flow will be weaker and of longer duration (Fig. 2.8b). There is still no net flow of water, but the peak current is stronger in one direction than the other. At a given place the amplitude of the half period tide varies approximately as the square of the amplitude of the principal tide. So, the inequality will be much greater at springs than at neaps. For this type of peak amplitude inequality the half period wave must be in phase or 180° out of phase with the principal wave. If the waves are 90° out of phase, the flow in one direction will last longer than in the other but the peak speeds will be equal (Fig. 2.8a). Since sand or gravel transport rate on the sea floor (Chapter 4) is approximately proportional to the cube of the difference between the current speed and the threshold speed for moving that material, any asymmetry in peak ebb and flood current speeds will have a significant effect on the net transport. Peak asymmetries, as discussed above, do not arise from a combination of a principal wave with a wave of one third of its period (e.g. M_2 and M_6) and so the latter is unimportant for net sediment transport.

2.4.3 NET SAND TRANSPORT BY TIDAL CURRENTS

As a general illustration consider a rectilinear current, Y, composed of a mean (tidal residual) A, say, and a tide of two frequencies, one twice the other, σ and 2σ with amplitudes B and C. The average value of Y^3 over a tidal cycle will be an approximation to the net sediment transport since, for simplicity, the movement threshold speed has been ignored. The resulting discrepancy will be small if the peak current speed is large compared to the movement threshold speed.

Now $Y = A + B \cos(\sigma t - b) + C \cos(2\sigma t - c)$

and the average value of Y^3 over a time $2\pi/\sigma$ is

$$A^3 + 1.5(AB^2 + AC^2) + 0.75B^2C \cos \phi$$

where $\phi = 2b - c$ and b is the phase of the wave with period $2\pi/\sigma$ and c is the phase of the wave with period π/σ. Special cases of this formula are of interest. First, if the mean current is zero ($A = 0$) the average value is proportional to $\cos \phi$, consistent with the above discussion of phase differences of 0°, 90° and 180°. Again for $A = 0$ the average value will be much larger if B is larger than C, showing that once-daily constituents will be less important than four times a day constituents for sediment transport in a region of twice-daily tides. Finally if A is of a similar magnitude to C and both are much smaller than B (a common occurrence for the M_2 tide around the British Isles, with the mean, A, generated by non-tidal forces) then both the mean and the higher harmonic will interact with the M_2 current to generate net sediment transport rates of similar magnitudes.

2.4.4 TIDAL CURRENT PATTERNS IN THE VICINITY OF SAND BANKS

Two more localized sediment transport/tidal current interactions can occur around sand banks (Sections 3.4.5 and 4.7). First, some patterns in the tidal flow are favourable for the formation of banks and second, once banks exist they can distort the flow field. Both these interactions can be represented by the generation of mean currents and higher harmonics.

One disturbance to tidal flow, which might lead to the formation of a sand bank, is a tidal eddy. The tidal eddy will occur in approximately the same position for each flood or each ebb current unlike a random, turbulent eddy. Generation of eddies by headlands is discussed by Tee (1977), and Maddock and Pingree (1978). The flow is increased at the tip of the headland and changes direction so that eddies form in the bays on either side. A similar process occurs when there is a sharp change in water depth (Zimmerman, 1978). The flow round Portland Bill and the supposed consequent formation of the Shambles Bank is discussed by Pingree (1978) and round Start Point with reference to the Skerries Bank by Pingree and Maddock (1979), as discussed in Section 3.4.5. Their arguments regarding the flow are applicable to other similar headlands. They suggest that the combination of circulation round the eddy, the Earth's rotation and bottom friction will generate a pressure gradient at the sea floor favourable to the formation of a sand bank near the eddy centre either if the circulation of the eddy has the same sense as the Earth's rotation or if the circulation of the eddy is strong enough. The increased flow at the tip of the headland is likely to lead to scour there. However, other factors could influence near-shore sand bank formation (Section 3.4.5) and sand banks well away from headlands definitely require other explanations, as discussed in later chapters.

If the axis of an existing sand bank is not aligned parallel to the peak tidal flow then higher harmonics and a mean flow will be generated by two mechanisms – the Earth's rotation and bottom friction. The Earth's rotation leads to clockwise circulation in the

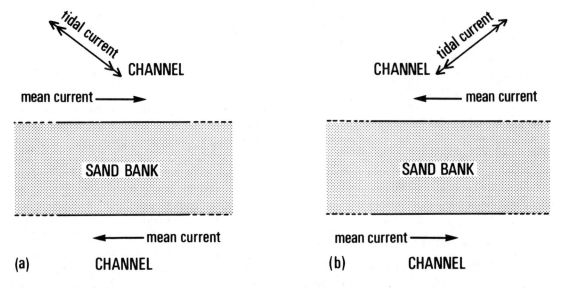

Fig. 2.9 Mean current (single headed arrows) generated by the action of bottom friction when a rectilinear oblique tidal current (double headed arrows) impinges on a long sand bank (after Huthnance, 1973). The mean currents reverse, cases (a) and (b) with the sense of obliquity of the tidal current (and would be zero for bank-parallel currents).

Northern Hemisphere and anticlockwise in the Southern, and bottom friction causes a circulation whose sense depends on the orientation of the undisturbed tidal current, assumed nearly rectilinear, relative to the bank (Fig. 2.9). For most of the Norfolk sand banks, both circulations are clockwise. Their magnitudes and those of the first harmonic agree reasonably well with Huthnance's (1973) theory.

2.5 Flow near the sea floor

The majority of measurements of tidal currents have been taken near the sea surface, with a few measurements at mid-depth and a very few near the sea floor (the relevant region for sediment transport calculations). However, near the sea floor the frictional and viscous drag forces become large and significantly affect the current velocity. Bowden (1978) has reviewed the present knowledge of this benthic boundary layer. Fig. 2.10 is a schematic diagram of the structure of this layer.

2.5.1 CONSTANT STRESS LAYER

For steady flow, there is a region next to the boundary where the shear stress is a constant, τ_0, equal to the mean stress applied to the sediment bed. From this stress, the friction velocity, u_*, is defined as

$$u_* = (\tau_0/\varrho)^{\frac{1}{2}}$$

where ϱ is the fluid density.

In the constant stress layer, the velocity of the fluid, u, at a height z can be expressed by

$$u = 2.5 u_* \ln(z/z_0)$$

where z_0 is a length scale associated with bed laminar boundary layer thickness (smooth flow) or boundary roughness (rough flow), with

$$z_0 \approx 0.1 \nu/u_* \text{ for smooth flow}$$

and

$$z_0 \approx d/30 \text{ for rough flow}$$

where ν is the kinematic viscosity and d is a characteristic dimension of the roughness elements, e.g. grain diameter for a flat bed of grains. The thickness of the constant stress layer is approximately $0.2 u_*^2/(fV)$, typically of order a metre for sand-moving currents, where V is the current speed away from the boundary and $f = 2\omega \sin\phi$, ω is the rate of

Tidal currents

about $0.1L$, it seems that the flow direction is constant and its speed conforms to a logarithmic profile.

2.5.3 EFFECTS OF THE OSCILLATORY NATURE OF TIDAL CURRENTS

Both constant stress and Ekman layers result from steady flows, whereas the tide oscillates. This leads to two complications. First, although the time scale required to set up a logarithmic layer is short, the time scale for an Ekman layer is comparable to tidal periods and so the layer will not be able to develop fully. (The layer is also limited in many cases because the water depth is insufficient.) Secondly, from a balance of the frictional forces, which are largest near the sea bed, and the pressure gradients due to tidal elevation gradients, it can be deduced that the current at the sea bed is in advance of that at the sea surface (Proudman, 1953, article 152). For twice-daily tidal currents the phase difference is of the order of 10°–20°, so that the current near the bottom will attain its maximum of the order of 20–40 minutes before the current near the surface. Since Ekman layers usually cannot develop fully in tidal motion, observations of the predicted direction change, clockwise with height in the Northern Hemisphere and anticlockwise in the Southern, are scattered, ranging from 0° to 40°. However, the change is usually only a few degrees in strong tidal currents, as also predicted theoretically (Nihoul, 1977). Estimated values of friction velocity lie between 0.5 and 5 cm/s with the logarithmic layer extending up to a few metres.

2.5.4 DRAG COEFFICIENT AND BOTTOM STRESS

An alternative method of calculating the bottom stress, τ_0, is to consider a drag coefficient, C, such that

$$\tau_0 = C\varrho u^2 (= \varrho u_*^2)$$
$$\text{i.e. } C = (u_*/u)^2$$

Measuring the current at 1 m above the sea floor gives a value of C as 3.1×10^{-3} (Sternberg, 1968, discussed in Section 4.2.2).

There are two complicating factors to these

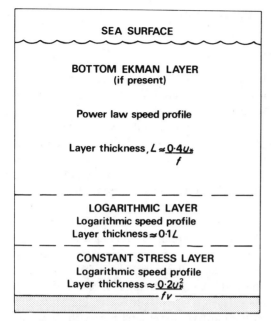

Fig. 2.10 Schematic diagram of the benthic boundary layer in a rectilinear tidal current. Relative layer thicknesses can vary greatly. The Ekman layer thickness stated is often not realized because of insufficient water depth.

rotation of the Earth and ϕ the latitude. The validity of the logarithmic profile for current speed seems to extend beyond the constant stress layer. Hence, by fitting a logarithmic profile to currents measured near the sea floor the stress, τ_0, applied to the sediments can be calculated.

2.5.2 EKMAN LAYERS

The Earth's rotation generates another boundary layer, called an Ekman layer, in which there is a balance between the force due to the Earth's rotation, the pressure gradient in the layer and the stress at the boundary. Ekman layers are generated at the sea surface, when there is wind stress, and at the sea floor, by bottom friction. The thickness, L, of a fully developed Ekman layer is approximately $0.4u_*/f$ which is at least 30 m for sand-moving currents and over 100 m in strong currents or in low latitudes (therefore f is small). A characteristic of the flow in an Ekman layer is that it changes direction with height, but in the lower part of the layer, within methods of calculating the bottom stress. First, the methods assume a sea floor with only small-scale roughness. If there are large sand waves or other bedforms which cannot be averaged over, not only will the current profile vary depending on where on a large bedform the measurements are taken (e.g. for large sand or gravel waves, Smith, 1970; Dyer, 1971) but also the shape of the sea bed will generate form drag. So the drag exerted by the sea bed on the water will consist of drag on the sedimentary particles and form drag from small-scale roughness and also form drag from the large bedforms (Taylor and Dyer, 1977). The second complication arises from recent measurements of the bottom stress by correlating the fluctuations of the horizontal velocity with the fluctuations of the vertical velocity (Gordon, 1974; Heathershaw, 1976). This method can sample the stress frequently and shows that it is intermittent with peak values 10 and occasionally 30 times the mean value in events lasting 5 to 10 seconds and recurring at intervals between 20 to 100 seconds. This will lead to scatter in mean stress calculations (Heathershaw and Simpson, 1978).

2.5.5 CURRENT PROFILES ABOVE THE LOGARITHMIC LAYER

By taking the average of the current speed over several minutes (common for many recording and direct reading current meters) the effect of this intermittency is much reduced. Such current measurements of the vertical profile on the continental shelf above the lowest few metres, where the speed profile is logarithmic, show that the tidal current can be approximated by a power law, some variations of which are:

$u = u_s(z/h)^{0.2}$ (van Veen, 1938)
$u = u_s(z/h)^{0.15}$ (Cartwright, 1961)
$u = u_s(z/h)^p$ $0.14 < p < 0.3$ (Dyer, 1970b)
$u = u_s[0.63 + 0.37((2z/h) - (z/h)^2)]$ (Bowden and Fairbairn, 1952)

where u is the speed at a height z, h is the water depth and u_s the surface speed.

2.6 Internal tides

2.6.1 NATURE

The only changes in a vertical section of the tidal current so far discussed have been generated by friction at the sea floor. However, if the sea water is stratified (i.e. its density decreases upwards) waves with tidal periods can occur within the water column which do not displace the sea surface and are associated with quite different vertical current profiles. These are called internal tides. Wunsch (1975) gives a recent review.

In some continental shelf seas, as for instance most of the Irish Sea, strong tidal mixing ensures that only small vertical density gradients exist and here internal tides will not be important. However, in many shelf seas tidal mixing is weaker and the near-surface layer is often appreciably lighter than the near-bottom layer for at least part of the year. (The causes of the density gradient include heating at the sea surface in summer and freshwater inflow from a river, since sea water density is determined by its temperature and salinity.) It is in the zone of large density gradient between the layers (the pycnocline) that internal tide amplitudes will be largest.

There are also small changes in density in both the near-surface and near-bottom layers. Because the vertical density profile is continuous the internal tides can be expressed as a series of vertical modes, the higher modes having a more complex structure, especially in the vicinity of the pycnocline. Usually for internal tides only the lowest few modes are important – the lowest mode represents bodily movement up and down of the pycnocline and at any one time current flows one way above the pycnocline and in the opposite direction below it, with maximum speeds near the sea surface and near the sea bed. In higher modes the currents are weak near the sea floor, so that only the lowest mode is likely to be significant for sand transport.

2.6.2 CAUSES

Since the wavelengths of internal tides are short (less than 200 km) they cannot be generated by the same mechanisms as are the usual surface tides (which are little affected by stratification). However, one mechanism for internal tide generation is the interaction between the currents due to the surface tide and variations in water depth, particularly near the top of the continental slope. From the shelf break the internal tides propagate both seawards and shorewards but then decay rapidly as they travel over the continental shelf. Observations show that internal tides are intermittent and that their phase can vary relative to the surface tide.

2.6.3 MEASURED CURRENTS OF INTERNAL TIDES

Some of the few measurements of near-bottom currents associated with internal tides on the continental shelf were made in a Norwegian fjord (Fjeldstad, 1964), in the central North Sea (Schott, 1971), off the coast of Nova Scotia (Petrie, 1975) and off the coast of Oregon (Hayes and Halpern, 1976). All of these recorded the presence of weak currents near to the sea floor which may be significant for sediment transport when they are added to other currents, as in tidal lee waves (Cartwright, 1959, discussed in Section 3.3.3). They may be important in localized and exceptional circumstances – for instance in breaking of the internal tide (Defant, 1961, Section 16.5), which can occur in shallow water on narrow continental shelves if the pycnocline intersects the sea floor, or if the wavelength and bathymetry are right for resonance.

2.7 Tides past

In the study of marine sedimentary deposits a knowledge of past, as well as present, tidal currents is essential. Changes to the ocean tides can arise either from changes to tide generating forces or from changes in the ocean's response to these forces.

2.7.1 EFFECTS OF TIDAL FRICTION

Referring to Section 2.2 the parameters involved in the tide generating forces are the masses of the Earth, Sun and Moon, the Earth's radius, the distances from the Earth to the Sun and from the Earth to the Moon and the Earth's gravitational acceleration, g. The evidence suggests that over the last 3000 million years only the distance from the Earth to the Moon is likely to have changed appreciably. The reason for this change is due to the tides themselves since, because of friction, the tides dissipate energy (at present at about 4×10^{12} watts) causing the Earth's rate of rotation to decrease (Cartwright, 1978). The total angular momentum of the Earth–Moon system is constant and one way to compensate for this decrease is for the Moon to recede from the Earth. Calculations based on eclipse data recorded over the past 3000 years support this argument with constant rates for slowing down of the Earth's rotation and for the Moon's recession (Muller and Stephenson, 1975). On a geological time scale it is supported by growth lines in fossils which show that there has been a decrease in the number of days per year and per month (Scrutton, 1978; Kahn and Pompea, 1978), although the estimated rates differ.

An extrapolation based on the astronomical data suggests that about 1000 million years ago the Moon would have been catastrophically close to the Earth (Gerstenkorn, 1955; MacDonald, 1966). However, no supporting geological evidence for this event has yet been obtained and it now seems that the Earth–Moon system has existed for 3800 million years with an appreciable separation for at least 2500 million years. The effects of tidal action can be seen in sediments as old as 3000 million years and although the evidence is fragmentary it indicates that tidal currents were stronger and tidal amplitudes possibly slightly higher in the distant past (Piper, 1978).

2.7.2 EFFECTS OF CHANGES OF BATHYMETRY

Probably of greater importance than the distance between the Earth and Moon will be changes in sea floor bathymetry, both local and global. For instance, the opening of Dover Strait less than 10 000 years ago will have had a profound impact on tidal currents in the Southern North Sea. Changes in bathymetry due to sediment deposition can affect tidal regimes especially if a degree of resonance is involved (Johnson and Belderson, 1969). The response of the ocean to the tide raising forces is dependent on the configuration of the continents and the bathymetry of the ocean floors which at present lead to an enhanced twice-daily response. The configuration of the continents is only known with any accuracy up to 300 million years ago and although the width of the continental shelf can be estimated over this time span the rest of the bathymetry is unknown. Sündermann and Brosche (1978) ran a numerical simulation for the twice-daily tide for 250 million years ago (Pangaea Ocean, Upper Permian). The model showed that the ocean tide was not significantly changed in general magnitude from today's. Unfortunately the model did not represent adequately the tides on the continental shelf and the authors state that it is now being refined to do so. From this chapter it is clear that tides, past and present, on the continental shelf are controlled by the local bathymetry, leading to tidal resonance if the dimensions are right, amplification from shoaling and to energy losses from bottom friction over the wide shelves that are thought to have existed at some stages in the geological past.

2.8 Main conclusions

1. Daily and twice-daily tides are generated by the relative gravitational attraction of the Moon and, to a lesser extent, of the Sun. The combination of their forces leads to spring–neap cycles, with periods of 13.66 and 14.77 days, which have in turn longer period modulations, caused by variations in the Earth's and Moon's orbits.

2. The response of the oceans to the tide generating forces is governed by the shape and depth of the oceans and continental shelf seas. Their present configuration leads to enhanced twice-daily tides.

3. The tides propagate around the world's oceans and into the continental shelf seas. They are amplified by reductions in depth and width, and by approaches to resonance, where the natural period of oscillation of a sea coincides with a tidal period.

4. For wider continental shelves than at present the current amplitudes will be maximum at distances of about 1/4, 3/4, 5/4, etc. tidal wavelengths from the coast.

5. Most of the tide's energy is dissipated in continental shelf seas working against bottom friction, and this leads to a stress at the sea floor which in places can move sediment.

6. Sand transport rate increases much more rapidly in proportion to the current speed. Hence asymmetry in peak tidal flows will cause net sand or gravel transport. Such asymmetries can occur when the tide is distorted by changes in bathymetry, for instance shallowing water depth and near to headlands and sand banks.

Chapter 3

Bedforms

3.1 Introduction

Plan views of the sea floor pictured by means of side-scan sonar (sonographs), together with the more generally available echo-sounder profiles as well as data from divers, photographs and television, opened the way to recognition, description and mapping over wide areas of bedforms made by tidal currents. Many of the views of sandy sea floors presented as sonographs invite overall comparison with the more familiar aerial photographs of sand-strewn deserts, although there are, of course, notable differences in the sizes and shapes of sand bedforms from the two environments (Section 3.6). Tidal current bedforms of sand and gravel are almost always either 'longitudinal' or 'transverse', by which is meant within 20° or so of parallel or normal to local peak tidal current direction. This direction of peak tidal flow is almost constant at any one site and varies little between sea surface and sea bed (Section 2.5). In contrast, in many desert regions winds strong enough to move sand have a wide spread of directions. Hence it is not surprising that many aeolian bedforms have patterns that are more complex than those occurring on tidal current swept sea floors.

The identification of some tidal current bedform types, whether longitudinal or transverse, is not always easily done. They may be readily identifiable when reasonably well developed, but are frequently poorly formed or sparsely distributed. A selection of sonographs indicating much of the variety of tidal current bedforms has been given by Belderson, Kenyon, Stride and Stubbs (1972). Further examples are given in Plates 3.1–3.20. For the sake of convenience, erosional features are here included under the definition of 'bedforms'.

The present chapter is primarily concerned with the nature of the main bedforms that are made by tidal currents on a continental shelf (Figs 3.1 and 3.2), with some mention of their probable modes of origin. It serves to set the scene for the use of these bedforms as indicators of sediment erosion, transport and deposition in later chapters.

Flume and river bedforms in relation to marine bedforms

The smaller subaqueous bedforms are most readily studied in laboratory flumes, in which detailed observations can be made under controlled conditions, whereas the largest ones can only be studied by acoustic and other methods in the much deeper waters of the sea. Intermediate sized features can sometimes be examined during low water river flow or during low tide in estuaries, but such bedforms

28   Offshore Tidal Sands

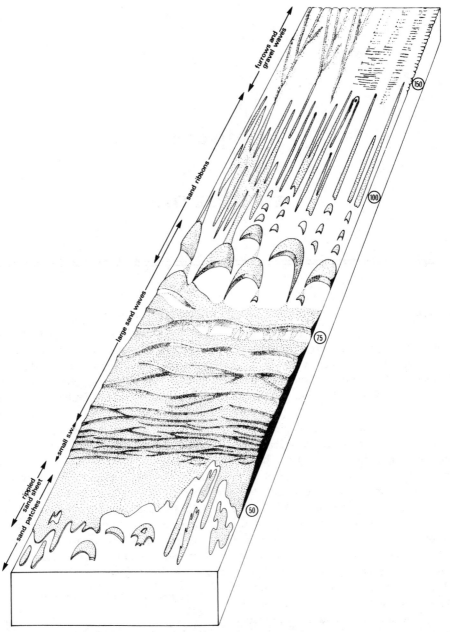

Fig. 3.1 Block diagram of the main lower flow regime bedforms made by tidal currents on the continental shelf, with the corresponding mean spring peak near-surface tidal currents in cm/s.

have been subject to some conditions that do not occur beneath the sea. The descriptive system here adopted for the offshore bedforms of the continental shelf is a combination of these types of observation, aimed at giving as much insight as possible into the significance of the marine bedforms. Allowance is made for differences between the flume, river and tidal current environments. In particular, the periodically reversing nature of the tidal flow, combined with irregular non-tidal currents and

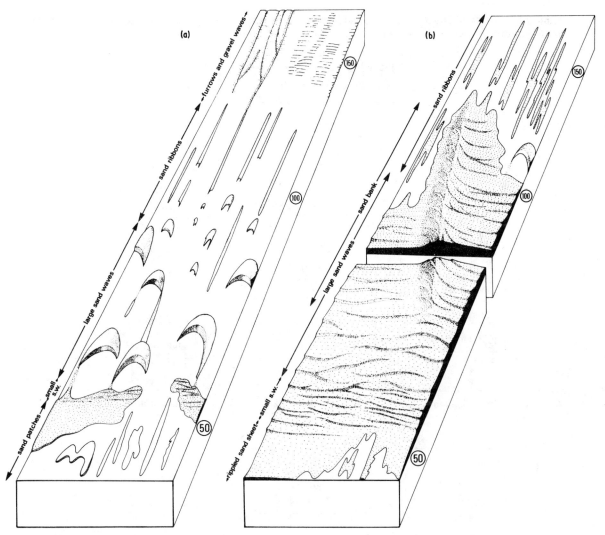

Fig. 3.2 Block diagrams of bedforms made by tidal currents on the continental shelf with low supply (a) or abundant supply (b) of sand, respectively.

waves, can produce variations from some purely unidirectional flow bedforms. Also the depth of flow on most of the continental shelf is much greater than in rivers. Furthermore, the depth of flow at any one spot varies in time by a much smaller percentage than in most rivers, except for intertidal and shallow subtidal areas. Finally, sand availability in flumes and rivers is commonly sufficient to cover the whole bed so as not to restrict transport rates. In Chapters 3 to 5 this will be assumed for the sea floor also, except where a statement on the context shows otherwise.

3.2 Relevant flume bedforms

Quantitative flume data for the bedforms made by a small steady unidirectional current are far more complete than data for the larger scale bedforms in the field. They therefore give useful guidance provided effects which can be shown to be peculiar to the small depth are not extrapolated up, and account is taken of possible modifications due to the time variations of larger-scale currents. Use is made of bedform state well away from the walls, entrance or exit of the flume.

Table 3.1 The classification of transverse and plane bed states in flumes using data for quartz density sand at about 17°C for median diameter less than and greater than 0.7 mm

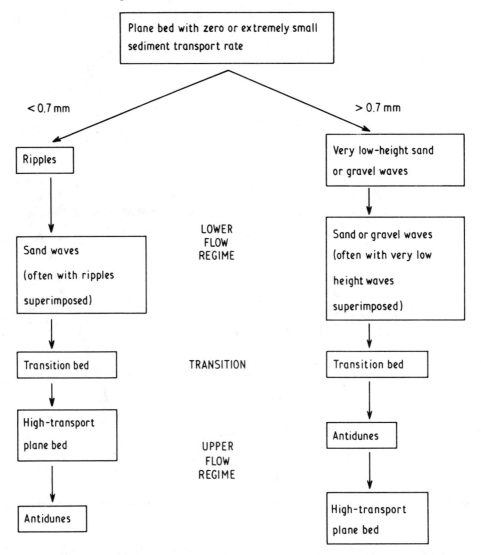

Bedforms of unidirectional steady flow observed in laboratory flumes, after equilibrium has been reached, have been given a number of classifications and nomenclatures. However, subsequent observational and theoretical studies do not suggest that significant changes are needed in the basic classification most commonly adopted in the literature (ASCE, 1966) which follows Simons and Richardson (1963). Also, its division of transverse bedforms into lower and upper regimes, separated by a transition, is useful and generally accepted (Table 3.1). The following discussion of flume results tries in addition to emphasize some little recognized, though relevant, matters.

3.2.1 LOWER FLOW REGIME FLUME BEDFORMS (SAND RIPPLES AND SAND WAVES)

The lower flow regime bedforms (Table 3.1) consist primarily of sand ripples and sand waves, the latter usually being called dunes by workers in hydraulics. Sand ripples are generally acknowledged to be genetically different from sand waves, in that their dimensions are not only smaller, but also approximately independent of water depth (for known bottom stress), provided that this exceeds several ripple heights (10 cm or so).

The ASCE classification specifies a height of 5 cm as separating sand ripples and sand waves. However, among both sand ripples and sand waves, wavelength varies more than height, and is therefore a potentially more useful basis for classification. Wavelengths of sand ripples do not exceed about 60 cm, while the wavelengths of very low-height sand waves (discussed shortly) are generally several times larger than this, and the wavelengths of sand waves with height greater than 5 cm are many times 50 cm (e.g. see data and descriptions of Guy, Simons and Richardson, 1966). These values refer to sand of roughly quartz density and laboratory temperatures, usually 10°–20°C. For much lower density material larger heights and wavelength would correspond.

Flume studies using different sizes of sand have found that there is some maximum median diameter above which current ripples do not form. For quartz-density sand at about 17°C this size was stated as about 0.7 mm by Southard and Boguchwal (1973). It is expected to be rather larger for lower sediment density or lower water temperature. For median diameters above about 0.6 mm they found a zone of lower-regime, flat-bed, between the zones of sand ripples and sand waves. In the ripple and flat-bed zones sand transport rates were much smaller than for sand ripples in smaller sand sizes, and the ripples in these coarse sand sizes formed and decayed much less quickly. Southard and Boguchwal's coarse sands were artificially well sorted (the main ones used lay within about 0.40–0.52 mm, 0.53–0.59 mm, 0.62–0.69 mm). In much more poorly sorted sands ripples can form in their finer constituents, whereas the coarser grains appear mainly in the ripple troughs. Southard and Boguchwal's sands were subrounded to rounded, but in very angular material (e.g. many shell sands) maximum median diameter for ripples will differ again, also depending on how 'grain size' is defined for such material.

Above the maximum grain size for ripples an initial plane bed with appreciable sand transport rate, such as would form ripples for smaller grain sizes, is followed, as current speed and bottom stress increase, by what may be called 'very low-height sand waves'. These have heights in the defined sand ripple range of less than 5 cm, but larger wavelengths than sand ripples. They are then followed by sand waves as defined by the ASCE classification, with heights exceeding 5 cm (and usually longer wavelengths again). 'Very low-height sand waves' can be recognized in the descriptions of bed states and tabulated bedform heights and wavelengths in individual flume 'runs' for very coarse sands given by Guy et al. (1966) and Williams (1967).

The conditions governing the change from sand ripples to sand waves have rarely been studied in flume work, as most flume studies make use of steady-flow conditions at given flow speeds. However, it is known that as current speed increases the sand ripples are smoothed out and longer-wavelength bedforms appear 'over a very narrow range of only a few cm/s' of current speed (Middleton and Southard, 1977). Initially these bed features cannot have reached the specified ASCE height to qualify as sand waves (>5 cm). Thus, they are called very low-height sand waves, as confirmed by Guy, Simons and Richardson's (1966) data for fine and medium sands. Very low-height sand waves seem to be included in Pratt's (1970) 'Phase 3' bedforms, which Costello (1974) includes with his 'bars'. However, both 'Phase 3' bedforms and 'bars' do not specify bedform heights and seem to include bedform (sand wave) heights exceeding 5 cm (e.g. data of Guy et al., 1966), which occur for slightly higher current speeds. Very low-height sand waves, and sand waves in coarser grain sizes are observed to be more nearly two-dimensional than sand waves in finer sizes.

Theory of dimensions (Yalin, 1972) predicts that data for unidirectional steady flows with an equilibrium bed state of sand ripples and such data for sand waves (including very low-height ones) should be separated by a 'boundary' on a 'Shields diagram' (as commonly used for plotting movement threshold data in terms of dimensionless bottom stress and grain-shear Reynolds number). In fact, such a plot of the flume data available to us

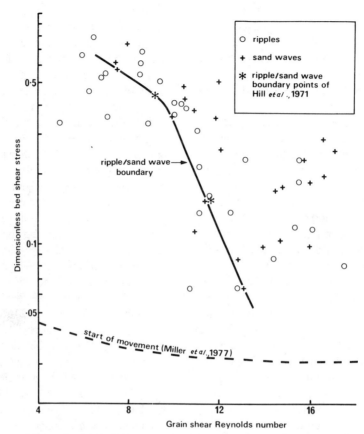

Fig. 3.3 Plot in terms of dimensionless bed shear stress and grain-shear Reynolds number, of flume data sets (Annambhotla, 1969; Barton and Lin, 1955; Chabert and Chauvin, 1963; Crickmore, 1967; Franco, 1968; Guy, Simons and Richardson, 1966; Hill, Robinson and Srinivasan, 1971; Pratt, 1970; Stein, 1965; Taylor, 1971; Vanoni and Brooks, 1957; Vanoni and Hwang, 1967) showing sand-ripples, or sand-waves with or without superimposed sand-ripples, together with two approximate data sets for observed ripple/sand-wave instability (Hill et al., 1971). The linear grain shear Reynolds number scale emphasizes the boundary area between the two bedforms. Many points well removed from it have been omitted. The boundary shown joins the sand-wave points with weakest flows, with one exception shown to be anomalous (after Johnson et al., 1981).

(Fig. 3.3) for which all the required values of the variables and bedform information were given and for which depth >12 cm shows considerable overlap of points for sand ripples and for sand waves (including very low-height sand waves). Where the bed state was given as sand ripples on sand waves it was counted as sand waves. Such sand ripples on the upstream, gentle slope of sand waves form there for higher values of total bottom stress and grain-shear Reynolds number than they do on a flat bed. This is presumably because the smaller local bottom stress, which is total bottom stress minus the 'form drag' or flow resistance of the sand waves, is within the range which would give sand ripples on a flat bed. However, in descriptions of some flume runs the sand ripples seem to have been more conspicuous than sand waves or very low-height sand waves under them. The bedform may have been thus identified as 'sand ripples' instead of 'sand waves' in such cases, so that on the above-mentioned data plot some 'ripple' points may really represent sand waves. Therefore the curve drawn in Fig. 3.3 is a 'lower sand wave boundary' drawn through or below all points (except one which is shown to be anomalous)

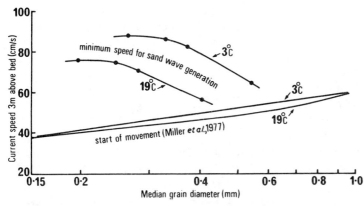

Fig. 3.4 The minimum current speed at 3 m above the bed required for generation of small sand waves from an initially rippled sand bed for two water temperatures (and for the start of sand movement on a flat bed). Salinity is 35 parts per thousand, immersed sand density is 1.62. (The curves are derived from Johnson et al., 1981.)

which correspond to reported sand waves. Thus the portion of Fig. 3.3 below this curve fairly certainly corresponds to beds with sand ripples only. The curve of Vanoni (1975, Fig. 16) representing the sand ripple/sand wave boundary aims to go midway through the overlap zone. His curve can be shown to correspond to a rather stronger current speed on it by about 10–20% for given grain size and water depth. Some of the width of the overlap zone must also result from differences between different flume sands that are not expressed through their median diameters, such as sorting and shape differences.

This 'lower sand wave boundary' corresponds to the curves in Fig. 3.4 for temperatures 3° and 19° (near-minimum winter and near-maximum summer for the Southern Bight of the North Sea) and a salinity of 35 parts per thousand, in terms of current speed at 3 m above the bed. This assumes a sand density of 2.65 (immersed density 1.62), and a near-bed logarithmic velocity profile (Section 2.5), with $z_0 = 0.2$ cm, which was found to be typical of a large number of flume runs with sand ripples or small sand waves. Application of these curves to the Southern North Sea is discussed in Section 4.6.

There do not seem to have been flume studies aimed specifically at finding a minimum grain size for the existence of sand waves, as there have been to determine the maximum grain size for current ripple formation. Sand waves do not seem to have been reported in quartz sand of less than 0.17 mm median grain size at laboratory temperatures. Any cross-stratification attributable to sand waves found in well sorted sandstones with median grain size less than say about 0.15 mm would suggest tropical water temperatures at the time of deposition.

3.2.2 TRANSITION BED CONDITIONS

'Transition' state bedforms may be regarded as largely smoothed out sand waves, which have much reduced height and often longer wavelength. This has been shown both by flume data and theoretical predictions (Engelund and Fredsøe, 1974). Transition state bedforms often change rapidly and are apparently unstable, with adjacent areas of almost plane bed sometimes simultaneously present. The 'sand waves' observed in flumes by Barton and Lin (1955) and later workers, in a shallow river (Culbertson and Scott, 1970) and in the Brahmaputra River (Coleman, 1969) may be classed with transition bedforms. They are bodies of sand with long, almost flat tops, and very long wavelengths that move downcurrent much more rapidly than usual sand waves. Supporting evidence for their transition bed status is provided by the high local Froude numbers on top of the 'sand waves'. These are over about 0.5 in each case. These criteria make these 'transition bed flat-topped sand waves' seem quite distinct from the lower-regime sand waves, and also quite irrelevant to continental shelf sand patches (Sections 3.3.4 and 3.4.4) whose shape is similar, because Froude numbers on top of continental shelf sand waves and sand patches are much smaller (<0.1). The term 'sand waves' will thus here refer to the usual lower flow regime features.

3.2.3 UPPER FLOW REGIME FLUME BED STATES

Upper flow regime bed states for fine and medium sand are theoretically possible only in very shallow water (1 m or less deep) on the continental shelf (Sections 3.3.1 and 3.3.3). For fine and medium sand the upper regime bed states in flumes start from an almost plane bed with strong sediment movement. With an increase of current speed this bed state gives way to antidunes. These are symmetrical sharp-crested sand bodies which form and disappear relatively rapidly, and remain stationary or move up or down-current very much more slowly than the current speed. They are linked to waves on the water surface approximately in phase with them. The sediment transport continues to be down-current, and the antidunes are merely waveforms in the bed with the moving grains passing through or over them. For coarse sand, an increase in current speed is observed to give antidunes before, rather than after, the upper regime plane bed state.

3.2.4 PAUCITY OF LONGITUDINAL BEDFORMS IN FLUMES

The flume bedforms described above are all approximately transverse (averaged over crest sinuosity) to the flow, and are independent of the flume walls. Most of the longitudinal flume bedforms seem to be fins or ridges extending down-current from sand waves or thin streaks on a flat floor composed of different fractions of the sediment, where it is poorly sorted. These streaks may be equivalent to parting-lineations found in sedimentary rocks. Many accounts of flume studies do not mention the ridges or streaks. This is perhaps because the 'streaks' are regarded as just part of the upper flow regime flat bed condition or because the sand was too well sorted for them to be clearly visible. However, a few flume studies have observed more obvious longitudinal features (Casey, 1935; Vanoni, 1944) which we would call sand ribbons. Casey observed them for two bed sediments. The first was poorly sorted, with 37% less than 0.25 mm and 41% between 1 mm and 2 mm, and the second had 84% in the coarse sand grade. Vanoni used a bed of fixed grains of 0.88 mm diameter with a suspended load in the flow of 0.16 mm median diameter (97% in the fine sand grade). Thus, flume formed sand ribbons seem to require poorly sorted moving sediment, or a different substrate.

Alternate bars are a boundary-tied bedform which sometimes form adjacent to vertical flume walls or steep river banks. They might just possibly have an equivalent in flow between long and parallel marine sand banks, although the slopes of these are far less steep.

3.2.5 NOTE ON THEORY OF TRANSVERSE BEDFORMS IN FLUMES

Despite their familiarity in flumes, rivers and the sea, there is not as yet a generally accepted quantitative theory of origin for sand ripples and sand waves. However, their morphology and general behaviour are reasonably well documented and the conditions for their occurrence seem fairly well established empirically. Thus, if flow conditions and median grain size lie in the respective ranges already discussed, then sand ripples or sand waves will be built (for sand waves, provided there is sufficient time and sand availability; sand ripples, of course, require little time or sand volume). Therefore, details of the existing theoretical work will not be discussed. Lag effects are an important aspect mentioned in Section 4.2.3. Dimensional theory indicates (Yalin, 1972) that molecular viscosity of the water is in some way important for sand ripple formation (probably through viscous forces in the steep velocity gradients adjacent to the bed and the sand grains) but not for sand waves (which depend in some way on turbulence in the whole depth of flow). This was used in Section 3.2.1.

For antidunes the theory gives very good agreement (Engelund, 1970; Engelund and Fredsøe, 1971) with the empirical curve of minimum Froude number versus the ratio of depth to wavelength (for two-dimensional antidunes). Reported depth/wavelength ratios for antidunes in rivers and flumes seem to be 0.2 or less, and correspond to minimum Froude numbers for antidunes of about 0.8 or more. Even higher values of minimum Froude number would apply for theoretical representations of three-dimensional antidunes (from the theory of Engelund and Fredsøe, 1971).

3.3 Transverse bedforms of the continental shelf

The prime distinction among transverse bedforms that is made in this book is between sand ripples

(Section 3.3.2), sand waves (Section 3.3.3) and transverse sand patches (Section 3.3.4). The distinction between sand ripples and sand waves is made because of their basic differences in size, water depth dependence and genesis. Sand patches differ again in that the floor between them is made mainly of gravel. Subdivisions of sand waves, as described in Section 3.3.3, have been made in the literature on the basis of morphology and size, especially in readily observable intertidal areas. These do not appear to reflect process differences that are comparable to those between sand ripples and sand waves.

As discussed in Chapter 4, tidal current speeds will be expressed in terms of mean spring 'peak' values, that occur during the approximate 12.5 hour tidal cycle for the predominantly twice-daily tides in the areas from which the data were obtained.

3.3.1 UNLIKELIHOOD OF ANTIDUNES OCCURRING ON THE CONTINENTAL SHELF

Froude numbers over 0.8 needed for antidune formation (Section 3.2.5) could only occur in the sea in very shallow water where the currents were also strong (e.g. Froude number is 0.8 for a speed of 250 cm/s and a depth of 1 m). Such values might occasionally occur on the tops of tidal current sand banks or in intertidal areas. Any antidunes that do form on the continental shelf are expected to be very transitory. This is because their wavelength changes with that of the sea surface standing waves to which they are linked, which in turn change with the rapid changes of current speed and water depth which occur in such situations. Antidunes have very small building or destruction times owing to their fairly small volume and the high sand transport rates associated with them. It is difficult to imagine their preservation unless by extremely rapid deposition on top of them.

Symmetrical transverse bedforms occurring in greater water depths and with weaker currents have occasionally been described as antidunes. These can quite reasonably be interpreted as wave-formed sand ripples, or when larger, as symmetrical sand waves in the case of tidal currents with equal ebb and flood peak speeds.

3.3.2 SAND RIPPLES

Sea floor sand ripples are not in general detectable by acoustic means, and observations of them are made mainly by divers or by underwater photography or television. However, since they only depend on near-bed flow their presence or absence should be predictable from near-bed current speed and grain size, using their existence conditions already discussed (Section 3.2.1, Fig. 3.3). Their dependence only on near-bed flow also means that their dimensions should have the same ranges as in flumes. The available quantitative observations support this. Thus underwater sand ripples, whether on the continental shelf or in flumes or rivers, may be distinguished from sand waves by having (a) a wavelength (averaged over, say, 10 successive ripples) usually less than about 1600 times and usually greater than 600 times the median diameter of the sand in which they have been constructed (see Yalin, 1972, Fig. 7.27), (b) a height not greater than about 300 times median diameter, and (c) both wavelength and height statistically independent of water depth (provided this exceeds 10 cm or so). Sand ripple heights are, in general, between 1 and 5 cm. Wavelengths are less than about 60 cm and usually between 5 and 12 times sand ripple height (Fig. 7.30 of Yalin, 1972). They can only occur for median grain sizes less than the values discussed in Section 3.2.1.

(a) *Current ripple shape*

Asymmetrical current ripples in sand have avalanche lee slopes (slip faces) over their full heights facing in their direction of travel and tend to be straight crested for a flow speed just above the threshold of grain movement (Guy *et al.*, 1966). For higher current speeds in the sand ripple existence range they generally become markedly three-dimensional. Average crest orientation remains transverse to flow, but otherwise sand ripple shape is very variable both in time at a fixed position, as well as between different fields of sand ripples (as comprehensively illustrated by Allen, 1968). This shape variation presumably indicates a variety of detailed modes of instability. Thus, the detailed shape of a given sand ripple in plan view may be determined mainly by the flow within a few ripple wavelengths horizontally or vertically of it, and perhaps also by grain size, so that the shape may be only of local significance.

The effective bed roughness height presented by sand ripples to the flow is also fairly constant in flumes. However, bed roughness height and ripple shape vary over tidal cycles (Dyer, 1980).

(b) *Effects of sea surface waves*

The bedform terms current-ripple and wave-ripple are used respectively for the purely current-formed asymmetrical bedform and the symmetrical sharp-crested bedform produced by the oscillatory water movements associated with sea waves generated by strong winds. In addition, there is the transitional (somewhat asymmetrical) current-wave ripple. This has rounded crests and its degree of asymmetry increases with the ratio of current speed to wave oscillatory water speed, as shown by flume studies (Harms, 1969; Hammond and Collins, 1979). In contrast to current ripples the height and wavelength of wave ripples in sand tend to increase rapidly with the grain size. The wave ripples made of very coarse sand and gravel may be as much as 25 cm high and have wavelengths of 1.25 m (e.g. Flemming and Stride, 1967 for gravel of 2.6 mm median diameter). These large gravel ripples in coarse material are frequently observed on high resolution side-scan sonar records (Plate 3.1). For a given grain size the wavelength first increases and then levels out, and can then decrease, with increasing water particle oscillation amplitude (e.g. flume data analysis by Mogridge and Kamphuis, 1972; and continental shelf observations, Inman, 1957). Sand ripples due to sea waves alone can be closely two-dimensional, with crests perpendicular to the wave travel direction.

Sea waves by themselves apparently do not produce bedforms larger than these sand and gravel ripples just described. However, by increasing sand transport rates, they could assist a current (which had higher near-bed speed than the wave oscillatory movements) in producing bedforms of types which could be formed by currents alone.

(c) *Occurrence of sand ripples*

Sand ripples due to sea waves, currents, or both together, are predicted to occur almost everywhere on the continental shelf around the British Isles where the bed is made of sand, except in sheltered hollows. Furthermore, most sandy floors of other continental shelves swept by tidal currents are expected to have sand ripples on them for much of the time. These will often be superimposed on other bedforms, but would be smoothed out by storm waves or currents which had sufficiently high near-bed speeds.

3.3.3 SAND WAVES

Sand waves have much larger wavelength than sand ripples. They also have a larger wavelength to height ratio, established by Yalin (1972, Fig. 7.29) from flume data to be generally greater than 15. Nearly all reported river and continental shelf sand wave trains satisfy this as well, the ratio sometimes reaching as much as 100.

Sand waves are found over wide areas of the continental shelf around the British Isles (Fig. 4.5). Large sand waves are associated empirically with near-surface mean spring peak tidal current speeds of more than about 65 cm/s on average (corresponding to about 50 cm/s at 3 m above the bed in 30 m water depth). Small sand waves require a slightly lower speed. Their relationship with other bedforms is shown in Fig. 3.1. Before describing their morphology, however, it is necessary to discuss some problems of nomenclature and origin.

(a) *Nomenclature*

The usage of terms such as giant ripple, megaripple, metaripple, pararipple, accretion ripple, transverse sand-ridge, bed undulation, dune and sand wave has become a morass of individual preference. However, only three of these terms are commonly used (Table 3.2). The term dune is used preferentially by hydraulicists for the sand waves in flumes and rivers, although a few flume workers in the past extended the term to include sand ripples. The term dune is also widely used for aeolian bedforms of a wide range of sizes, including some that are longitudinal bedforms. In contrast the terms sand wave and megaripple are the main ones used by marine geologists

Table 3.2 A few examples to show some of the terms used for transverse bedforms of sand made in the lower flow regime.

Present book	Sand ripples	Small sand waves	Large sand waves
Allen, 1980a	Sand ripples		Dunes if flow unidirectional Sand waves if flow tidal
Hydraulics workers	Sand ripples		Dunes
Middleton and Southard, 1977	Sand ripples	Dunes	Sand waves
Yalin, 1972; Cornish, 1914		Sand waves (including antidunes)	
Coastal Res. Gp., Massachusetts University, 1969; Boothroyd and Hubbard, 1975; Dalrymple, Knight and Lambiase, 1978	Sand ripples	Megaripples	Sand waves

(the term megaripple being especially used by workers studying tidal flats and shallow subtidal sand waves). Unfortunately, though perhaps logically, Cornish (1914) and Yalin (1972, p. 204) used the term sand wave as the synonym of all wave-like bodies of granular material, including sand ripples, 'bars' in rivers, and upper-flow regime bedforms.

The present authors consider the term megaripple to be unsatisfactory as it implicitly ignores the generally acknowledged genetic difference between sand ripples and sand waves. Megaripples are, in fact, small sand waves, formed in flows too strong for near-bed viscosity effects to be significant, and so should not be considered simply as large sand ripples. Thus, the threefold classification into ripples, megaripples and sand waves, with increasing size that is used by some workers (e.g. Coastal Research Group, University of Massachusetts, 1969; Boothroyd and Hubbard, 1975, and many subsequent workers such as Dalrymple *et al.*, 1978) should be discarded. Confusion also arises when the term dune is used for small sand waves, but not for the large ones (e.g. Middleton and Southard, 1977).

The present authors propose to use the term sand wave to refer to the subaqueous, lower flow regime, transverse bedforms of sand that have larger wavelengths than sand ripples. Where two or more sizes of sand wave occur together it is convenient to refer to them as small sand waves and large sand waves, without implying any genetic difference. This is because there seems to be a complete gradation in size, plan view and lee slope angle between sand waves at varying locations. They are all attributable to the turbulence in the flow. The term sand wave had appeared by 1882, when Hider described such features from the bed of the Mississippi River. This could be considered to give precedence to the term. Of course, similar forms in gravel will be called gravel waves.

Some workers (e.g. Middleton and Southard, 1977) have made a distinction in shape as well as size between what have been called megaripples or dunes and sand waves. We suggest that the frequently more complex forms of intertidal and shallow subtidal sand waves, where there is a relatively great water depth variation, are not a reliable guide to sand wave morphology in water greater than about 10 m depth. Furthermore, we would disagree with the proposition that small and large sand waves are different in major formation processes; instead, their morphological differences may be associated with their great building and lag time differences.

(b) *Tidal current versus unidirectional current sand waves*

Allen (1979, 1980a and b) classed small sand waves in reversing tidal currents with small sand waves in unidirectional currents on the basis that their asymmetry can be reversed between each ebb and flood current. However, he subdivided 'giant' sand waves into unidirectional-flow sand waves, called 'dunes', and tidal-current sand waves. Allen considered the tidal currents to give an oscillatory boundary layer analogous to that under surface water waves, even though he acknowledged that 'there is no theoretical or experimental evidence one way or the other' for carrying over the properties of the wave boundary layer to the tidal current one.

It seems to us that large sand waves formed by tidal currents can be considered simply as unidirectional current formed ones which have been modified, to a lesser or greater extent, by the reversing current. The two need not be differentiated for the following reasons: (a) Sand transport rate, especially of bedload, lags much less than a tidal period behind current speed (Section 4.2.3). If a fairly constant lag time is assumed the development in time of any size of sand wave can, in principle, be obtained by integrating sand transport rate as a function of current speed, whatever the form of time variation of that speed. Thus treatment as a boundary layer may not be necessary. (b) However, if one does consider boundary layers, the tidal shelf one is less purely oscillatory than an offshore wave boundary layer. This is because of its stronger harmonic constituents (not mentioned by Allen), especially M_4 (Section 2.4). Also, frequently stronger non-tidal currents can be superimposed on any tidal residual (mass transport) current (prediction of whose magnitude is in any case uncertain). In some areas with comparable non-tidal and tidal currents the latter vary with spring–neap cycles, etc., and the total current can be sometimes unidirectional and sometimes reversing. Thus, all gradations from unidirectional to exactly oscillatory currents are possible, as Allen (1980a) in fact recognizes. Furthermore, non-tidal unidirectional currents can themselves reverse at random, or long regular (e.g. seasonal) intervals. For example, (1) in the entrance to the Baltic Sea there is normally a bottom inflow, with occasional outflow observed to reverse sand wave asymmetry (Kuijpers, 1980); (2) the boundary between a strong oceanic current and associated nearer-coast counter-current can migrate laterally (e.g. the Agulhas Current and Gulf Stream); and (3) predominantly tidal currents can have their peak direction reversed by strong seasonal non-tidal currents (e.g. in the Torres Strait, Section 4.5.3). (4) the period of a tidal 'wave' is several thousand times larger than those of surface waves, and its boundary layer is much thicker than the wave boundary layer, and fully turbulent. In the wave boundary layer the accelerations and retardations of water movement speed are relatively important. In fact suspension load can be greatest when the flow is just reversing (Kennedy and Locher, 1972). In contrast, in tidal oscillations the bottom stress is nearly in phase with current speed (Yalin and Russell, 1966) and the acceleration forces have little effect on sand transport, corresponding to the small time lag already mentioned.

(c) *Relationships of sand wave height, wavelength and water depth*

Wavelengths and heights of sand waves correlate very poorly with water depth in mass plots. Reasonable correlations may be obtained if (a) mean values of height and wavelength over ten or so sand waves in a train are used, excluding near-end or bifurcation portions, so as to reduce the effects of the standard deviations of up to 40% of individual heights and wavelengths; (b) only the larger sand waves in a group are considered in situations where two orders of size of sand wave are present. Then the smaller sand waves seem to be a function of local flow conditions such as on up-current slopes of larger sand waves, rather than involving the whole water depth; (c) the transition bed state (Section 3.2.2) is excluded; (d) sufficient time has elapsed since the present water depth and current regime became established; (e) sufficient sand is available for the sand waves to build up to equilibrium size – this may take many years for large sand waves, especially if current speeds well above the minimum for sand waves are infrequent; and (f) water depth exceeds 10 m say at low tide to exclude shallow subtidal and intertidal sand waves. If (a) to (f) are satisfied, then the wavelength of sand waves is predicted from Yalin's (1972, Fig. 7.28) analysis of flume and some river data to average about 6 times the water depth (with scatter by a factor of up to 2 or 3 either way) for a water depth greater than about 30 000 times grain diameter (i.e. 6 m water depth for 0.2 mm sand). This is consistent with observed mean wavelength to depth ratios, usually from 2 to 18 for the continental shelf. This factor of about 6 is also about what might be expected from the size of the larger turbulent 'eddies' (Yalin, 1972). It can therefore be argued, as by Yalin and Price (1976), that the oscillatory nature of tidal currents has little effect on sand wave heights and wavelengths. The maximum height/depth ratio is observed to be about 1/3. In deeper shelf waters this ratio will be much smaller.

At locations on the continental shelf where assumptions (a) to (f) are satisfied, current speed

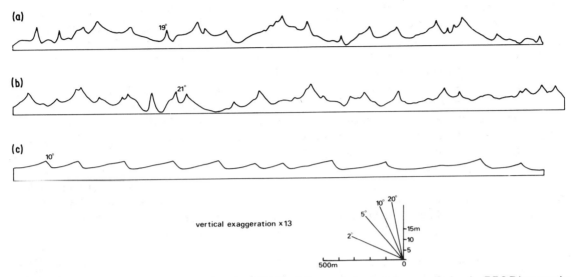

Fig. 3.5 Cross sections of sand waves to the west of Holland from an echo-sounder record taken by RRS Discovery in 1972. Profiles (a) and (b) are from the Southern Bight Bed-Load Parting, showing some sand waves with symmetrical cross-section. In profile (a) the sand waves are located on top of an undulating floor. Section (c) shows sand waves with asymmetrical profile with long flat gentle slopes and tangential bases to the steep slopes.

seems to have little effect on sand wave height and wavelength. However, near the lower end of the speed range for which large sand waves are the equilibrium bedform (about 60 cm/s mean spring peak near-surface tidal current speed) the height/depth and wavelength/depth ratios are expected to be smaller. If assumption (c) is removed, as the transition bed state is approached, the height/depth ratio decreases and the wavelength/depth ratio increases.

Another qualification needs to be considered, although it does not clearly affect mass plots. Theory predicts that, if the water depth does not vary, the height of two-dimensional sand waves decreases as the relative importance of suspended-load to bed-load sand transport rate increases. This may explain why the large sand waves of the Southern Bight of the North Sea show marked progressive decrease in height along the sand transport-deposition path (Fig. 5.6) without any obvious correlation with water depth. The median grain size certainly decreases northwards along this path and it was argued that the relative importance of suspension transport, for maximum water movements, increases northwards (McCave, 1971b). However, it seems more likely that the northerly decrease in the height of the large sand waves is due to the progressive northward decrease in the opportunity for building them (Section 4.6).

(d) *Sand wave slope angles in the sea*

Remarkably little attention has been paid to the steepness of sand wave slopes in marine studies. A compilation has now been made of the 'lee' slope angles given in 25 references or measured in many echo-sounder profiles available to the authors from tidal and unidirectional current-dominated seas. Subsidiary near-crest slip faces, often reversing with the ebb and flow of the tidal current, are excluded from mention. Where the remainder of the lee slope is not of uniform angle, the angle of its steepest third or so is used. This means that the angle is not reduced by including a concave trough. The data available to the authors indicate that for small asymmetrical sand waves (up to about 2 m high) the lee slope angles vary from about 17° to 35° but are most commonly greater than 20°. In contrast, the large asymmetrical tidal sand waves have lee slope angles of about 4° to 30°, but nearly always less than 20°. The angles of the gentle slopes range from about 0.5° to about 4° for strongly asymmetrical sand waves, to 14° or so for almost symmetrical ones. Symmetrical sand waves might therefore be considered in effect to have steep slopes on both sides. The few available measurements (Harvey, 1966; Jordan, 1962; Ulrich and Pasenau, 1973, and our unpublished Southern North Sea measurements) for symmetrical marine sand waves formed by tidal currents suggest that the slope angles (on both sides of a sand wave) usually range from 10° to 20° (Fig. 3.5). However, symmetrical sand waves formed by tidal lee waves have small angles (<4° for the La Chapelle Bank examples from Cartwright, 1959).

Our lee slope measurements for the two available sets of large asymmetrical sand waves in very coarse sand in estuaries (Ballade, 1953; Dyer, 1970b) give slopes, respectively, of 11° to 14° and 14° to 17°. These are presumed to refer to mobile sand waves (confirmed in Ballade's case by repeated surveys during a period of months that proved net displacements). Thus they seem consistent with the slopes of other sand waves which are mostly in medium or fine sand, i.e. the absence of suspension transport does not seem relevant, at least for large sand waves.

The indications, therefore, are that small sand waves of tidal flows generally have steeper lee slopes than large tidal sand waves, and that avalanche faces are much more likely to occur on the small sand waves. Allen (1970) gives laboratory values of residual angle after avalanching in still water as 31.4° to 33.9° for quartz sands (medium to coarse grained, either very well or poorly sorted) and 36.5° for coralline algal beach 'sand' (median diameter 3.17 mm, and very well sorted). He found values of 25° for polished and 34° for rusty-surfaced equal diameter steel spheres. A smaller surface-texture effect is expected for well-rounded, well-sorted natural grains with values unlikely to be appreciably below 30° in still water.

The angle of lee slopes in unidirectional currents in a flume with fine to very coarse sands have been reported (Guy, Simons and Richardson, 1966) to be 20° to 32° depending on the degree of counter-circulation (due to flow separation) in the sand wave trough. Unfortunately we have not found other flume evidence on this point. Certainly with high resolution echo-sounders and side-scan sonar small sand waves can be observed not only upon the more gentle slopes of large marine sand waves but also sometimes upon their steeper lee slopes (Plate 3.2), showing that these were at much less than avalanche values.

Large sand waves formed by the unidirectional Agulhas current off the east coast of South Africa (Flemming, 1978, 1980) vary from sharp-crested, high-angle (30° or more) active forms to rounded, low-angle (as little as 2°) degenerate forms, as measured from his published profiles. The high-angle forms, thus, have lee slopes steep enough for avalanching. This interpretation of angles is in keeping with some large sand waves in rivers. Except at very high current speeds for which they are partly smoothed (Section 4.6), they usually have high lee slope angles, such as the 'dunes', up to 7.5 m high, in the Brahmaputra River (Coleman, 1969). These have large flow separation zones implying steep lee slopes, as shown in published echograms. Unfortunately, in most studies the lee slope angles have not been recorded. However, under unidirectional flow conditions avalanche face development is predicted theoretically (Bagnold, 1956; Smith, 1969) for sand waves which have built up sufficiently to produce flow separation. This probably commences at lee slope angles of 5° to 9° (Allen, 1980a). Flow separation causes the bed-load sand transport rate to decrease abruptly to near-zero just beyond the crest, so resulting in a high deposition rate there and rapid steepening of the slope below to avalanche values. Many fluviatile sand waves are, however, also partly smoothed in consequence of transitional flow (cf. Section 3.2.2). Examples of this are some large sand waves in the Mississippi (Lane and Eden, 1940, p. 287) and probably Imbrie and Buchanan's (1965) 'accretion ripples'. This situation is uncommon in the sea.

Available evidence suggests that the spread in the slope values for each type of sand wave in the sea is due to the variation between highly active forms, with relatively steep lee slopes, and temporarily moribund, smoothed off forms with gentle slopes. The latter occur where there has been a decline in tidal current speed, and where wave oscillatory movements and burrowing organisms have helped to reduce the slope angles since the sand wave was last active. (A large sand wave is more likely to be active at present if it has active small sand waves or sand ripples on its slopes.) Reduced slope angles are also observed (Section 3.4.5) for moribund as compared to active sand banks, which are also largely due to transverse sand transport.

A tentative generalization for a sea with predominantly tidal currents with speeds which at least occasionally reach threshold for sand wave generation or maintenance is as follows. Where the minimum speed is exceeded frequently and there is sufficient net transport of sand the slopes of large sand waves will be high, commonly 10° to 30°. This will apply to the lee slope with respect to the more competent of the ebb or flood current. It will also apply to both slopes of symmetrical sand waves caused by equal ebb and flood currents. For predominantly tidal currents which only rarely exceed the threshold speed for sand wave generation and where there is little sand transport, the large sand waves will have lower angles. These will probably be less than 15° for the lee slope, with similar values for both slopes of symmetrical forms. Slope angle values will of course change where there are appreciable non-tidal water movements associated with the tidal currents.

Allen's (1980a) predicted numerical values of lee slope angles (β in his notation) for different degrees of tidal current asymmetry are not considered reliable. His predicted low slope values (about 3°) for symmetrical sand waves are far removed from observed values in the sea (given above). The data sets used include the La Chapelle Bank lee-wave sand waves with angles <4°. His angles are based on a plot of a rather small number of sets of values, with scope for drawing different straight lines giving a similar degree of agreement. (However, use of such reasonable-looking alternatives only gives a predicted β range of 2.8°–3.3° for the symmetrical case and 4.2°–5.4° for peak speeds of 120 and 100 cm/s). Furthermore, his 'symmetry index' is defined by mean slopes from trough to crest, i.e. reduced by any trough rounding, and his current-asymmetry values neglect distortions (e.g. the M_4 constituent) from sinusoidal of the tidal current speed/time variation. Thus, the steady current component is assumed equal to half the peak ebb/flood difference, and Allen states that additionally the ebb and flood peak speeds he used may not have been representative of the sand wave fields concerned.

(e) *Sand waves in plan view*

Individual sand waves within a continuous field of offshore sand waves (Plates 3.2 to 3.4 and Belderson et al., 1972, Figs 47 to 57) are generally more or less sinuous crested and orientated approximately normal to the peak tidal flow. The alternate lingoid and barchanoid units along the crests are generally not strongly developed, so that more extreme three-dimensional forms, with associated scour-pits, are rare. This implies that most of these large offshore sand waves have fairly constant elevations of crests and troughs. There is some evidence from sonographs (Terwindt, 1971b) and from Seasat side-scan radar images of sea-surface disturbance set up by turbulence generated by the tidal current flowing over sand waves in the outer Bristol Channel that large sand waves can be orientated somewhat obliquely to the direction of the regional peak tidal flow. However, the trend is not constant even for a given region (Fig. 3.6). Houbolt (1968) showed this oblique orientation for parts of the Southern Bight of the North Sea. However, we consider that this might be erroneous as his courses were oblique to tidal flow, thus requiring some correction for the ship crabbing.

In some extensive fields of small sand waves the bedforms are arranged in lanes that extend parallel to the peak tidal flow. The wavelength of the sand waves in adjacent lanes can vary slightly (Plate 3.4), possibly reflecting grain size differences or variations in sand thickness. There may be some relationship between these lanes and sand ribbons (Section 3.4.4), which are frequently composed of trains of small sand waves.

Isolated, sand-starved sand waves moving over gravel pavement are generally barchan-shaped, with the horns pointing in the direction of net sand transport (Fig. 3.2a). There are many examples of a horn that terminates in the middle of the convex side of another barchan located further 'downstream' from it (Plate 3.5 and, Belderson and Kenyon, 1969), implying that sand passes from one to the other. Thus,

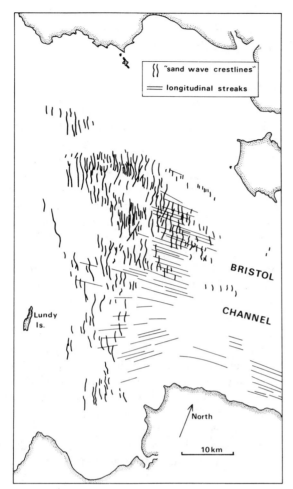

Fig. 3.6 Orientation of sea surface features in the outer part of the Bristol Channel, England, taken from a Seasat side-scan radar image. The approximately transverse features correspond with sand wave crests some 50 m below the surface. The sand wave crest orientations vary from about normal to a maximum angle of about 15° to the direction of peak tidal currents, as indicated by the longitudinal streaks on the water surface. Where not normal to the tidal currents, the sand wave crests are rotated in an anticlockwise sense relative to the tidal current. The (flood) tide was flowing at near-peak strength towards the east at the time of observation.

the barchans are interdependent, each helping to nourish the next in line.

Skewed or 'hooked' barchans in which one of the horns is preferentially developed have been observed, most notably over a wide area around the western entrance to the English Channel. Here they are predominantly skewed such that the northern horn of the westward facing sand wave is by far the longer. Possibly this indicates that there is some net sand transport transverse to peak tidal flow (Section 4.4.3).

(f) *Outer limits of sand waves*

Around the British Isles small sand waves (with wavelengths of only a few metres) extend into regions of slightly weaker tidal currents than would give by themselves the bottom stress required to form small sand waves in flumes (Section 3.2) (and considerably less than would be needed to form large sand waves in the sea). This is because the outer limits of both small and large sand waves are related to the sum total of water movements, including storm-induced currents and storm-wave effects that are added to the tidal current strength (Sections 4.5.2 and 4.6). At least in the eastern half of the Southern Bight of the North Sea, the small size of these sand waves near their outer limit is not due to any lack of sufficient sand to form large ones.

(g) *Small sand waves on large ones*

Small sand waves (of heights exceeding 5 cm) often occur on both the upstream and lee slopes of the relatively larger sand waves in estuaries and on the continental shelf. Either slope of a large sand wave that is inclined at much less than the residual angle after avalanching may reasonably be considered as a portion of sand bed in its own right. Thus, formation on it of smaller sand waves or sand ripples may be expected for flows similar to those giving these bedforms on a flat sand floor. The current speeds for which they can form will be increased from the flat floor case by effects of non-uniformity of flow over the large sand wave and of local bed slope. However, such effects do not appear to be important here because of the often gentle lee slope angles, while upstream facing slope angles are even smaller.

More important is that for such sand ripples or small sand waves the effective bottom stress is reduced from the large-scale bottom stress by the form drag of the large sand wave (as noted for sand ripples on small sand waves in Section 3.2.1).

Small sand waves can frequently be orientated obliquely to at least parts of the crests of large sand waves (Plate 3.2; Terwindt, 1971b; Belderson et al., 1972, Figs 49 and 52). This is to be expected, as the orientation of the small sand waves will be determined by the local flow direction, which is inevitably deflected from the regional flow direction by the obstruction presented by any sinuosity of the underlying sand waves. There are also some indications that small sand waves may in places have a mean orientation not parallel to the mean crest orientation of the large sand waves, which themselves are not always perpendicular to the regional peak current direction. However, it should not be argued, as has been done (Eisma, Jansen and Van Weering, 1979), that the presence of small sand waves on top of the larger ones shows that the latter were formed under older tidal conditions. Quite the reverse is likely. Thus the presence of small sand waves indicates active sand transport, which, if continued for long enough, will result in the building of large sand waves.

(h) *Sand wave smoothing*

Upper flow regime conditions of high transport rate (but not high enough to give antidunes, Section 3.3.1) may occur on the shallow tops of sand banks and in very shallow water elsewhere. This is expected to cause smoothing out of existing sand waves in sand with median diameter finer than about 1 to 2 mm. For very coarse sand and gravel in flumes this condition appears to require even higher bed shear stress than that for antidunes, so that for the sea only the smaller sizes of sand need be considered. Sand wave smoothing is most likely to occur when tidal currents are reinforced by wind-driven currents and sea waves. The near-bed oscillatory speeds of sea waves increase with decreasing water depths and so could be very important in shallow water. They are locally further increased by wave refraction near the tops of sand banks. Partial smoothing, corresponding to the transition bed state in flumes, is expected for slightly less strong water movements. Fuller discussion will be found in Section 4.6, as it is related to high sand transport rates.

(i) *Sand waves formed by tidal lee waves*

On the outermost part of the continental shelf west of Brittany in a water depth of up to about 200 m there is a group of exceptionally large, almost symmetrical sand waves with rounded crests. They are up to 5 km long, 1 km apart and 12 m high (Stride, 1963a), with slopes of only up to about 4°.

These sand waves have been attributed to the action of tidal lee waves that develop at the summer thermocline when the tidal current flowing landwards 'feels' the edge of the continental shelf (Cartwright, 1959). This mechanism is not likely to be of widespread importance, because of the combination of conditions required. A well-marked thermocline will usually be present in low latitudes, but is expected to occur only during the summer and autumn in middle latitudes. Salinity increase downwards could provide or enhance stratification, but at the present time there are probably not many salinity-stratified shelf-edge areas. Also, portions of continental slope with progression of the tidal wave up and on to the shelf are in the minority, as shown by co-phase lines on charts of ocean tides. Another requirement is that, in spite of fairly large water depth near the shelf edge, near-bed current speeds should be sufficient to move the considerable volumes of sand needed to form the large sand waves.

3.3.4 TRANSVERSE SAND PATCHES

In widespread regions where near-surface mean spring peak speeds are relatively weak (about 50 cm/s or less) and there is no continuous sand cover (equivalent to a sand sheet of Section 5.2) any available sand is found as sand patches of various sizes. The long axes of these patches are generally aligned either more or less transverse or longitudinal to the peak tidal flow. The longitudinal variety are described in Section 3.4.4.

The most obvious characteristics of the transverse sand patches, when well-developed, are their usual thickness of 2 to 3 m (with a maximum of 4 m) and their sharply defined edges. Where first recognized in the Celtic Sea (Kenyon, 1970), they tend to be crescentic to ragged in plan view (e.g. Plate 3.6; and

Belderson et al., 1972, Figs 59 to 64) and tabular in profile, with steep edges and a flat top. A photograph of a steep edge of one of these tabular sand patches is shown by Delanoë and Pinot (1980), although these authors mistakenly term the feature a 'sand ribbon'. In addition, examples have been found that have a steep edge on one side only, which suggests that these sand patches are at least occasionally moving in a preferred direction. From this sense of asymmetry it is inferred that they are barchanoid (or lunate) in plan, rather than linguoid as was previously thought. Perhaps they should, therefore, be considered as ill-formed or quasi-sand waves.

Sand movement associated with the sand patches is thought to be due generally to the combined effect of tidal and other water movements, such as storm-induced currents and wave effects. Sand grains stirred up by the greater bed roughness, and hence turbulence and bottom stress, over the intervening gravel areas are probably gathered into the sand patches by a to and fro sweeping action. This is because the travel speed of the sand grains is greater over the gravel due to the greater bed roughness and bottom stress (Bagnold, 1941). The 2 to 3 m thickness of the patches may, thus, represent the limiting height to which sand may be moved both as bed-load and in suspension at the steep, sharply defined edge of the sand patch. Beyond this edge there is the less turbulent environment over the sand patch itself. The 'transition-bed sand waves' of Section 3.2.2 bear some similarity to transverse sand patches, but require very much higher Froude numbers and current speeds. They are not thought to be relevant here.

3.4 Longitudinal bedforms of the continental shelf

There are six main types of quasi-longitudinal bedforms that are developed more or less parallel with the main tidal flow. These are (here including erosional features as bedforms) scour hollows, longitudinal furrows, obstacle marks, sand ribbons, longitudinal sand patches and sand banks (Figs 3.1 and 3.2). As with transverse bedforms these can be associated empirically with peak tidal current speeds falling within restricted ranges. Unfortunately little guidance about the nature or origin of such bedforms can be derived from rivers and flumes, since longitudinal forms not closely related to the steep channel sides are uncommon in these situations (Section 3.2.4). The exceptions are some studies of very small-scale ribbons and lineations in flumes (e.g. Casey, 1935; Bagnold, 1941; Vanoni, 1944; Werner, Unsöld, Koopmann and Stefanon, 1980) and flash floods (e.g. Karcz, 1967).

3.4.1 SCOUR HOLLOWS

Intensively scoured floors are confined within areas where near-surface mean spring peak tidal currents reach speeds of about 150 cm/s (about 3 knots) or more. These scour hollows are often associated with tide-rips and 'boils' on the sea surface and tend to be elongated in the directions of the peak tidal currents. Convincing examples up to 28 m deep have been recognized in the North Channel of the Irish Sea, where they occur around and adjacent to more resistant dykes and larger igneous intrusions (G. F. Caston, 1975). These Irish Sea examples have an implied maximum erosion rate averaging 3 mm per year. Other scour hollows associated with strong present day tidal currents are found in various parts of the world such as the entrance to San Francisco Harbour, and the Bisan and Hayasui Straits off Japan (Mogi, 1979), where they are cut as much as 260 m below surrounding ground (Fig. 3.7).

Tidal scour at times of lower sea level may also have played a part in the formation of at least some of the large elongate deeps found in areas now swept by less than 150 cm/s (3 knots) near-surface mean spring peak tidal currents. Examples of such may be the 150 km long and up to 5 km wide Hurd Deep in the English Channel which reaches down to about 150 m below the neighbouring floor (Hamilton and Smith, 1972) and the 40 km long and 3 km wide Inner Silver Pit in the south-western part of the North Sea, which reaches down to 70 m below the surrounding floor (Donovan, 1973). These tidal scour hollows are probably analogous to the erosional hollows found in deserts and snow fields that are aligned with the prevailing winds.

In addition to the large tidal scour hollows which can be elongated in the direction of peak tidal flow

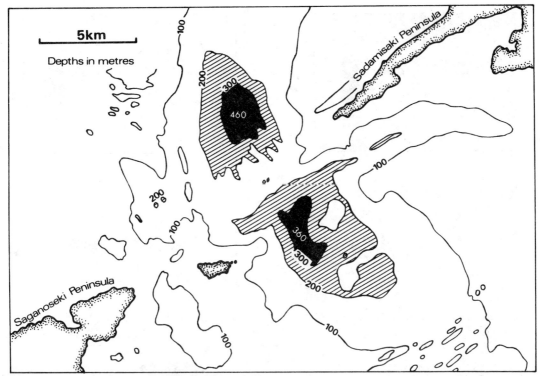

Fig. 3.7 Scour hollows more than 260 and 160 m deep below surrounding floor situated on both sides of a rocky sill in the strongly tidal Hayasui Strait, Japan (after Mogi, 1979).

there are other areas of rock out-crops etched into sharp relief because of the variable resistance to erosion of adjacent layers. Differential erosion of the rock floor of the Bristol Channel and English Channel was described by Donovan and Stride (1961b). This has picked out the strike and fault lines most dramatically over wide areas (Belderson *et al.*, 1972, Figs 6 to 18).

3.4.2 LONGITUDINAL FURROWS

Groups of furrows elongated almost parallel to peak tidal currents have been recognized around the British Isles on floors of gravel, sand and mud (Dyer, 1970a; Stride, Belderson and Kenyon, 1972; Flood, 1981) and on a rock floor in the Bay of Fundy (Klein, 1970a). Similar bedforms are found associated with unidirectional currents on the deep sea floor. In contrast, longitudinal erosional features unaccompanied by depositional features do not seem to have been reported for unidirectional currents in rivers and laboratory flumes. This is, perhaps, because these flows are already carrying their full capacity sediment load.

The longitudinal furrows cut in gravel floors of the continental shelf are associated with peak near-surface tidal currents that reach 150 cm/s (3 knots) or more at mean springs. They are up to 8 km long, 30 m wide and 1 m deep, with a cross section that is sometimes asymmetrical. The furrows in mud are associated with a tidal current which at a height of 7 m above the floor reaches a speed of about 70 cm/s during mean springs. These furrows are up to 1.5 km long, 15 m wide and 1 m deep.

In plan view the furrows on both gravel and mud floors vary from straight to slightly sinuous, with a tendency for individual furrows to join in a preferred direction (Plates 3.7 to 3.9, and Belderson *et al.*, 1972, Figs 27 to 30). In each case this direction corresponds to that of the stronger of the peak ebb or flood tidal current. For the examples so far described from gravel floors these directions of joining also

46   Offshore Tidal Sands

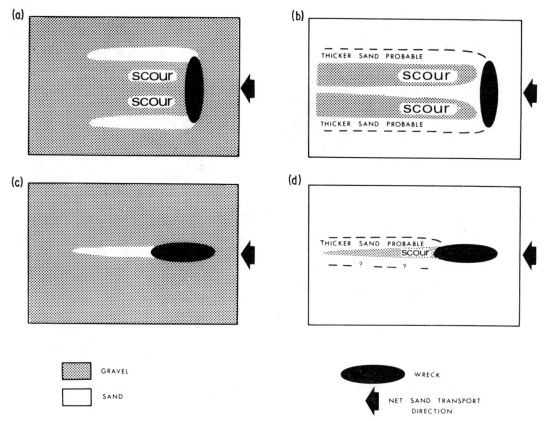

Fig. 3.8 Diagram showing the varied form of wreck marks, depending on the orientation of the wreck with respect to the tidal net sand transport direction (after G. F. Caston, 1979).

correspond with the longitudinal directions in which the peak tidal flows are decreasing. They have been used as indicators of the net transport directions for sand and gravel (Section 4.4.2(c)). Ancient examples of such furrows may have sometimes been referred to by geologists as 'channels' in descriptions of limited rock exposures on land (Section 7.3.4).

3.4.3 OBSTACLE MARKS

The flow of tidal currents over the sea floor is often disturbed by natural and man-made obstacles. As a result there are local areas of scour and deposition. The larger obstacle marks are shown by sonographs, and in shallow water, by aerial photography, while the smallest have been seen by divers or underwater photographs and television. The largest obstacle marks known are the banner banks, which are sometimes several kilometres long and tens of metres high (Section 3.4.5). The term 'sand shadow' can also be applied to deposits in the lee of obstacles.

Numerous wreck marks have been revealed by sonographs taken in the outer part of the Thames Estuary (G. F. Caston, 1979). Adjacent to the wreck there is generally a scour hollow which can be up to 15 m deep, together with a variety of erosional or depositional marks. These can be up to at least 1 km long and extend parallel to peak tidal flow on the 'downstream' (stronger of the peak ebb or flood current) side of the wreck (Plate 3.10). Where such bedforms also occur on the net 'upstream' side of the wreck they are much smaller.

The wreck marks vary in form according to whether the wreck lies transverse, oblique or parallel to the peak tidal flow and whether it is sited on a sand or gravel floor (Fig. 3.8). Varieties include (a) two sand ridges separated by a scour hollow; (b) two scour hollows emanating from the ends of the wreck and separated by a ridge; (c) a single sand ridge; or (d) a single scour hollow, sometimes with a ridge to one side of it.

Patterns of scour and deposition associated with submarine pipe-lines have also been observed. In places the features consist of remarkably regular groups of spurs and furrows 'attached' to the pipe-line (Plates 3.11 and 3.12).

Small-scale examples of obstacle marks (comet marks and sand shadows) have been seen on high resolution sonographs. There are numerous examples of these in the entrance to the Baltic Sea for instance (Werner and Newton, 1975; Werner et al., 1980). Here occasional strong unidirectional currents are superimposed upon much weaker tidal currents. The comet marks are longitudinal features up to 100 m long and several metres wide that have been eroded into coarse sediment lying between areas of finer grain sizes. They originate from obstacles such as isolated boulders, and develop down-current from them. On coarse, sand-starved floors they can be flanked by thin streaks of sand. Comet marks are also present in tidal seas (G. F. Caston, 1976; and Plate 3.13).

The sand shadows appear as elongated, current parallel, depositional bodies of sand which accumulate on the down-current side of natural obstacles. In this case the generating obstacles (as described by Werner and Newton, 1975) are larger (ridges up to 2 m or more high and up to 20 m long) than those responsible for comet marks. The sand shadow itself ranges from 10 m to 100 m or more in length. Such features are closely related to sand ribbons (Section 3.4.4), as fully developed sand ribbons have been observed to originate from various obstacles, including rocks, wrecks and the tips of barchanoid sand waves.

Divers have reported that a small accumulation of sand in the lee of a pebble can change sides when the tidal current reverses in direction.

In general, the dimensions of the various obstacle marks seem to be related to the size of the obstacle, the strength of the current and the nature and abundance of suitable materials. The fact that in a tidal sea the obstacle marks may be developed in both the ebb and flood directions of sand transport (though usually unequally) away from the obstacle is a potential (though probably not very fruitful) means of distinction between the effects of tidal and unidirectional currents in ancient shallow marine deposits.

3.4.4 SAND RIBBONS AND LONGITUDINAL SAND PATCHES

Over large areas of continental shelf swept by tidal currents there are longitudinal bodies of thinly spread sand that lie on, and are separated by, floors of shell or gravel. The bedforms found with strong tidal currents are called sand ribbons and those associated with weak tidal currents are called longitudinal sand patches (Fig. 3.1). The two bedforms may be inherently different types of features or merely the two end-members of a series. Both features are also made by unidirectional currents (e.g. Flemming, 1980). The association between sand ribbons and obstacle marks is clearly a close one, as some sand ribbons develop from an obstacle 'source'. The 'longitudinal megaripples' of van Straaten (1953) found in estuaries, and the large-scale current lineations of Imbrie and Buchanan (1965) on the Bahaman Banks are not sand ribbons in that they have a ripple-like profile and are not separated by strips of gravel.

The fact that a large proportion of the continental shelf is bare of sand, except for grains sheltering among gravel or shells, is explained by the higher transport speed of any sand grains in transit over bare floor than over a sand bed. Thus, sand grains tend to join the sand-bed areas, as they do in subaerial flow (Bagnold, 1941). In addition, helical circulation with longitudinal axes may play a part in forming sand ribbons and longitudinal sand patches (as seen for sand ribbons in flumes). These circulations near the sea floor would be driven by the bottom stress in a similar way to Langmuir circulations produced by wind shear stress on the land or the sea surface.

(a) *Sand ribbons*

The sand ribbons (*sensu stricto*) of the continental shelf vary greatly in size (Plates 3.13 to 3.15; and

Belderson *et al.*, 1972, Figs 31 to 39). Some are known to reach up to 15 km long, and exceptionally they can be 200 m wide. Their length to breadth ratio is generally in excess of 40:1. Their edges are parallel or almost parallel, and are often fairly sharply defined, sometimes on one side only, implying slightly thicker sand on that side. Their thickness is always relatively small, probably from a few grains thick (not enough to totally mask the gravel floor) up to about a metre.

The most typical sand ribbons (Kenyon, 1970) are found associated with near-surface mean spring peak tidal or non-tidal currents of about 100 cm/s (2 knots). At higher current speeds of about 130 cm/s (2.5 knots) the sand ribbons are made up of narrow trains of more or less straight-crested sand waves. (Trains of waves in fine gravel may also possibly occur on a floor of coarser gravel but it is often difficult from sonographs alone to draw a distinction even between small sand and gravel waves. Specific sampling is also necessary.) At weaker sea-surface tidal current speeds of about 90 cm/s (1.75 knots) ribbon-like trains of often sinuous or barchanoid small sand waves are found (Plate 3.14). At still lower speeds of about 77 cm/s (1.5 knots) generally smaller (10 m or so wide) sand ribbons are occasionally found on a gravel floor between isolated large sand waves.

Present attempts to relate sand ribbon spacing and water depth in the sea (e.g. Flemming, 1976) do not appear to be successful. Indeed, for a given water depth individual sand ribbons tend to broaden down the longitudinal velocity gradient of the peak tidal flow (Kenyon, 1970), which might indicate an inverse relationship with peak water speed.

Junctions of sand ribbons are relatively infrequent, but when they do join the ribbons usually merge down the longitudinal velocity gradient. Usually this direction is also that of the net sand transport, but rare exceptions may occur where the direction of net sand transport is up the longitudinal velocity gradient of the peak tidal flow.

(b) *Longitudinal sand patches*

At mean spring peak tidal current speeds of less than 50 cm/s or so at the sea surface (Fig. 3.2) the length to breadth ratio of the longitudinal sand bodies on a gravel floor is considerably smaller than for sand ribbons and their edges are much more irregular and non-parallel (e.g. Belderson *et al.*, 1972, Figs 68 to 72). For convenience, the present authors call these features longitudinal sand patches. However, the name may have to change should the bedform be shown subsequently to have the same basic origin as sand ribbons.

The longitudinal sand patches vary greatly in their size and shape, often having ragged edges (Plate 3.16). Their spacing is variable, perhaps being dependent on sand availability. When lying on a flat floor they sometimes attain a thickness of 2 to 3 m. Some longitudinal sand patches show a well developed asymmetry, with one edge sharply defined and the other one a feather-edge (Plate 3.17). Their asymmetry can be matched in sand ribbons, but in the sand patch case the associated tidal currents are much weaker. Thus storm waves and unidirectional currents can play a relatively more important role, perhaps imparting a lateral component to any shift of the features.

Normally a zone of sand ribbons (associated with strong tidal currents) is separated from a zone of longitudinal sand patches (associated with weaker tidal currents) by a sand wave zone. However, there are areas where there appear to be transitional-type longitudinal bedforms where sand waves might normally be expected. Two such areas lie to the south-west and north-west of Cornwall, England. An example of this bedform is shown in Plate 3.18. Other variants include rhythmic repetitions of bands of multiple patches within which the individual sand strips systematically vary in width laterally (e.g. Belderson *et al.*, 1972, Fig. 34). They are more regular and parallel-sided than many sand patches, but less so than genuine sand ribbons. They also tend to be thicker than sand ribbons, and are associated with near-surface mean spring peak tidal currents of about 80 cm/s (stronger than for sand patches but weaker than for genuine sand ribbons).

Longitudinal sand patches (together with the transverse ones) are likely to be found in the many areas around the world's continental shelves where the near-surface tidal and other currents do not often exceed 50 cm/s, but where there are occasional

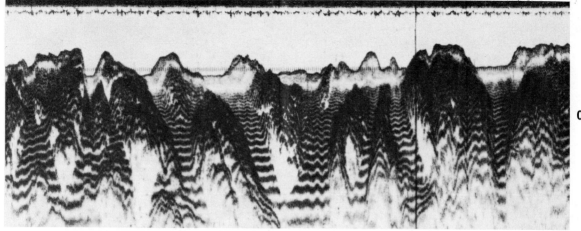

2 km

Plate 1.1 Sonograph showing a plan view of rough outcrops of Precambrian rock separated by flat-surfaced sediment ponds, located at about 56°51′N, 07°44′W, to the west of the Outer Hebrides, Scotland. The multi-beam method of presentation in this particular sonograph and Plate 4.1 allows variations in height of relief features to be observed and measured on much of the sonograph (Stubbs, McCartney and Legg, 1974; Kenyon and Pelton, 1979).

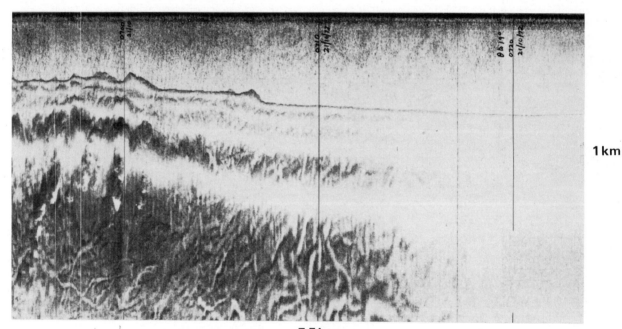

7.5 km

Plate 1.2 Sonograph showing iceberg plough marks and some rock outcrops in plan view on the continental shelf at 58°15′N, 06°05′E, off southern Norway. The sediment cover increases in thickness to the right, progressively masking the relief.

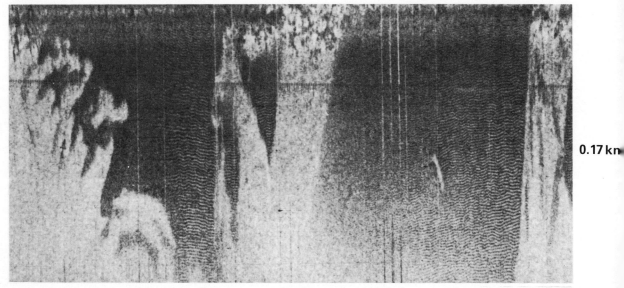

Plate 3.1 Sonograph showing patches of gravel floor (dark tone) covered by large wave-formed gravel ripples of about 1.5 m wavelength, separated by patches of sand (light tone). The ripples are well seen on the gravel because they trend almost parallel to the ship's course. Any wave or current-formed ripples on the sand would have shorter wavelength (less than the resolution of the 250 kHz sonar used). If current-formed, they may in any case have a different orientation. Celtic Sea, 50°42′N, 04°58′W.

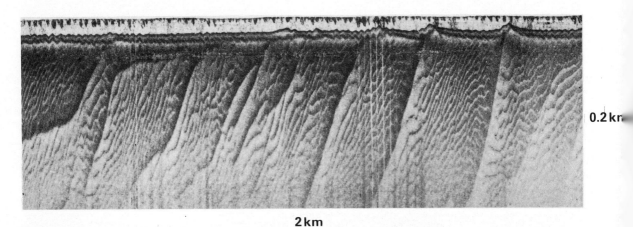

Plate 3.2 Sonograph showing sinuous-crested large (5–7 m high) asymmetrical sand waves with superimposed small sand waves oriented obliquely both in the troughs and on the slopes of the large sand waves. Southern North Sea, 53°22′N, 01°38′E.

Plate 3.3 Sonograph showing sand waves of different wavelengths (about 8 m–60 m) in adjacent areas. Southern North Sea, 53°22′N, 01°35′E.

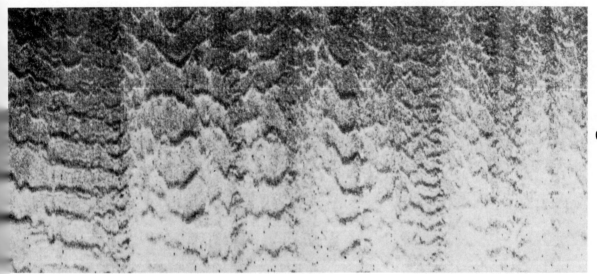

Plate 3.4 Sonograph showing sand waves of slightly different wavelength (about 15 m, 7 m and 4 m) arranged in adjacent 'lanes'. Southern Bight North Sea, 52°43′N, 02°21′E.

Plate 3.5 Sonograph showing a series of barchan-shaped sand waves on a gravel floor. Their horns point in the direction of net sand transport and shed sand from their tips in that direction (left) to 'feed' the backs of 'downstream' barchans. English Channel, 50°17'N, 02°30'W.

Plate 3.6 Sonograph showing crescentic to ragged shaped transverse sand patches (light tone) on a gravel floor (dark tone). They are tabular, having relatively steep sides about 2 m high, and flat tops. Celtic Sea, 51°29'N, 06°56'W.

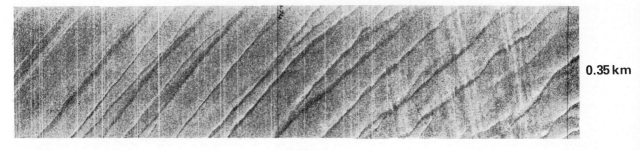

Plate 3.7 Sonograph showing longitudinal furrows in gravel; in this example the furrows are fairly straight, with relatively few downstream 'tuning fork' junctions. English Channel, 50°23′N, 02°27′W.

Plate 3.8 Sonograph showing longitudinal furrows in gravel. These furrows are fairly sinuous, with a marked dendritic pattern. English Channel, 50°24′N, 01°14′W.

Plate 3.9 Sonograph showing longitudinal furrows in gravel. The slight relief of the furrows is well brought out by the bands of shadow (light tone) and strong reflection (dark tone) representing the two sides of each furrow. English Channel, 50°24′N, 01°10′W.

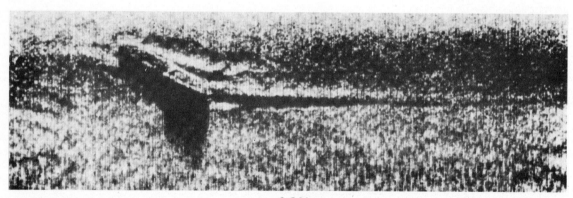

Plate 3.10 Sonograph showing a wreck (55 m long and up to 14 m high) with ridges and scour hollows emanating from each end. The unequal development of the obstacle marks in each direction indicates net sand transport from left to right, while the unequal development of the marks on the same side of the wreck is due to the oblique alignment of the wreck relative to the peak tidal flow (left to right). In this sonograph the shadows cast by the wreck and obstacle marks appear black (whereas any shadows in the accompanying sonographs appear white). Sonograph by courtesy of Kelvin Hughes (Smiths Industries) Ltd. Dover Strait.

Plate 3.11 Sonograph showing a pipeline exposed on the sea floor, with a regularly spaced series of obstacle marks extending away from one side of it. Southern North Sea, 52°53'N, 01°41'E.

Plate 3.12 Sonograph showing a pipeline in a field of sand waves, with indications of short obstacle marks (spurs and furrows) associated with it. Some sand waves are starting to rebuild across the pipeline. Southern North Sea, 53°09'N, 02°23'E.

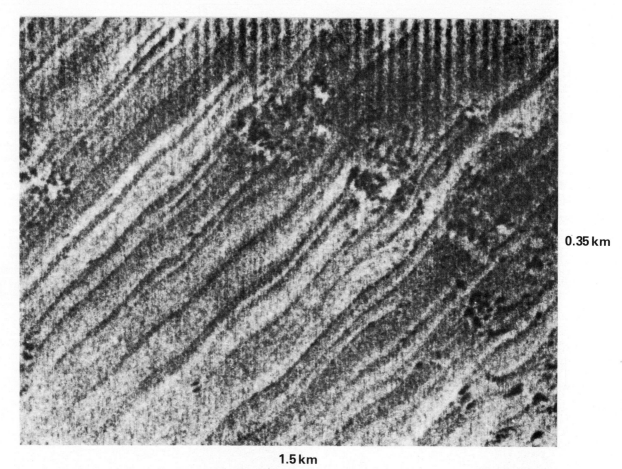

Plate 3.13 Sonograph showing large comet marks and associated sand ribbons (light tone) developed downcurrent (in terms of the stronger of the peak ebb and flood currents) from isolated small outcrops of rock (black). Bristol Channel, 51°15′N, 04°43′W.

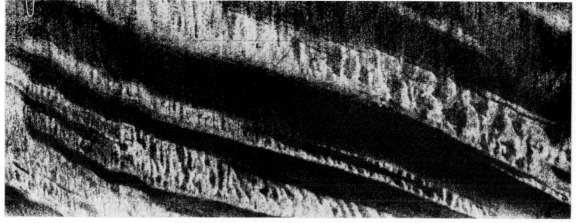

0.4 km

5.1 km

Plate 3.14 Sonograph showing sand ribbons (light tone) of variable width in which the sparse sand present is organized into trains of small sinuous or barchanoid sand waves. English Channel, 50°38′N, 00°32′E.

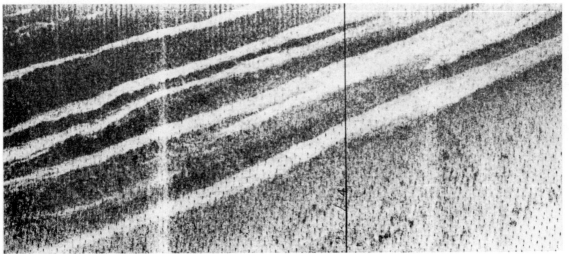

0.5 km

2.7 km

Plate 3.15 Sonograph showing sand ribbons (light tone) bifurcating down a transport path (which in this rare case is up the longitudinal velocity gradient of the tidal currents). The regular pattern of dots, particularly evident at the bottom right hand corner, is due to echo-sounder interference. Bristol Channel, 51°15′N, 04°35′W.

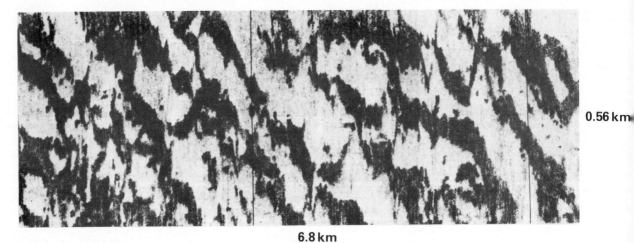

Plate 3.16 Sonograph showing ragged-edged longitudinal sand patches (light tone) with relief of about 2 m on a gravel floor (dark tone). West of Ireland, 52°30′N, 11°28′W.

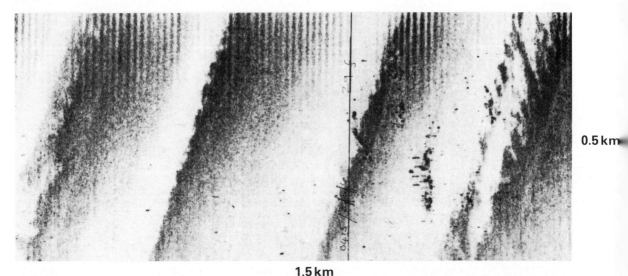

Plate 3.17 Sonograph showing longitudinal sand patches (light tone) with a fairly sharp boundary on one side and a gradational one on the other. The progressive decrease in thickness of sand across the patches (from a maximum of about 2 m) probably indicates a slight lateral component of shift of the basically longitudinal sand patches over the gravel floor. Celtic Sea, 50°38′N, 05°14′W.

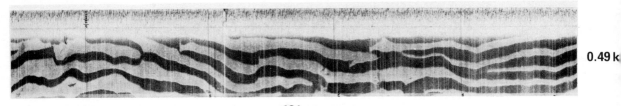

Plate 3.18 Sonograph showing a bedform thought to be transitional between sand ribbons and longitudinal sand patches. The features have a smaller length to breadth ratio, more wavy edges and a thicker sand cover than most sand ribbons. English Channel, 49°48′N, 05°50′W.

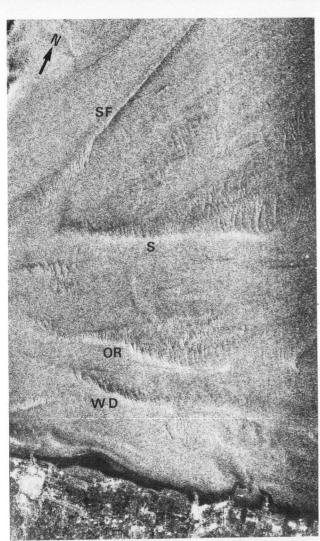

Plate 3.19 (left) SEASAT side-scan radar image of patterns on the sea surface due to sand banks and associated sand waves in the Southern Bight of the North Sea. The sand wave crests make an oblique angle with the crests of the sand banks as do longitudinal streaks, believed to be current-parallel features. South Falls (SF), Sandettie (S), Outer Ruytingen (OR) and West Dyck (WD). Photograph by courtesy of the European Space Agency.

Plate 3.20 (below) Air photograph of an island in Torres Strait (between Australia and New Guinea) from which banner banks are seen originating in a preferred direction (westwards) away from the ends of the island. Photography by courtesy of Division of National Mapping, Australia.

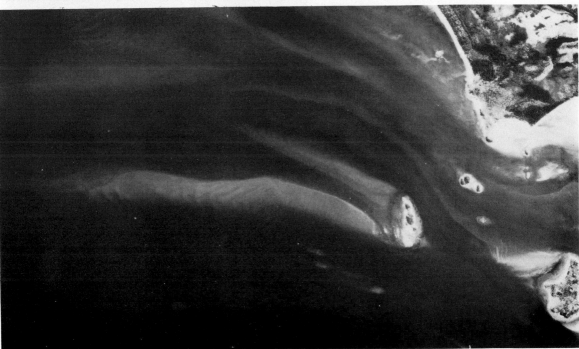

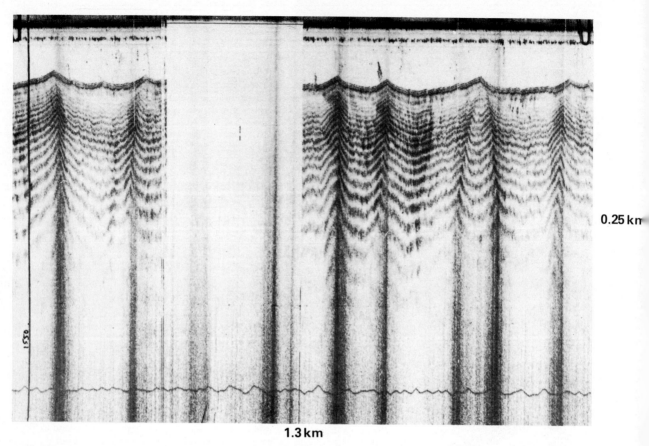

Plate 4.1 Multi-beam side-scan sonar (250 kHz) record of slightly asymmetrical sand waves, 8 m high. The dark strips running up and down the sonograph are produced by the noise of sand in motion on the crests of the sand waves (as demonstrated in the section of record where transmission was switched off and only the noise received). The multi-beam method of presentation allows variations in height along individual sand wave crests to be readily observed and measured (Stubbs, McCartney and Legg, 1974). Southern North Sea, 51°38′N, 02°15′E.

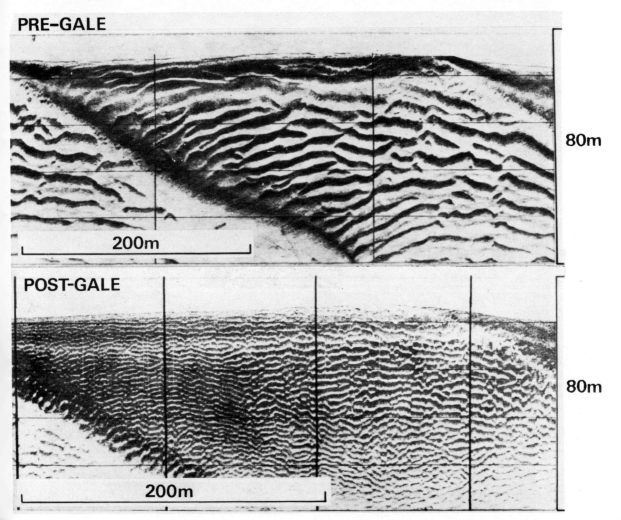

Plate 4.2 Two sonographs showing small sand waves before and after a gale; the wavelength of the small sand waves (situated on the flanks of large sand waves) has been markedly reduced (after Langhorne, 1976). Outer Thames Estuary, 51°48′N, 01°37′E.

Plate 6.1 Coarse calcareous algal gravel or 'Maërl'. In addition to the nodular and branched fragments of calcareous algae, the gravel contains fragments from regular echinoids and bivalves including *Circomphalus casina* (*Venus casina*) and *Chlamys distorta*. Note encrusted rock fragments. Western English Channel, 50°18.31'N, 04°42.33'W, water depth 18 m. Width of photograph is equivalent to 11.5 cm.

Plate 6.2 Part of the surface of a boulder supporting a rich carbonate producing epifauna of serpulid polychaetes (*Pomatoceros triqueter*, *Serpula vermicularis* and *Hydroides norvegica*) bivalves including *Anomia* sp., alcyonarians and sponges. Fair Isle Channel, north of the Orkney Islands, 59°39.66'N, 02°29.30'W, water depth 90 m. Width of photograph is equivalent to 19 cm.

Plate 6.3 A selection of molluscs of the gravel sheet fauna. (a) *Bulbus islandicus* (*Amauropsis islandica*); (b) *Lunatia montagui* (*Natica montagui*); (c) *Paphia rhomboides* (*Venerupis rhomboides*); (d) *Circomphalus casina* (*Venus casina*); (e) *Clausinella fasciata* (*Venus fasciata*); (f) *Timoclea ovata* (*Venus ovata*); (g) *Spisula elliptica*; (h) *Gouldia minima* (*Gafrarium minimum*); (i) *Gari costulata* (note *Lunatia* boring); (j) *Palliolum tigerinum* (*Chlamys tigrina*); (k) *Glycymeris glycymeris*; (l) *Modiolus modiolus*. The shell supports an epifauna including the barnacles *Balanus balanus* and *Verruca stroemia* and the serpulid *Pomatoceros triqueter*. The specimens photographed came from several locations in the Fair Isle Channel between the Orkney and Shetland Islands. Scale: the white bar represents 5 mm, except for (k) and (l) where it represents 30 mm.

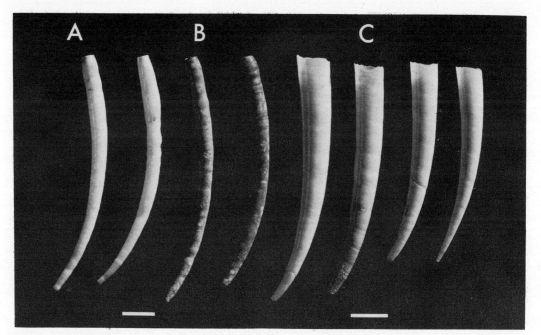

Plate 6.4 Tubes of the polychaete *Ditrupa arietina* contrasted with shells of the scaphopod *Antalis entalis* (*Dentalium entalis*). (a) Fresh *Ditrupa* tubes. (b) *Ditrupa* tubes discoloured by algal and fungal boring. Note the degree of expansion of the *Ditrupa* tube, the curvature and the narrowing of the tube at its anterior end. (c) *Antalis entalis* (*Dentalium entalis*) shells. Note the more rapid expansion of the tube and the absence of any narrowing of the tube at the anterior end. Scale: the white bar represents 5 mm.

Plate 6.5 Coarse fraction (>1 mm) of the rippled sands on the continental shelf west of Scotland consisting almost entirely of fragments of *Ditrupa* tubes. Some unbroken tubes are present. Width of photograph is equivalent to 7.5 cm

Plate 6.6 A selection of molluscs of the zone of rippled sand from several locations on the continental shelf west of Scotland. Other species present are listed in Section 6.12.2. (a) *Aporrhais pespelecani*; (b) *Bela nebula* (*Mangelia nebula*); (c) *Acteon tornatilis*; (d) *Turritella communis* (note *Lunatia* boring); (e) *Cylichna cylindracea*; (f) *Acanthocardia echinata* (*Cardium echinatum*); (g) *Arctica islandica* (*Cyprina islandica*); (h) *Lucinoma borealis*; (i) *Aequipecten opercularis* (*Chlamys opercularis*); (j) *Mysia undata*; (k) *Dosinia lupinus*; (l) *Pandora pinna*; (m) *Gari fervensis*. Scale: the white bar represents 10 mm except in (b), (c) and (e) where it represents 5 mm.

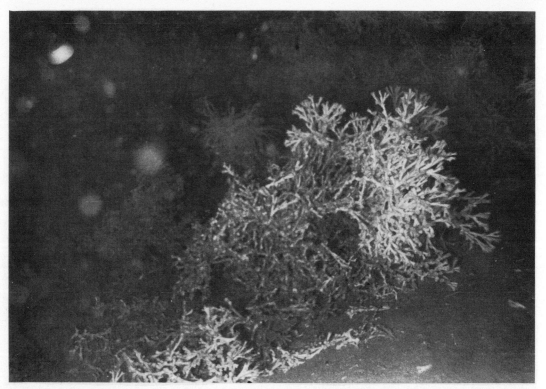

Plate 6.7 Edge of 'patch' of the deep water coral *Lophelia pertusa* (*Lophelia prolifera*). The living colonies are up to 1 m in height. The dead coral debris is colonized by an epifauna of serpulids, bryozoans and large anemones. Approximate width of picture in the foreground is 180 cm. Rockall Bank, 57°36.640′N, 14°29.415′W, water depth 256 m (after J. B. Wilson, 1979b).

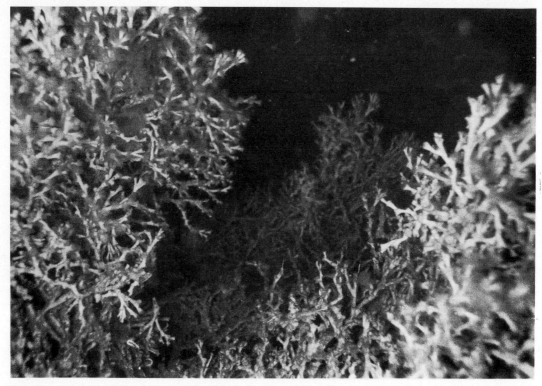

Plate 6.8 Living colonies of the coral *Lophelia pertusa* (*Lophelia prolifera*). Note the delicate nature of many of the branches and the expanded polyps on the colony in the right foreground. Approximate width of picture in the foreground is 190 cm. Rockall Bank, 57°36.637′N, 14°29.412′W, water depth 256 m (after J. B. Wilson, 1979b).

Plate 6.9 Characteristic molluscs from the edge of the continental shelf west and north of Scotland. (a) *Colus howsei*; (b) *Typhlomangelia nivalis*; (c) *Pyrene haliaeeti*; (d) *Troschelia berniciensis*; (e) *Volutomitra groenlandica*; (f) *Solariella amabilis*; (g) *Cylichna alba*; (h) *Acropagia balaustina* (*Tellina balaustina*); (i) *Astarte sulcata*; (j) *Propeamussium hoskynsi* (*Chlamys hoskynsi*); (k) *Chlamys sulcata*; (l) *Manupecten alicei* (*Chlamys alicei*); (m) *Parvicardium minimum*; (n) *Cardiomya costellata* (*Cuspidaria costellata*); (o) *Bentharca nodulosa*; (p) *Limopsis aurita*. The specimens were taken from several locations. Scale: the white bar represents 3 mm.

Plate 6.10 >2 mm fraction of shell gravel from the continental shelf west of Scotland (56°45.16′N, 07°54.34′W). Note valves of the inarticulate brachiopod *Crania anomala*, the bivalves *Arca tetragona* and *Palliolum tigerinum* (*Chlamys tigrina*) and bryozoan fragments including *Sertella* sp. Water depth 99 m. Width of photograph is equivalent to 10.5 cm.

Plate 6.11 >2 mm fraction of late Pleistocene and early Holocene shell gravel from the continental shelf west of Scotland (55°49.04′N, 08°00.48′W). Note opercular and rostral plates of the barnacle *Balanus balanus*, worn valves from the bivalves *Chlamys* sp. and *Nucula* sp. and tubes of the polychaete *Ditrupa arietina*. Water depth 130 m. Width of photograph is equivalent to 14.5 cm.

storms to provide additional energy to help erode and transport sand. Sand ripples are expected to be a normal feature on the surface of the sand patches.

(c) *Potential preservation of sand ribbons and sand patches*

Since the whole of the thin sand layer of sand ribbons is subject to intermittent transport they have a poor preservation potential as deposits. Thus, if after formation there is a permanent but gradual decrease in the sand transport rate then their fairly small thickness of sand could readily be reworked into sand waves or even sand patches that are in equilibrium with the new flow conditions. Nevertheless it seems likely that in exceptional circumstances they could be preserved. For example, they could be preserved if they were formed by a temporary strong current that ceased abruptly, and the usual currents were too weak to move much sand.

The longitudinal sand patches, however, are probably incipient deposits. Given a supply of sand and sufficient time for it to be distributed they are expected to grow and amalgamate into a much more extensive sand sheet (Belderson and Stride, 1966). Likewise, the 'sand streams' of Dobson, Evans and James (1971) may, in areas of somewhat stronger currents, probably represent the lateral amalgamation of sand ribbons where the sand is locally more abundant. In such occurrences, however, any substantial accumulation of sand is likely to be fashioned into a longitudinal belt of large sand waves or possibly even into a sand bank (Section 3.5 and Fig. 3.2b).

3.4.5 TIDAL SAND BANKS

Tidal sand banks (tidal current ridges of some workers) are the largest bedform of seas with strong tidal currents (Plate 3.19). Those examples that are active in European seas at the present time can be up to about 50 km long, 6 km wide and 40 m high (Table 3.3), although none reaches the extreme size in all dimensions. Their use in sand transport studies is summarized in Section 4.2(b), and their texture, composition, internal structure and potential preservation are considered in Section 5.3.

Table 3.3 Maximum known dimensions of longitudinal sand banks of the modern tidal current environment of north-west European seas

Location	Length (km)	Breadth (km)	Thickness (m)	Angle of steeper slope degrees
Offshore	52	3	43	6
Estuarine	27	6	20	4
Banner	33	1	25	1

The tidal sand banks should not be confused, as sometimes in the literature, with the shoreface-connected sand ridges of continental shelves with weak tidal currents where most of the sand transport is caused by wind-induced currents of occasional severe storms, or with nearshore bars related to the zone of breaking waves (or even with the sand ribbons or sand waves described above). A group of large sand bank-like features in the North Sea to the east of Flamborough Head were termed 'sand hills' by Dingle (1965), who thought from the limited data then available that their long axes were at right angles to the peak tidal flow. However, these features are now known from Institute of Oceanographic Sciences, Hydrographic Department, and SEASAT data to be aligned at only the expected small angle to the direction of peak tidal flow, as in the case of the other sand banks of tidal seas.

The sand banks of tidal seas either tend to occur in groups, both offshore and in estuaries, or else as solitary near-coastal and banner banks in the lee of headlands, islands or submerged rock shoals (Figs 5.14 and 5.15). The modern sand banks are commonly separated by a floor of gravel, and the tidal currents passing over them are usually stronger than those associated with the modern off-shore sand sheet facies (Section 5.3). Since the grain size of the sand within the banks is finer than would be expected with the existing current strength, the sand must therefore either be trapped in an estuary after being transported from seawards, or lie in the lee of an obstacle, or be supplied to one side of a region faster than it can be removed from the other side, or have been originally so abundant that there has not yet been time for all of it to be carried away.

Some tidal sand banks, particularly near the coast and in estuaries, are strongly 'V' or 'S' shaped in plan view. The shapes of these have usually been related to mutually evasive ebb or flood dominant channel systems. In some cases they may represent forms that are transitional between a single linear sand bank and three adjacent linear sand banks resulting from its split up (V. N. D. Caston, 1972). Many offshore sand banks, however, are linear or only slightly sinuous-crested. These are sometimes relatively broad and round at one end and rather pointed at the other, the broader end being at the 'upstream' end of the bank in terms of the net sand transport direction (G. F. Caston, 1981). The linear sand banks were originally thought to extend parallel to the peak tidal currents (e.g. Off, 1963), but have now been shown to have long axes that are generally oblique to the regional peak tidal flow by as much as 20° (Kenyon, Belderson, Stride and Johnson, 1981).

The spacing within groups of sand banks tends to increase with increasing depth of tidal flow between the banks (Off, 1963; Allen, 1968), although there is much scatter in this proposed relationship. The scatter may be an indication that some of the sand banks are not adapted to the present water depth because they were formed at lower sea levels. Their spacing is likely to be determined by initial water depth and current regime and also perhaps by relative abundance of sediment and its grain size (Section 5.3.1).

(a) Actively maintained and moribund sand banks

The tidal sand banks of north European seas may be divided into those that are actively maintained by the modern (late Holocene) tidal current regime, and those formed at times of lower sea level which are now in a moribund state (Figs 5.14 and 5.15, and Kenyon *et al.*, 1981).

Various criteria have been used to distinguish between these. The actively maintained ones are found where the near-surface mean spring peak tidal currents generally attain well over 50 cm/s (1 knot); they have large sand waves present upon them; the bank crests (when not shallow enough to be flattened by wave action) tend to be sharp; they are generally asymmetrical in cross section, with the steeper of their two slopes approaching a maximum angle of 6°; their crests are relatively shallow, often approaching low-tide sea level, where the effects of wave action may plane them off; and they are usually separated by gravel floors. When confined in estuaries, with a plentiful supply of sand and smaller, shorter period sea waves, they have broad, flat tops, many of which are exposed at some low tides.

In contrast, the moribund sand banks are found where the tidal currents now reach a near-surface mean spring peak speed of less than about 50 cm/s (1 knot); they are not expected to have large sand waves upon them; their crests have rounded profiles and their slopes are only 1° or so; they are separated by sandy or muddy floors; and their crests are in relatively deep water.

No absolute demarcation can be drawn between the actively maintained and the moribund tidal sand banks. The term moribund itself implies a possible small amount of sand movement. The moribund phase is followed by the burial phase. For example, in the outer Celtic Sea there is an extensive group of moribund sand banks (Fig. 5.14) presumably first formed at lower sea levels (Stride, 1963b; Bouysse, Horn, Lapierre and Le Lann, 1976) which now have gently dipping slopes. Those in the region of weakest modern tidal currents (the northern-most ones) have rounded crests and an absence of sand waves. However, along the length of each sand bank and laterally south-eastwards from bank to bank their crests become sharper and sand waves occur upon them as they come increasingly under the influence of stronger modern tidal currents. Thus, there is both a longitudinal and a lateral transition from a predominant moribund state to a more active state within this group of sand banks.

(b) Axial obliquity of sand banks

The long axes of the great majority of the sand banks (other than those tied to headlands) are oriented at a small oblique angle relative to the peak tidal flow direction, generally in an anticlockwise sense. (It must be remembered that the observed sand banks are in the northern hemisphere.) The usual angle between the long axis of the sand bank and the regional peak tidal flow direction is between about 7° and 15°, but with extreme values ranging from 0° to 20° or more (Kenyon *et al.*, 1981).

Most tidal sand banks have asymmetrical cross-sectional profiles. Where such is the case the steeper slope is observed to be on the side of the bank that faces obliquely down the general direction of regional net sand transport (Section 4.4.2). Thus, for the purpose of predicting regional net sand transport paths the asymmetrical cross-section of a sand bank may, in general, be used as a directional indicator in much the same way as if it were a sand wave aligned at a highly oblique angle to the peak flow, except that one must also know whether the long axis of the sand bank is oriented clockwise or anticlockwise relative to that flow. The bank should therefore tend to migrate bodily in the net 'downstream' sand transport direction, unless tied to an obstacle. This is consistent with available data on the internal structure of sand banks (Section 5.3.3), where large-scale growth or migration surfaces lie parallel to the steeper face of the bank (e.g. Houbolt, 1968). The long-term migration of some of the Norfolk sand banks in the direction faced by their steeper side was demonstrated by Houbolt (1968) and V. N. D. Caston (1972).

(c) *Sand wave orientation on sand banks*

The modern, actively maintained, sand banks all have large sand waves upon them. The crests of the sand waves on the upper slopes of the sand bank are usually oriented obliquely to the bank crest and have their steeper slopes facing towards the crest from either side of the bank (e.g. Houbolt, 1968; Caston and Stride, 1970; V. N. D. Caston, 1972). The sharp crestlines of such sand banks are thus maintained by sand waves driven obliquely towards the crest from either side. So long as the sand banks were thought to be parallel to the regional peak tidal flow, it was necessary, in order to explain this sand wave 'circulation' system on the banks, to have recourse to mutually evasive ebb–flood channel systems (as occur in some sand-choked estuaries) or to specify ebb-dominance on one side of a channel and flood-dominance along its other side. But, as described above, the sand banks generally lie slightly oblique to the peak tidal flow. Thus, the explanation for the opposed sense of sand wave migration on either side of the crest is found in the relatively greater exposure of one side and protection of the other side of the bank (together with a narrow band of adjacent floor), during peak flood flow, and vice versa during peak ebb flow. The degree of obstruction of the sand bank to the tidal flow is sufficient to allow the local net sand transport direction on the lee (steeper) face of the bank to be opposite to the net regional direction.

In rather more detail the observations show that the sand waves, as they approach the crest of the sand bank, do not maintain their crestlines at approximately normal to the regional peak tidal flow, but bend around to become more parallel to the crest of the bank. In the vicinity of the British Isles the sense of this veering over banks offset in an anticlockwise sense from the peak tidal flow is towards the right as the sand wave approaches the bank crest (Fig. 4.25, Model a). On the few offshore sand banks with axes offset in a clockwise sense from the peak tidal flow the sand wave crests are found to veer towards the left on approaching the bank crest (Fig. 4.25, Model b).

At least part of this veering of the sand wave crests towards the sand bank crest is attributable to variations in the cross-bank component of the current during the tidal cycle. During much of the tidal cycle the cross-bank component will represent a greater proportion of the total flow than it does at peak flow. Moreover the bank itself will also present an increasingly greater obstruction to the flow the greater the angle of approach of the flow towards the crest. This is due to the much reduced depth of flow over the obstruction. Furthermore, the wider the tidal envelope, the greater will be this effect, so that overfalls and much sand transport can take place across the crest of the bank (e.g. Stewart and Jordan, 1964). This may account, for instance, for the observed parallelism of sand wave crests to the bank crest on Cultivator Shoal, Georges Bank, USA (Uchupi, 1968), where the tidal envelope is wide. Despite such potentially destructive action, the sand banks have frequently grown and been maintained near enough to the sea-surface for their crests to be strongly affected by wave action, or even to be exposed at low tide. This shows how strong is the instability which caused the banks to be built and/or how great is the volume of sand which has to be arranged into bedforms in the neighbourhood.

(d) *Origin of linear offshore sand banks*

Offshore sand banks have frequently been ascribed to pairs of helical circulations rotating in opposite senses with axes parallel to the peak tidal current. However, the actual obliquity of bank axes to peak flow would require the same obliquity in helical circulations if such circulations originally determined the bank orientation. Near-surface Langmuir circulations in the sea are well-known. However, there appears to be no observational evidence of helical circulations over the whole depth of water on the offshore parts of the continental shelf. Thus, if such flow is used to explain sand bank formation it is necessary to invoke laboratory evidence. Here instabilities of a rotating fluid in the form of helical circulations (roll vortices) were shown to have axes inclined at angles either $0°$ to $8°$ or $14.6°$ (in the opposite sense) to the mean current (Greenspan, 1968). The spacings were, respectively, 25 to 33 and 11.8 times the boundary layer thickness. Theory predicts similar instabilities (Gammelsrød, 1975, giving earlier references). Theory also appears to give an adequate explanation of Langmuir circulations near the sea surface and longitudinal helical circulations/roll vortices in the lowest few kilometres of the atmosphere (of which there are some direct observations, Hanna, 1969). However, this work for the lower atmosphere, laboratory and near-surface boundary layers is not directly applicable to the whole depth of currents on the continental shelf. One difference is that the ratio of spacing of circulation axes to boundary layer thickness in these boundary layers in the air is one or more orders of magnitude less than what would be the corresponding ratio for sand bank formation. Also convection frequently plays a part in the lower atmosphere circulations.

Whatever their origin, once the offshore sand banks have formed helical circulations may develop in conformity with bank axes. These will assist in building up the banks further. However, such circulations are probably weak in view of the small slope angles of the sand banks (Huthnance, 1981).

Smith (1969) suggested that cross-bank components of flow are important in sand bank building and maintenance. Even if the sand bank axes were aligned parallel to the peak current, cross-bank components could arise if the tidal current envelope was not too narrow, or because of non-tidal currents. Cross-bank current components are amplified near the sand bank crest due to the reduced depth there, corresponding to the turning of resultant current. Of course, the more recent discovery of the general obliquity of sand bank axes to peak tidal current direction provides another source of cross-bank current components. The problem has been analysed theoretically by Huthnance (1981) for a rectilinear tidal flow far from sand banks. This gives a mean flow (tidal residual) due to the effects of earth rotation and bottom friction (the latter being stronger the shallower the water, Huthnance, 1973). First, it is necessary to assume the presence of an 'initial bed waviness' of small amplitude with contours making angle α say with peak current direction. The theory then shows that a wide range of initial wavelengths would grow in amplitude. Growth rate would be a function of wavelength. It would have a broad maximum for a wavelength about 250 times the water depth (e.g. 7.5 km for 30 m depth), and be approximately 30% less for a wavelength of twice or half that. The growth rate is insensitive to the assumed contour angle α. For example, for a wavelength 250 times the water depth the growth rate is predicted as only 10% less for $\alpha = 15°$ than for the bank orientation of $\alpha = 27.8°$ (giving maximum growth rate), and only 20% less for $\alpha = 42°$. Huthnance thus suggests that the sand bank orientation which eventually predominates may be susceptible to 'external' influences, such as the trend of a coastline. In principle, the developing sand banks would have separation and orientation equal to those which are predicted to give fastest-growing amplitude.

(e) *Banner banks*

The banner banks are generally the smallest of the sand banks (Table 3.3). They can be developed in various situations such as in the lee of headlands, islands, submerged rock shoals and gaps in rock ridges. They are sometimes paired, with one larger than the other indicating the net direction of sand transport. They have a different origin to the sand banks discussed above, since they are caused by the

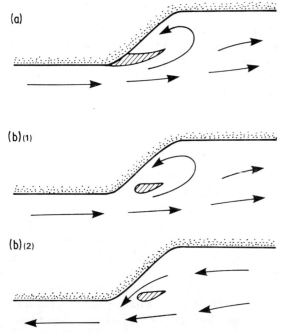

Fig. 3.9 Hypothetical explanation for the formation of banner banks. A sand spit, normally attached to a headland under unidirectional current conditions (a), is kept detached from the headland by a reversing tidal flow (b(2)) which flows strongly from the bay past the headland.

presence of fixed obstacles. They may therefore be considered as giant sand shadows or obstacle marks. In plan the smaller ones are almost pear shaped, with the broader end of the bank nearest the associated headland or other obstruction. The larger ones tend to be more linear.

Skerries Bank, near Start Point, England, is a pear shaped banner bank that has been studied in some detail. The ebb tidal current is dominant on rather more than the western half of this sand bank while the flood tidal current is dominant on its eastern side (Acton and Dyer, 1975). This information is confirmed by sand wave asymmetry (Kelland and Bailey, 1975). The sand bank would appear to have grown as a result of the separation of ebb and flood paths resulting from the projection of the coast at Start Point. The flood tidal current flows more strongly outside Start Point while the ebb flow is confined by the bay. The pear shaped Warts Bank, off the Isle of Man in the Irish Sea, is in a similar situation, with one side swept preferentially by the ebb and the other by the flood tidal current, the two portions sloping up to a median ridge that is oblique to main tidal flow (Jones, Kain and Stride, 1965).

A further situation, where an island obstructs the tidal flow, can lead to two sand banks being developed. The longer sand bank extends outwards from the island on the side facing down the regional net sand transport direction. Where a long island lies athwart the peak tidal flow directions (such as Lundy, in the outer part of the Bristol Channel) it is even possible for such a pair to be developed at either end of the island, although only two of them are large in the case of Lundy (see also Plate 3.20 for an example from Torres Strait).

In contrast, when a rock ridge is well submerged the associated sand bank is developed on the side facing down the direction of net sand transport, with only a minimal amount of sand on the 'net upstream' side. Such an example is the Shambles Bank whose east end is tied to a rock shoal rather than being tied to Portland Bill itself, as Cornish (1914) thought. It has been argued (Pingree, 1978; Pingree and Maddock, 1979) that this sand bank is due to an eddy (shown by numerical modelling of the tidal currents) developed on the east-going tide. However, it was too simplistic to argue as did Pingree (1978) that it was due to an eddy in the tidal residual current. For instance, the bank does not lie in the centre of the supposed eddy but to one side of it, so that the bank may still owe its exact location to the rock shoal at its eastern end.

An alternative simple explanation for the formation of banner banks is proposed in Fig. 3.9. This supposes that a sand spit which would normally be attached to a headland under a unidirectional current or longshore drift regime, is, in a tidal current regime, detached from the headland by the locally concentrated reverse tidal flow adjacent to the headland. Pingree's model should apply to an eddy produced by a unidirectional current. If this is correct, then examples of such banner banks (as opposed to spits) should be evident from strong unidirectional currents. The authors are unaware of any such examples.

Sand banks can also be associated with gaps in a high rock ridge that extends across the tidal flow. In the case of the Chaussée de Seine, west of Brittany, there is a sand bank occurring both to north and south of such a gap (Hinschberger, 1963). The growth of sand banks associated with submerged rock shoals is likely to be limited both in height and width by the size of the shoal itself. With a meagre sand supply the resulting bedform may only be thin and have some resemblance to a sand ribbon (Belderson and Kenyon, 1969).

Similar bodies of sand have been seen in deserts where they have been described either as sand shadows or sand drifts (Bagnold, 1941). The main difference is that the tidal current bed features are sometimes paired, although the larger deposit points from the obstacle in the direction of net sand transport. This paired development would seem to be a useful criterion for recognition of formation by a tidal, as compared with a unidirectional current.

3.5 Relationship between bedforms

Gradation between different bedforms can occur both with changing peak current speed, and with constant peak current speed but increasing sand availability (Figs 3.1 and 3.2). Furthermore, the development of some bedforms may be triggered by the presence of others. Thus, although it is useful to delineate bedform 'zones' as typified by the presence of a particular bedform, this will not mean that that bedform is spread continuously throughout the zone, or is indeed the sole bedform type found in that zone. Some examples of these relationships are given below.

(a) *Effect of a longitudinal velocity gradient*

Sonographs show that narrow sand ribbons can evolve from longitudinal furrows so that the two bedforms appear to grade one into another down a longitudinal velocity gradient of tidal flow. This might simply imply that the furrow is acting as a conduit for any available sand, which progressively becomes concentrated by the 'bunching' effect of the 'downstream' decrease in current speed. If such is the case then it is perhaps an argument in favour of Flood's (1981) proposal that both furrows and ribbons are situated beneath converging pairs of helical circulations, which would require to be very long.

(b) *Effect of increasing sand availability*

Sand ribbons, some longitudinal trains of sand waves, and sand banks can all occur at the same peak current speed. They possibly represent a sequence of bedforms developed in response to increasing sand availability, as is the transition from isolated barchans to fields of sinuous-crested sand waves.

(c) *Localized concentrations of sand*

The wing-tips of barchan-type sand waves are frequently associated with sand ribbons, which implies the loss of sand from them down the direction of net sand transport. Likewise the obstacle marks behind boulders, rock-outcrops, wrecks and pipelines often develop into clearly defined sand ribbons. Here the features are acting as foci for the sand ribbon development (possibly in the same way as mentioned above for furrows) by diverting and concentrating sand flow. They may, at the same time, also trigger off longitudinal helical circulations which maintain the sand ribbons.

(d) *Co-existent bedforms*

Sand waves and sand ribbons can occur as co-existent partners. The ribbons may either be considered to be trains of small sand waves or the sand waves to be secondary co-existent partners upon the sand ribbons. The superimposition of bedforms, one upon another is a well-known phenomenon in deserts, where sand ripples are nearly always found on the flanks of aeolian dunes and dunes on the flanks of larger features, such as the draa of the Sahara. Similarly, in the case of bedforms made by tidal currents the active small sand waves commonly have sand ripples upon them (for medium and fine sand), while the active large sand waves commonly have small sand waves upon them and perhaps ripples on these (Section 3.3.3(g)). Should the smaller of these bedforms be absent from the larger one then the latter is likely to be in a more or less moribund state (Section 3.4.5), depending on whether the absence of the smaller bedform is permanent or only temporary. Additional support for this view was provided by Jackson (1976) who doubted the validity of Allen and

Collinson's (1974) statement that in such cases where several orders of bedform are superimposed, only one order is active.

3.6 Aeolian equivalents

Desert bedforms are widely enough known and similar enough to need some mention here since they provide useful analogues without being exactly similar to many continental shelf bedforms. Desert bedforms of sand attain a much greater maximum size than their submarine equivalents. This is because they are not restricted by the very large effective depth of the air flow. In particular, the transverse aeolian barchan dunes may be the equivalent of small marine sand waves, while large marine sand waves may find their aeolian equivalent in the slow-moving, regularly spaced draa of North Africa. The draa sometimes have slip faces, but more commonly are covered on both windward and leeward slopes by dunes. In contrast, the aeolian sand ripples and barchan dunes generally have slip faces inclined at the residual angle after avalanching.

Aeolian sand ripples tend to have straighter crests than underwater current ripples of sand. This has been attributed to saltation being the predominant bed-load sand transport mechanism rather than rolling as in underwater bed-load movement of sand (Bagnold, 1941). Also suspension transport only applies to grain sizes less than about 0.08 mm ('dust').

The shapes and patterns of aeolian bedforms larger than sand ripples tend to be more variable than those of tidal-current bedforms larger than sand ripples. This may be due to the presence of thermal convection cells and eddies during hot days when winds are light, as well as because the winds usually have a wider spread of directions than those of most tidal flows at the times when they are moving sand. Exceptions can occur over sand bank crests and in shallow estuarine waters where interference patterns of small sand waves have been seen.

In deserts, barchan dunes are associated with uni-directional winds (varying up to 20° about a mean direction) blowing over hard surfaces with sparse supplies of sand. On a pavement over which sand is being rapidly moved, aeolian dunes can only develop at the convergent nodes of the flow pattern. These will be in the form of barchans. In the divergent nodes the flow is too fast for stable linguoid forms to develop, which should otherwise be present there (Cooke and Warren, 1973). Barchans skewed in plan view to a varying degree as observed on the sea floor are also found in deserts, and have been attributed to asymmetry in wind pattern or in sand supply.

The longitudinal tidal current bedforms likewise have their aeolian equivalents. Scour hollows and longitudinal furrows have been observed (e.g. Cooke and Warren, 1973, p. 252), while sand shadows, obstacle marks and sand ribbons are all common aeolian features (e.g. Bagnold, 1941). The linear seif dunes may also be analogous in some way to tidal sand banks. One point of difference is that the tidal sand banks rarely have marked 'Y' or tuning fork shaped junctions of the type which occur with some seif dunes. Sets of linear seif dunes cover great areas of desert, whereas offshore tidal sand banks have relatively few in a set. The reason is that tidal current speeds vary over much shorter distances than do the prevailing winds over continents.

3.7 Main conclusions

1. Bedforms in flumes can be a useful guide to bedforms made by tidal currents on the continental shelf provided the effects peculiar to the small flume depth are not extrapolated up, and account is taken of modifications due to the reversing flow.

2. Each member of the suite of distinctive bedforms found on the continental shelf around the British Isles can be related empirically to the near-surface mean spring peak tidal current speed in the vicinity.

3. Thus, on travelling down a longitudinal velocity gradient of a tidal current, the various bedforms can be assigned to a sequence of zones placed in relation to the local strength of the tidal current.

4. Although a bedform may typify a particular zone, this does not mean that the bedform necessarily occupies the whole of that zone, or indeed that it is the sole bedform to be found there. This is because the supply of sand determines the degree of development of various bedforms.

5. Much theoretical and observational work remains to be done concerning the origin and morphology of tidal current bedforms. For example, the wide variations in morphology (such as slope angles) of sand waves in relation to local conditions are little understood.

6. Tidal current bedforms are summarized in Table 3.4.

Table 3.4 Summary properties of continental shelf tidal-current bedforms. Numerical values for predominantly clastic quartz-density sands.

Bedform: sections of Chapter 3 with main discussion	Observed wavelength or separation	Observed height, crest to trough	Observed slope of steeper side, excluding subsidiary crests and near-trough rounding	Median grain size ranges, observed or predicted	Mean springs near-surface peak speed range for which bedform observed (cm/s)	Probable origin
Transverse bedforms, i.e. with long axes within about 20° of perpendicular to peak tidal current direction							
Sand ripples 3.2.1, 3.3.2	<60 cm	<5 cm	Avalanche	<0.7 mm at 15°–20°C	Less than sand wave minimum, also higher speeds for ripples on top of larger bedforms	Turbulent 'bursts' reaching the bed, viscous forces important
Predicted very low-height sand waves (or gravel waves) 3.2.1. Not yet recognized from the sea	50–c.200 cm	<5 cm (defined)	Avalanche	<0.7 mm at 15°–20°C, also gravel	Predicted speed range the minimum for sand or gravel waves down to about ¼ of it (from flume data)	Turbulence within a metre or so of the bed (viscous forces unimportant)
Small and large sand waves, also gravel waves 3.2.1, 3.3.3	Up to ×20 water depth	5 cm to 1/3 water depth	Small sand or gravel waves: Avalanche slopes. Large sand or gravel waves: usually 4°–20° (asymmetrical), 10°–20° (symmetrical)	Sand > ~0.15 mm, gravel	> about 65 for large sand waves, more for gravel waves, slightly less for small sand waves with much less building time	Turbulence in whole current, or the lowest 10 m or so for small sand waves on top of other bedforms
Tidal lee-wave sand waves 3.3.3	Determined by the lee waves, probable range 100 m to several km	Up to 12 m for the Celtic Sea examples	1°–4°	> ~0.15 mm	> ~70	Tidal-period internal-waves in temperature- or salinity-stratified water
Transverse sand patches 3.3.4	Variable	<4 m	Up to avalanche at edges	Sand	<50	Accumulations of sand 'swept' from surrounding gravel floor

Table 3.4—cont.

Bedform: sections of Chapter 3 with main discussion	Observed wavelength or separation	Observed height, crest to trough	Observed slope of steeper side, excluding subsidiary crests and near-trough rounding	Median grain size ranges, observed or predicted	Mean springs near-surface peak speed range for which bedform observed (cm/s)	Probable origin
Longitudinal bedforms, with long axes within about 20° of peak tidal current direction						
Scour hollows 3.4.1	Individual features	Max. hollow depth found, 260 m	Depends on local sediment consolidation	Not relevant	>150	Local erosion by strong currents: longitudinal for uniform material
Furrows in gravel 3.4.2	<100 m (width up to 30 m)	Max. furrow depth found, 1 m	Varies	Gravel	>150	Helical circulations
Furrows in mud 3.4.2	<50 m (width up to 15 m)	Max. furrow depth found, 1 m	Varies	Mud	70 (for the one set observed)	Helical circulations
Obstacle marks 3.4.3	Individual features	Up to tens of metres (banner banks): scour down to 15 m (wreck marks)	Varies	Sand	>75 for large obstacle marks	Obstacle changes the current and turbulence intensity patterns
Sand ribbons 3.4.4	Typically 30–1000 m	<1 m	Sharply defined to feather edge	Sand, with gravel floor between	75–150	(1) As for patches (2) Helical circulations, with axes parallel to peak current
Longitudinal sand patches 3.4.4	Variable	<3 m	Sharply defined to feather edge	Sand, with gravel floor between	<50	Accumulations of sand 'swept' from surrounding gravel floor
Active sand banks 3.4.5	5–20 km	Up to low-tide water surface: max. known height 43 m	Up to 6°	Sand, usually with gravel floor between	>60	Similarly to sand waves, by the cross bank components of sand transport
Moribund sand banks 3.4.5	5–20 km	<55 m	c.1°	Sand, with sand or mud floor between	<50 (now)	Ditto

Chapter 4

Sand transport

4.1 Introduction

Tidal currents acting on their own are competent to move sand on large parts of the continental shelf around the British Isles and other strongly tidal shelves at appreciable rates for appreciable percentages of the total time. Furthermore, when aided by other water movements, sand can be moved over most of the areas of such continental shelves. Gravel can also be moved in strong-current areas.

The continually changing strength of the tidal currents (over periods ranging from hours to millenia) complicates the description of their effects on sand transport, especially as the added effect of other water movements must be considered as well.

The chosen approach begins with a discussion of the relation of sand and gravel transport rates to tidal current speed and points to the difficulty of giving absolute rates of sand and gravel transport. The peak speed (ebb or flood) of the mean spring tidal currents (Fig. 2.6) is then used as a good general guide to the regional importance of the tidal currents, and is compared with the observed distribution of bedforms.

The case is then made that the direction of the stronger of the peak ebb and flood currents can generally be used to indicate net sand transport direction (Fig. 4.1), exceptions (Figs 4.3 and 4.6) being rare. These directions are available for many positions. They are compared with net sand transport directions on the bed revealed by widespread bedform and other data. In historical terms it was these data that provided the links between sand transport and tidal currents.

Finally, the effects of the other stages of the spring–neap and other tidal cycles and the combined effect of the tidal currents and other water movements are considered in general terms and are related to the overall sedimentation pattern. By such a progressive approach it is possible to establish principles and gradually lead up to situations that resemble those existing in the sea, without presenting all the necessary details at the start.

In addition to inference from the widespread bedforms, actual sand transport is manifest from direct observations such as those made by divers, or by means of television, submersibles, the use of various tracers, sand traps, burial or exposure of man-made objects, measured sand wave and sand bank movement and changes in shape, and from the noise of sand movement on the crests of sand waves observed by acoustic means (Plate 4.1). The conformity of the several lines of evidence provides a satisfactory measure of proof that the overall sedimentation pattern that is inferred is essentially correct.

Sand transport

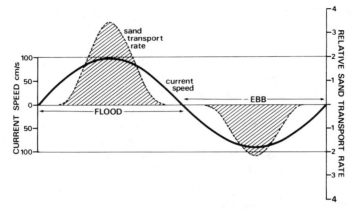

Fig. 4.1 Rectilinear tidal currents of equal ebb and flood duration but different peak speeds. This is the usual situation in offshore tidal seas. It is responsible for a greater sand transport in the direction of the stronger current and therefore for causing net sand transport in that direction. The relative sand transport rate at different current speeds is taken as the cube of current speed minus a threshold current speed of 30 cm/s, times a constant that will give convenient numerical values. Lag effects have been neglected. They would displace the sand transport rate curve somewhat to the right, but would have little effect on the total transport.

4.1.1 AVAILABILITY OF SAND FOR OFFSHORE TRANSPORT

Sand occurring on continental shelves at the present time has a number of different origins. Each has provided some material that was or still is out of equilibrium with modern water movements and so has been redistributed by them or will be distributed by them in the future. Most of it comes from land deposits that were encroached on during the early Holocene rise of sea level. Since then additional sand has reached the continental shelf via rivers, or as a result of coast erosion, wind and ice transport, or has been derived from the break-up of shelly faunas or grown as ooliths, for example.

An oft-repeated statement in the past was that little, if any, sand is now being lost from the land for distribution offshore. While not wishing to overemphasize such loss, it is worth mentioning some possible mechanisms, although strictly the discussion of them is outside the scope of the present book. For instance, the delta sediments of the Fraser River in western Canada are being reworked and removed by the strong tidal currents flowing across the delta front. On many coasts which have heavy swell or waves, these cause strong local seaward flows (rip currents) which aid the passage of sand outwards from the coast to ground beyond the breaker zone.

Sand held in suspension by wave activity at high water will also move offshore as the tide level falls to low water, so providing an opportunity for some of the finer sand to stay in suspension and be transported further offshore. An implied loss of fine sand to seaward is shown on the East Anglian coast of the British Isles by the coarsening of the sands in the southward direction of sand transport along the beach (McCave, 1978).

Movement of sand to the offshore zone can also be more or less aided by the width of the sand-transporting part of the tidal envelope and by lag in picking up, transporting and putting down sediment (Section 4.2.3). Sand carried along the coast to a headland could be streamed away from its tip. Some of it may then be moved further offshore by the lag effects acting in association with the usual current speed decrease away from the headland.

4.2 Relation of sand transport rate to tidal current speed

4.2.1 SAND TRANSPORT RATE IN FLUMES AND RIVERS

Flume studies for steady uniform currents over a sand bed have provided numerous data sets which include the total sand transport rate, and in many cases its approximate bed-load and suspension-load components. River and irrigation canal studies have provided data sets for greater water depths, usually with most of the suspension-load measured and bed-load estimated from it. However, some of the data sets contain largely unknown lag effects (Section 4.2.3), since current speed was varying along the flow path or changing in time. Data sets for the rate of gravel transport (virtually all as bed-load) are also available. These transport rates have not usually been analysed in terms of current speed but rather in terms of bottom stress. The analysis can use non-dimensionalized transport rates and non-dimensionalized bottom stress (Shields parameter) to cater for different sand densities, water temperatures and viscosities. Theory of dimensions shows (Yalin, 1972) that non-dimensionalized bottom stress cannot be replaced in the analysis by a (non-dimensionalized) current speed, but that a friction factor (often called a drag coefficient) would be required as well. Bottom stress in flumes, rivers and irrigation canals can usually be calculated to adequate accuracy from measurements of water-surface slope (times mean depth, water density and gravity; for flume data, an estimate of the frictional drag on the side walls is subtracted).

4.2.2 RELATIVE SAND TRANSPORT RATE OVER THE SEA BED

For sea-floor applications the bottom stress is difficult to determine, as discussed later, whereas current velocity, even if not measured at the required position and time, can often be numerically modelled or its tidal part predicted from tidal analysis of past measurements. Secondly, bed-load transport rate is likely to be associated mainly with the surface drag ('skin friction') part of the bottom stress (produced by the sediment grains) rather than with the bedform drag part, due to the disturbance to flow caused by the bedforms. 'Bottom stress' will be used to denote the sum of surface drag and bedform drag. The surface drag can be estimated from the logarithmic velocity profile formula (Section 2.5), and is thus determined by current speed. Suspension-load transport rate of sand depends on current speed as well as on bottom stress. Therefore, there will be no discussion of transport-rate formulae in terms of bottom stress for bed-load, suspension-load or their sum derived from flume, river and irrigation canal studies. The well-known divergences in their predictions for any specific case arise largely because many of the formulae are based on data sets covering only parts of the ranges of the variables, especially depth. A few of the formulae are much less unreliable (White, Milli and Crabbe, 1975).

In the sea, water-surface slope is only known closely enough to estimate bottom stress when special determinations of the slope have been made (e.g. Wolf, 1980). Without accurate determinations of water surface slope the sea-bottom stress has to be estimated in some other way. Three methods (Section 2.5) have principally been employed: (1) measurement and correlation of near-bed longitudinal and vertical turbulent velocity components, giving the 'Reynolds stress'; (2) use of near-bed current velocity profiles fitted to logarithmic forms, and (3) by taking current speed squared times water density times a drag coefficient. However, for methods 1 and 2 the derived stress is a small-scale value which will vary over large sand waves (sand ripples, and probably small sand waves would be averaged over). To obtain the bottom stress it would have to be averaged over large sand waves and augmented by their form drag contribution. Also the derived values of friction velocity (square root of small-scale stress divided by density) are less reliable than measurements of near-bed speed. This is because they have a ratio of standard deviation to mean which is several times greater (Heathershaw and Simpson, 1978; Dyer, 1980).

Method (3), in fact, starts from current speed to estimate bottom stress using a drag coefficient. The values of drag coefficient in different locations and its dependence on bedforms are uncertain. Drag coefficient (friction factor) prediction is still one of the main problems in river hydraulics, so that recourse must be had to sea floor data. Many published values for the sea floor (e.g. by Sternberg, 1968; giving a mean value of 0.0031 in terms of current speed 1 m above the bed, Section 2.5) are small-scale values obtained by the methods 1 and 2 given above (dividing bottom stress values by current speed squared). These have high standard deviations, as discussed above, and also illustrated by published individual values of drag coefficient. They refer to surface drag plus the form drag from sand ripples. It is not yet possible to estimate the form drag of individual sand ripples or sand waves reliably, partly owing to lag effects and the changes in the ripples during the tidal cycle (Dyer, 1980), and the greater lag in sand waves (Section 4.2.3). Therefore, it is not yet possible to estimate large-scale drag coefficients by combining contributions from individual sand ripples and sand waves together with the surface drag from the grains, or by combining small-scale drag coefficients just mentioned with estimates of drag from sand waves.

Numerical modelling techniques have included attempts to optimize drag coefficient values to give the best agreement between observed and computed, depth-averaged, M_2 tidal currents. However, the latter do not vary much over a range of drag coefficient values, so that this is not a sensitive way of estimating drag coefficient. Computed M_4 tidal currents or tidal ranges at the coast are more sensitive to drag coefficient value, but are also affected by the detailed differences between different authors' computational techniques, and it is not clear which of the latter is the most accurate. Tee (1977) found that use of a drag coefficient of 0.003 instead of 0.001 did not appreciably affect the direction of tidal residual current (in the Minas Channel and Basin, Bay of Fundy), while decreasing its magnitude. Commonly, a constant drag coefficient (usually 0.0025 with current speed averaged over depth) has been used over the whole area of interest. Exceptions are made in some cases with an increased coefficient in shallow water, e.g. inverse proportionality to depth for depths <30 m (Pingree and Griffiths, 1979).

It is considered that the above discussion justifies use of a formula for sand transport rate in terms of current speed for the discussion of tidal-current sedimentation in this book. A few estimates of relative sand transport will be made in Figs 4.1, 4.2, 4.3, 4.6 and 4.23. For these, the total sand transport rate is assumed proportional to a coefficient (actually a function of grain size and sorting) times the cube of the amount by which current speed at 3 m above the bed exceeds a threshold value. Since only relative transport rates for different current speeds are required, a numerical value of this coefficient is not needed (although one is assigned in some of the diagrams to give convenient numbers). Such a relationship is well-established for aeolian bed-load sand transport rate (Bagnold, 1941) for which it can be slightly modified to cater for unusual degrees of sand sorting (Chiu, 1972).

For underwater sand transport, proportionality of bed-load rate to the cube of current speed without subtracting a threshold would, with assumed constant drag coefficient, correspond to the proportionality of bed-load rate to fluid power (bottom stress times flow speed) derived theoretically by Bagnold (1966). Suspension-load sediment transport rate was derived by Bagnold (1963, 1966) as proportional to fluid power times flow speed divided by mean settling speed (fall velocity) of the grains in suspension. That is, it is proportional to the fourth power of current speed. Often overlooked is Bagnold's emphasis that his arguments that led to the above formulae assumed negligible flow resistance (form drag) from bedforms. For flow over sand ripples or sand waves this form drag must impose lower transport rates, particularly for bed-load (averaged over the ripples or sand waves). To find expressions for this reduction would require extensive analysis of flume and river data. In the meantime, the procedure of subtracting a threshold value from current speed, before cubing it, is adopted here. This means of effecting a reduction has been shown to agree with flume data of Guy, Simons and Richardson (1966) – using data for 0.19 and 0.45 mm sands (Gadd, Lavelle and Swift, 1978). It has commonly been used, as by Allen (1980a) and Heathershaw and Hammond (1980). The empirical curves given by Colby (1964), based on extensive flume and North American river data and shown to give reliable estimates of irrigation canal sand transport (Chaudry, Smith and Vigil, 1970) can be shown to agree well with this form. All of these comparisons were with observed total sand transport rates.

The 'threshold' values subtracted from current speed which give best agreement between the cubic formula and observed transport rates are not necessarily the same as 'movement threshold' values, for the grain size concerned, determined from flume experiments done specifically to investigate the beginning of movement. Also flume definitions of movement threshold have differed. Most require considerable continuing movement of grains whereas a few low density or initially exposed grains can move at current speeds well below most threshold values. Estimates from the above cubic transport rate formula (or most other prediction formulae) for the relatively low current speeds that give sand ripples or very low height sand waves can thus be in error by larger factors than for higher current speeds. Fortunately, however, the absolute transport rates are extremely small, and several orders of magnitude smaller than transport rates over sand wave beds. Thus, they are not significant for total sand transport. Also, the two threshold values (16 and 19 cm/s) found to give best agreement between the (current minus threshold) cubed formula and the flume transport rates of Guy, Simons and Richardson (1966), used by Gadd, Lavelle and Swift (1978) referred to the flume depths of 9–33 cm. These would correspond to considerably larger values of current speed at the level of 3 m above the bed adopted here for tidal currents. A threshold value of 30 cm/s will be used for illustrations in this book (20 cm/s in Fig. 4.11). In view of the above discussion this is not intended as an exact movement threshold value.

The same cubic formula also gives reasonable estimates of relative gravel transport rate, using a much higher threshold current speed appropriate for gravel. The sand and gravel transport rates and bottom stress discussed above are moderately large scale values averaged over sand waves. Local transport rates over different parts of a sand wave, or sand bank, vary from cubic formula rates and depend on local slope angle and turbulence intensity as well as local current speed.

Most of the available observations of tidal currents refer to their strength fairly near the sea surface. Furthermore, most numerical models are two-dimensional and give average current speeds over the water column. Such values need to be reduced to obtain values at 3 m (or some other small distance) above the bed. Section 2.5 lists some formulae for velocity profiles. Most of the recent current-meter recordings were made either much closer to or further from the bed than 3 m, which was therefore chosen as a compromise. For the Southern Bight of the North Sea a value of 0.75 will be used for the ratio of 3 m above-bed to near-surface tidal current speed. This is an approximate average for recorded current values at these levels at near peak ebb or flood (Deutsches Hydrographisches Institut, 1958). It happens to correspond to a one-seventh power law (as in the second profile formula listed in Section 2.5). The tidal-current periodicity makes little difference (Section 2.5) to the velocity profile at or near peak ebb or flood. Areas where much of the total current is non-tidal can have different current speed profiles.

4.2.3 LAG EFFECTS IN TIDAL CURRENT SAND TRANSPORT

Currents which vary in speed with time, as do tidal currents, produce bed-load and suspension-load sand transport rates which are determined by previous water movements as well as those at the time. Such 'lag' effects can significantly alter the direction of net sand transport and mean that bedforms are not generally in equilibrium with the current.

Bedforms also show a lag with respect to sand transport rates, because of the finite time it takes to move the sand that builds or otherwise changes them. Thus, previous sand transport rates are involved. However, bedform lag itself affects lag in bed-load and suspension-load transport rates since the bedform dimensions have a major influence on bottom drag coefficient and bottom stress.

Sand grains that are moving adjust within a few seconds to a change in local water movement direction so that it will be assumed that sand transport rates, which involve the large number of individual grain velocities, are directed parallel to the current. (Transport rates and current vector are here assumed to be averaged over the same length of time, say a few minutes, to average out shorter-duration turbulent fluctuations). For rotary tidal currents this neglects any systematic directional lag in the mean square values of the turbulent velocity components and the correlation between longitudinal and vertical components which leads to the Reynolds stress. No such directional lag of these mean squares and correlations seems to have been reported, unlike their lag in magnitude, mentioned presently. Then the bed-load

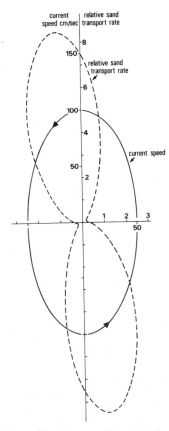

Fig. 4.2 Schematic diagram to illustrate the principle of sand transport rate lagging behind a change of current speed during a tidal cycle. The lag time will depend on grain size and any sand ripples and sand waves present. It is here assumed to be 1/24 of the predominant tidal period, i.e. about half an hour throughout a twice-daily cycle. This makes the direction of net sand transport 7.5° anticlockwise (for a tidal current vector that rotates in the anticlockwise sense around an elliptic tidal envelope).

transport rate vector may be taken as having the magnitude calculated from the current vector at a time earlier by the bed-load lag time. The same would apply for suspension-load and the total transport rates (illustrated in Fig. 4.2 for an idealized situation of an elliptic tidal envelope and an assumed half hour lag time throughout a twice-daily tidal cycle). This is one mechanism causing net sand transport to be oblique to peak current direction. Thus, by zigzag motion of grains, sand can be moved away from a shore, as mentioned in Section 4.1.1. This lag is also expected to be one factor involved in the oblique alignment of tidal sand bank axes to peak tidal current direction. Probably sand bank alignment is determined at an early stage of bank development by the tidal currents at that time.

Individual sand grains, that are moving as bedload, react in a few seconds to a change in the speed of local water movements, as they do to a change in its direction. However, the bed-load transport rate, involving the large number of individual grain velocities, does have a lag in magnitude behind a change in current speed, since there is a lag in turbulence intensity. In a flume (Anwar and Atkins, 1980) and in most continental shelf investigations (e.g. Bohlen, 1976; McCave, 1979) the mean squares of the turbulent velocity components and their correlation giving Reynolds stress have been found to be considerably greater on a decelerating than an accelerating current for the same current speed. This corresponds to an hour's lag for a twice-daily tidal oscillation. This lag in the turbulence intensity is, however, apparently reduced near to the bed (Bowden and Ferguson, 1980). A more direct indication of lag in bed-load transport rate is given by sand ripple displacement. Kachel and Sternberg (1971) found this to continue at an unaltered speed for the 10 minutes for which observations continued after a sharp decrease in current speed.

The suspension transport rate can lag significantly behind a change in current speed because (1) the speed necessary to put a grain into suspension is greater than the speed necessary for keeping it in suspension ('scour lag' of van Straaten and Kuenen, 1957); (2) it takes some time for the turbulence to raise some of the eroded grains well above the bed and for such grains to settle downwards to the bed again after a weakening of the current ('settling lag'); (3) of the already mentioned lag of the mean square turbulent velocity components (particularly the vertical velocity component), since suspension requires that upward turbulent velocity components are sufficiently often greater than grain settling speed.

Confirmation of suspension lag is given by measurements of suspended sand loads by frequent multi-depth sampling during twice-daily tidal cycles in the Thames Estuary (Thorn, 1975). These show lag times of suspension-load of about 40 minutes on the ebb and 10 minutes on the flood current for a fraction 0.1 to 0.15 mm; 70 and 10 minutes, respectively, for a fraction of 0.075 to 0.1 mm; and 60 and 40 minutes, respectively, for 0.06 to 0.075 mm sand. Sand ripples were present but not sand waves, as these would not be expected in such fine-grained (0.14 mm) bed sand.

Fortunately the effects of time lag in the bed-load transport rate will largely cancel out in magnitude when the rate is integrated to obtain the bed-load transport over a complete ebb or flood current. This will also hold for suspension-load transport or total transport. Lag has, therefore, been ignored in schematic diagrams shown in Figs 4.1, 4.3, 4.6 and 4.23. The change that lag imposes on net sand transport direction will vary greatly, however, and be due not only to variation and uncertainty of lag time values. Inspection of tidal-current envelopes shows that current direction at some positions remains constant for one or two hours about the time of peak current speed, while varying at other times, and in other cases changes around the time of peak speed by 10° in an hour, even though the tidal envelope as a whole is fairly elongated.

Lag of bedforms behind a change in current speed and direction

The preservation or part preservation of large bedforms is made possible through the lag in their response to decreases in current speed. Moreover, the operation of such a lag means that in some cases bedforms do not have time to reach their equilibrium dimensions. This is an important qualification for interpreting marine bedforms on a basis of flume data for steady uniform flow (Allen, 1973).

Sand ripple shape and size have been observed to adapt to changes in current speed within a few minutes in flumes and less than an hour in tidal currents, although in strong currents the ripple changes are complex and can correspond to a greater lag of equivalent bed roughness z_0 (Dyer, 1980). The time taken is expected to be longer for current speeds not much above the minimum for sand ripple formation or for coarse sand with a median grain size that is only just less than the maximum size in which sand ripples can form (Section 3.2.1).

The complete reversal of asymmetry of small sand waves, following a reversal of tidal current, has a greater lag time. This is about 1–2 hours for sand waves about 0.5–1 m high in fine sand and for sand waves 2 m high in coarse sand for correspondingly stronger currents (Terwindt, 1970; Smith, 1969). Lag time would be shorter for smaller sand waves, but could not be much longer in a tidal current that reverses every six hours, so that only the upper half of sand waves 3 m high in very coarse sand was reversed in the Loire estuary (Ballade, 1953). Several days were required for rebuilding, after destruction by storm waves, of small sand waves on the gentle slope of large sand waves in the Thames Estuary (Langhorne, 1976), as shown in Plate 4.2.

Large sand waves in wide deep rivers commonly react to the more extreme flow changes in a few days, as is well documented for the Mississippi River. Current speeds commonly exceed 200 cm/s and sand transport rates are extremely high. On the sea floor, except perhaps in bed-load convergences, there are unlikely to be large sand waves for such current speeds. Large marine sand waves are commonly associated (Section 3.3.3) with mean spring peak speeds under 100 cm/s. Thus, even the largest water movements in the sea, including the effects of storm waves, would hardly produce the extreme river sand transport rates, and even then for only a few hours at a time. The building and lag times of marine sand waves, except intertidal and shallow subtidal ones, will in general therefore be much longer than for sand waves in rivers. There are also expected to be considerable differences between fields of marine sand waves, since sand grain size varies and since sand waves are found for a considerable range of peak tidal current speeds.

Sand banks contain very large volumes of sand. Because of this they will respond only very slowly to changes in tidal current speed. This lag ensures that they should have a high preservation potential, particularly when a change in tidal regime brings a relatively sharp decrease in peak current speed.

4.2.4 TRANSPORT OF SEDIMENTS WITH TWO OR MORE MODES

On continental shelves with strong tidal currents, if the grain sizes available for transport are strongly

Sand transport 65

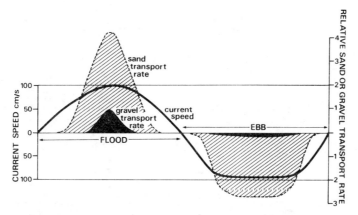

Fig. 4.3 Under exceptional conditions the net transport direction for sand could be opposite to that for gravel during extreme spring tides (when the currents are strong enough to move gravel). The figure shows one possible form of marked distortion of the tidal current speed-time curve from sinusoidal. The relative transport rate for gravel at different current speeds is taken as the cube of the current speed minus a threshold speed of 80 cm/s, and is drawn on an enhanced scale. Lag effects have been neglected.

bimodal (e.g. mainly fine sand and gravel) it may be necessary to consider the net transport of the modes separately. Their transport rates on ebb and flood can differ greatly, of course, and their net transport directions could perhaps also differ. This is less likely than for clay and fine silt sizes and perhaps the very fine sand grade, as these remain in suspension down to lower current speeds, perhaps throughout a period of several tidal cycles. In that case they are displaced in the direction of the net (residual) current for that period, rather than with the direction of the peak tidal current flow. However, the fine sand mode will have a much lower movement threshold, in terms of current speed, than will the gravel mode. It will respond (except for some grains sheltered among gravel particles) to the moderate as well as to the strong current speeds, whereas any gravel moved (except for a few exposed grains) will only respond to the latter, especially at extreme spring tides. Thus opposing directions of net transport of gravel and sand could in principle occur at some of the locations at which the speed of tidal current over a cycle is distorted very markedly from a simple sinusoidal variation, as is most likely for strong currents in shallow or constricted seas. One form of distortion which could give opposing net transport directions would be that shown in Fig. 4.3, with the ebb current having a wide flat peak and the flood current a much narrower peak reaching a stronger maximum. The latter would determine the gravel transport direction.

Opposite net transport directions of 'modes' might occur in a similar way if one 'mode' had much higher or lower density or was made of platey carbonate material.

These differences in direction of net transport of modes would tend in time to remove the bimodal character. So will the differences in speed of net transport of grains between modes which will occur even if direction is the same. Thus, the bimodal character of a sediment seems most likely to occur when one mode has been locally or recently supplied or produced.

Similar consequences are expected when the bed sediment has just a few per cent of a different constituent, but not enough to qualify as a 'mode'. Thus, mineral constituents cannot necessarily be assumed to be tracers of the movement of the rest of the bed sediment. Care must be taken that a specially introduced tracer material has the same density, shape and size characteristics as the part of the bed sediment it is required to trace. In the absence of information on the subsequent movement of such a tracer, the most reliable way of predicting the transport of a sediment with several constituents is to observe its behaviour in a flume over a range of current speeds. Such studies are particularly required for shelly material when it is far more angular or platey than quartz sand.

66 Offshore Tidal Sands

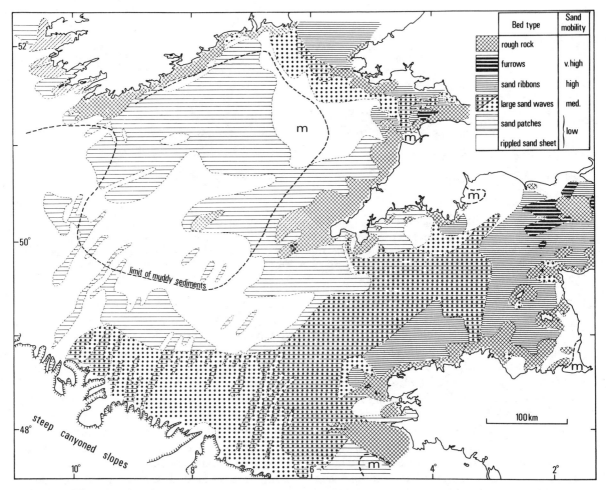

Fig. 4.4 The suite of bedforms, that are widespread on the floor of the Celtic Sea, Bristol Channel and western half of the English Channel, provide a good first impression of relative sand transport rates by tidal currents. Small sand waves will be present on the large sand waves, but the full extent of the outer zone where the small sand waves occur on their own is not known in enough detail to be shown. All the sand of quartz density for median diameters up to about 0.7 mm (and rather more for poor sorting and low density) can have current ripple marks on it for at least part of the time, except perhaps in some sheltered locations. Horizontal lines imply gravel floor, such as between some large sand waves.

4.3 Geographical variation in sand transport rate

4.3.1 RELATIVE SAND TRANSPORT RATE SHOWN BY MEAN SPRING PEAK TIDAL CURRENT SPEED

An indication of the importance of strong tidal currents around the world at the present time is shown by the extent of regions with relatively high mean spring tidal ranges (Fig. 2.5). Of more immediate use, however, are the mean spring peak speeds of the near-surface currents that are provided by tidal atlases for some regions, although those so far published for the north-west European continental shelf do not take account of most of the large number of current-meter recordings made since about 1970. However, those available and tidally analysed by about 1979 for the Irish Sea, Celtic Sea and Southern North Sea have been taken into account in drawing Fig. 2.6. These near-surface values can be converted into near-bed values by using a reduction factor

(Section 4.2.2). The mean spring peak tidal current speeds serve as a valuable indicator of relative importance of tidal current sand transport between different locations in a region with predominantly twice-daily tidal currents; they are approximately proportional to extreme tidal current speeds (e.g. about 80% in the Western English Channel, Section 2.2). Enhanced rates and more widespread sand transport that are due to the combination of tidal with the occasional non-tidal water movements are discussed in Section 4.5.

4.3.2 RELATIVE SAND TRANSPORT RATES SHOWN BY BEDFORMS

The relative location of different zones characterized by distinctive bedform types described in Chapter 3 provides a good first impression of the geographical distribution of zones of different current speed. This correlation was first established for the continental shelf around the British Isles (e.g. Belderson and Stride, 1966) and can be used elsewhere. On a local scale it is seen roughly by comparison of the small area of Fig. 4.4 with Fig. 2.6. Comparison around the whole of the British Isles shows that longitudinal furrows in gravel are found in association with near-surface mean spring peak tidal currents of more than about 150 cm/s; sand ribbons are associated with values above about 100 cm/s; large sand waves, often with small sand waves or sand ripples on them, with values of more than about 60 cm/s (Fig. 4.5); and sand patches with speeds of about 50 cm/s or less. These tidal current speeds are lower than water movement speeds necessary by themselves to produce the bedforms, as extreme tidal currents could be up to about 30% stronger and there will also be a variable contribution from the aperiodic (non-tidal) water movements. The minimum mean spring speeds for furrows, sand ribbons and sand waves in the Bristol Channel that are given by Warwick and Uncles (1980) are about 50 cm/s greater than the values given above. This can probably be explained by the weaker aperiodic water movements hereabouts, relative to the tidal currents, as compared with elsewhere around the British Isles. A fuller discussion of sand waves and sand ripples is delayed until Section 4.5 and especially Section 4.6.

4.4 Net sand transport by tidal currents

It can be deduced that on some continental shelves there is much long term net transport of sand along certain paths, together with some gravel transport as well. As by far the greatest bulk of bed-load transport is of sand-sized material, the term 'sand transport path' will be used for convenience, but with the recognition that in the regions of strongest tidal flow some gravel can also be transported. On continental shelves with strong tidal flow it is in general these currents that determine the net sand transport directions. The net transport paths around the British Isles are described first, as these are better known than elsewhere, while some other regions of continental shelf are referred to in Section 4.4.4. The enhanced sand transport rates and more widespread sand transport (and occasional reversals) that are due to the occasional non-tidal water movements are discussed in Sections 4.5.2 and 4.5.3.

4.4.1 NET SAND TRANSPORT DIRECTION PREDICTED FROM MEAN SPRING PEAK TIDAL CURRENTS

On the continental shelf around the British Isles it is usual for nearly every ebb peak tidal current at any point to exceed the immediately preceding and following flood peak tidal current or vice-versa. This arises mainly because of the interaction of the twice-daily constituent M_2 and its first harmonic M_4 (Section 2.4 and Fig. 2.8). The predominance of one direction of tidal flow is also expected to apply at most positions on continental shelves affected by strong tidal currents, whose once-daily constituents are small compared to twice-daily constituents. However, it could be affected by local enhancement of the relative magnitude of a constituent caused by its local resonance. The peak current direction at springs should be preferred to that at neaps if they differ, as sand transport rates at neaps will be far smaller.

The tidal current data indicate sizeable regions where the peak ebb tidal current is the stronger: these are situated adjacent to other sizeable regions wherein the peak flood tidal current is the stronger (Sager and Sammler, 1975). Around the British Isles

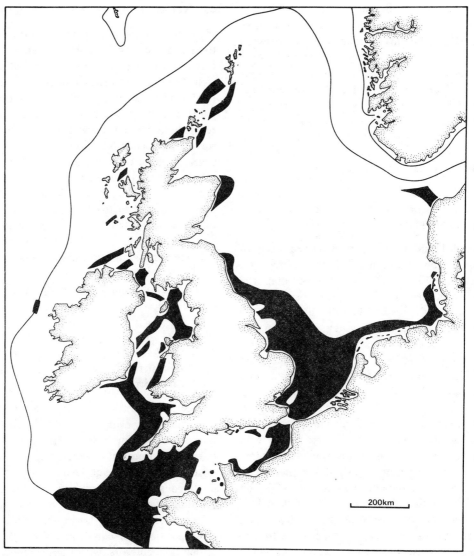

Fig. 4.5 The distribution of sand waves around the British Isles, based on widespread sonograph coverage and echo-sounder profiles.

the peak ebb and flood mean spring currents may differ from one another by as much as 60 cm/s (1.1 knot). However, as little as 5 cm/s (0.1 knot) difference is usually found to be enough to determine a definite direction of long-term net sand transport (as described in Section 4.4.2). This empirical finding is consistent with the steep increase of sand transport rate with increasing current speed (cf. Fig. 4.1). Thus, most of the sand transport occurs within about $1\frac{1}{2}$ hours of the times of peak ebb and peak flood twice-daily tidal current, depending on the ratio of peak speed to the threshold speed. This can be illustrated quantitatively by applying formulae (such as the cubic one discussed above) for sand transport rate as a function of current speed to the known variations of tidal current speed (such as the hourly values of mean spring twice-daily tidal currents tabulated by the Hydrographic Department, Admiralty, 1948 for numerous positions). Fig. 4.1 gives a common example.

Sand transport

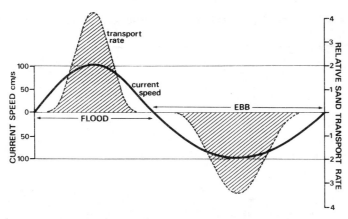

Fig. 4.6 An unusual situation with ebb and flood tidal currents of markedly unequal durations. The sand transport is greater during the slightly weaker (by 5%) but much longer duration (by 50%) ebb current, as shown by the relative areas between the ebb and flood sand transport curves and the horizontal axis. Lag effects have been neglected.

The direction of net sand transport is verified (by reference to field data in Section 4.4.2) to be that of the stronger of the peak ebb or flood currents for all such positions so far tried. There may be apparent exceptions for situations where the peak ebb and flood currents are tabulated as of equal speed (in tenths of a knot), or where they are nearly equal and the directions of the peak currents are not quite opposite. In addition, exceptions to the rule are theoretically possible for a few positions with strong tidal currents and shallow water, even without ebb/flood depth differences. The sinusoidal current speed-time variation could be greatly distorted in such a way that either the ebb or flood flow, although of slightly weaker strength, runs for a much longer period than the opposite flow. It can, thus, give a slightly greater total sand transport (Fig. 4.6) and therefore cause net sand transport in that direction. However, in most cases of such unequal ebb and flood durations (usually in estuaries or in a regular eddy) the shorter duration current has a higher peak speed. Even if this were only 10–15% greater it could be enough to determine the net sand transport direction.

The tidal current components are accompanied by turbulent velocity components of shorter periods and having a wide spread of directions. The instantaneous resultant water and grain movement velocities therefore, will have some spread of direction. However, this will be much less than the 180° of opposing tidal directions.

The empirical proof that the net sand transport direction is generally indicated by the stronger of the ebb or flood peak tidal current directions comes from the widespread and abundant bedform data (Section 4.4.2) for regions with dominant twice-daily tides. This conclusion would be expected to apply also to regions with dominant once-daily tides.

Areas with 'mixed' tidal currents are more complicated. These have twice-daily and once-daily tidal current constituents that are comparable in speeds. For example, this pattern is found on parts of the continental shelves off western Scotland (Cartwright, 1969) and eastern USA where tidal resonance enhances the once-daily currents. In some parts of the spring–neap–spring cycle the peaks of ebb current are stronger than those of the flood currents both immediately preceding and following, while in other parts of the cycle they are weaker. Examples of this are shown in tidal current predictions contained in the annual Tide Tables for USA and Australia. Sand transport summed over individual ebb and flood currents will correspondingly sometimes be greater in the ebb direction and sometimes in the flood direction. Therefore, in such cases summation over a whole spring–neap–spring sequence would be essential to establish a net sand transport direction from tidal current data alone. Longer term summation would be required where there are strong seasonal differences in non-tidal currents, as in Torres Strait (Section 4.5.3).

70 Offshore Tidal Sands

Table 4.1 Some methods of determining net sand transport directions in tidal seas and the probable periods for which data are available.

Method	Data Availability						
	Hours	Days	Months	Years	10's years	100's years	1000's years
Grain movement							
Sand impact on transducer	+	.	.	.	.	.	.
Sand traps	+	+	.	.	.	.	.
Tracers	.	+	+	.	.	.	.
Bedform movement							
Sand ripple migration	+	+	.	.	.	.	.
Sand wave migration	+	+	+	.	.	.	.
Sand bank migration	.	.	+	+	+	.	.
Morphology							
Sand wave polarity	.	+	+	+	+	.	.
Sand bank polarity	.	.	+	+	+	.	.
Sand patch polarity	.	.	+	+	.	.	.
Deposit							
Age and structure	+	+	+	+	+	+	+

A number of authors have used the term residual current to describe the difference in strength between peak ebb and peak flood tidal current at a given site. This usage is misleading, and should be discontinued, for the term residual is already more commonly used to cover the net movement of water which remains after all periodic tidal contributions have, in principle, been removed by filtering procedures. It includes the tidal residual mean current generated by the same forces that give the harmonics (Section 2.4.1).

4.4.2 FIELD EVIDENCE OF NET SAND TRANSPORT DIRECTIONS

Net sand transport directions in tidal seas are indicated by many complementary lines of evidence (Table 4.1). The direct observations of grain movement and bedform displacement, in association with observations of water movements, are well suited to increasing understanding of the processes in quantitative terms. They can also provide the net transport directions, but for practical reasons such observations can only refer to a few localities and are generally made for relatively short periods of time.

The regionally more useful indications of net sand and gravel transport directions can be inferred from the morphology of the widespread suite of bedforms of sand and gravel. These directions refer to substantial periods of time. They indicate remarkably consistent net transport directions during the varied periods of observation (Table 4.1). Predicted sand transport directions from tidal current data are referred to in Section 4.4.1 and computer simulations are referred to in Section 4.4.3. The longest term evidence of net sand transport directions are provided by the nature and location of the deposits (Section 5.2).

(a) *Sand wave movements and asymmetry*

For a unidirectional current in flumes, irrigation canals or rivers a sand wave advances downcurrent in the direction faced by the steeper of its two slopes. Thus, its asymmetrical profile indicates the net sand transport direction. For tidal seas sand waves have also been observed to migrate in the direction faced by their relatively steeper slope. These observations apply to a few sand waves at a number of localities. They were made in the tidal estuary of the River Loire, France (Ballade, 1953) and within a tidal channel off Florida, USA (Harrison, Byrne, Boon

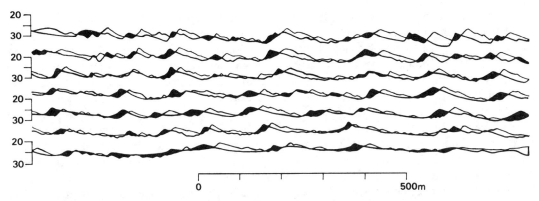

Fig. 4.7 Seven pairs of parallel echo-sounder profiles from Lister Tief (near the island of Sylt, north west Germany). These show that sand waves were moved by the dominant ebb tidal current in the direction faced by their steeper slopes during the period between two surveys 9 months apart. Areas shown in black represent forward movement of sand waves (after Ulrich and Pasenau, 1973).

and Moncure, 1970); in the open waters of the Irish Sea (Sly, 1966; Jones, Kain and Stride, 1965); in the North Sea (Fig. 4.7) off the Island of Sylt (Ulrich and Pasenau, 1973), amongst the sand banks off Germany (Samu, 1968), in the Southern Bight of the North Sea with respect to bottom markers (Harden-Jones and Mitson, in press), and by successive surveys (Cloet, 1980; Langeraar, 1966; and Terwindt, 1971b). Although in the last two examples the shifts did not exceed expected navigational errors they implied sand wave migration to be consistently in the direction indicated by sand wave asymmetry and by predicted net sand transport (Sections 4.4.1 and 4.4.3), so this is a reasonable interpretation of the data. Oscillations in the position of an asymmetrical sand wave with respect to bottom markers were nevertheless associated with a net progression in the direction indicated by its asymmetry (Langhorne, in press).

The profile of sand waves can thus be used to indicate the sand transport direction. This can be due to a unidirectional current as discussed in Section 4.5.3, or it can result from net sand transport, where the tidal currents are dominant and the ebb or flood is the more competent (Section 4.4.1). For the tidal case this direction is generally that of the ebb or flood tidal current that reaches the higher peak speed. An example of this concordance was found in the northeastern part of the Irish Sea, where long-term, near-bottom current observations were made in the vicinity of sand waves. The direction of the stronger flow was found to be in agreement with the asymmetry of the sand waves (Belderson and Stride, 1969).

The widespread occurrence of sand waves around the British Isles (Fig. 4.5) and the abundant data about the direction faced by their relatively steeper slope allows a broad regional comparison to be made of sand waves and tidal current data. This is assembled along with other lines of evidence in Section 4.4.3.

Displacement directions of sand waves which seem to be opposite to the directions of the local peak tidal currents have sometimes been obtained from just two or three successive surveys over a given patch of sand waves and/or over short periods. However, these observations may not give true long-term sand wave displacement directions for several possible reasons: navigational inaccuracies which are a particular nuisance when sand wave crests are short or sinuous; to-and-fro displacement of the crest and near-crest during the $12\frac{1}{2}$ hour tidal cycle; abnormal non-tidal current components between the surveys (e.g. due to occasional strong winds as described in Section 4.5.3); wrong re-identification of individual sand waves between successive surveys, perhaps due to near-crest shape changes between ebb and flood; or use of the crest positions of small sand waves moving over large ones.

The finding of apparently zero, or statistically insignificant differences in position of sand waves between successive surveys cannot automatically be taken (as it has been by some workers) to indicate that they are static or even 'relict' features. They may be considered as such only if the present-day water movements have been measured and analysed over a long enough period to show that peak tidal current speeds corresponding to main tidal constituents are below the threshold speed for sand waves, even when augmented by reasonable estimates of other non-tidal peak water movements. In such cases a mud-cover, deposited from suspension, could be present, although it is not proof of sand wave 'death', as mud can be deposited at quite high current speeds if its concentration in suspension is sufficient (McCave, 1970). Commonly, the uncertainty in the location of sand wave crests between successive surveys is greater than the magnitude of any predicted migration that is based on known water movements and formulae for estimating sand transport rate that are consistent with river sand wave displacement rates. These are typically 10–100 times greater owing to the higher current speeds and its unidirectional nature.

For unidirectional flow the horizontal displacement rates of small sand waves are, in general, much greater than those of large ones for the same current speed and sand transport rate. The same conclusion applies for tidal currents using net displacement and net sand transport rates. This is to be expected, owing to the much smaller volumes of sand contained in the small sand waves. Where there are two sizes of sand waves present in unidirectional currents, observations show that the larger sand wave is displaced when the smaller sand waves tip their load down its lee slope on reaching its crest. In this case the sand waves have steep lee slopes near the angle of repose. Those large marine sand waves whose steeper slopes are everywhere only between 5 and 15 degrees can still be slowly displaced in the direction of the stronger peak current, probably by differential rates of displacement of small sand waves up and down their slopes (Ludwick, 1971).

In those few localities where the long term sand wave asymmetry and the short term observations of near-surface tidal currents seem to be in conflict, the former is taken as the preferred indicator so long as the sand waves are not due to unidirectional currents (Section 4.5.3). This is because the cross-sectional shape of the sand wave is governed by the actual net sand transport near the floor. Thus, the asymmetry of large sand waves may well be a more reliable indicator than some of the tidal currents given on navigational charts which were based on early observations that lasted only a day or so. The direction faced by the steeper slopes of the largest sand waves present should be used, even if the angle of slope is small. This should refer to their whole height, not just near the crest where the asymmetry can reverse between successive ebb and flood tides. The asymmetry of the whole bedform can reverse between ebb and flood for sand ripples and some small sand waves for which the required volume of sand can be moved in the half tidal period. The maximum height of such reversing sand waves will depend on grain size and on ebb and flood peak speeds (Section 4.2.3). The asymmetry of these small sand waves, whether on a flat floor or upon large bedforms, represents more local or transitory conditions and is of less value as an indicator of net sand transport.

(b) *Sand bank polarity*

The majority of offshore and estuarine linear tidal sand banks are aligned at a small angle to the strongest tidal flow (Section 3.4.5). The steeper slope is then located on the side facing away from the most competent current (e.g. Fig. 4.8), in the same way that the steeper slope of a sand wave faces 'downstream' and indicates the net sand transport direction (Kenyon, Belderson, Stride and Johnson, 1981). Sand waves on sand banks merely show the associated local sand transport directions (Section 4.7). In contrast, however, the outline of a sand bank in plan view shows the regional net sand transport direction. Thus, the end pointing in the direction of net sand transport can be narrow and pointed, while the opposite end can be broader and more rounded. This is emphasized when aprons of sand waves can be recognized on flat floor immediately beyond the ends of the sand banks (G. F. Caston, 1981).

(c) *Longitudinal furrow polarity*

The preferred 'downstream' junction of longitudinal furrows (Section 3.4.2 and Plates 3.7 to 3.9) can be a useful indicator of net transport direction of sand

Sand transport

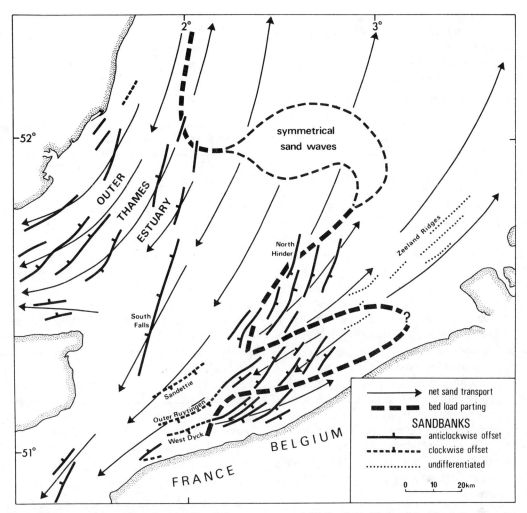

Fig. 4.8 Net sand transport directions in the southernmost part of the Southern Bight of the North Sea, based on sand wave, sand bank and tidal current asymmetries. The long axes of most of the sand banks are rotated in an anticlockwise sense with respect to the main tidal flow and have a clockwise sense of local sand transport on opposite sides (after Kenyon, Belderson, Stride and Johnson, 1981).

(and possibly some gravel) on ground where sand waves are absent (Stride, Belderson and Kenyon, 1972). Although the furrows are confined to areas of very strong currents and so are limited in value in geographical terms, they can serve to confirm and refine the position of bed-load partings, such as the English Channel Bed-Load Parting.

(d) *Sand ribbon polarity*

Individual sand ribbons (Section 3.4.4) tend to broaden and to merge together when followed in the direction in which the peak tidal current speed is weakening (Kenyon, 1970). This direction is usually also the net sand transport direction.

(e) *Obstacle mark polarity*

Obstacle marks (Section 3.4.3) can give a valuable indication of the net sand transport direction by tidal currents and can be especially useful when sand waves are absent. The features may extend outwards

74   Offshore Tidal Sands

from the obstacle in both the direction of the ebb and flood currents. However, the longer obstacle mark is generally developed on the side of the obstacle facing in the direction of net sand transport (Fig. 3.8 and Plates 3.10 to 3.13). This method was developed and used for ground lying north of Dover Strait (G. F. Caston, 1979).

(f) *Gradients of decreasing grain size*

In well-sorted clastic deposits a gradient of decreasing grain size can be used to infer the long term, net sand transport direction on a path which is associated with a gradient of decreasing peak current strength. There are good examples of these gradients on several transport-deposition paths in the Irish Sea and Bristol Channel (Belderson, 1964; Belderson and Stride, 1966; G. F. Caston, 1976; Cronan, 1969) and on the northerly transport-deposition path of the Southern Bight of the North Sea (Figs 4.10 and 5.3). They provide ultimate proof that the net transport directions have been effective during the period of the late Holocene (Section 5.2.1). In some regions there is net transport of sand up a longitudinal velocity gradient of the tidal currents. This should give rise to a lag deposit which will be increasingly coarse in the direction of transport. Such a situation can occur off a headland or near to an island. But the sand may also become trapped in a bed-load convergence where any deposit should be ill-sorted, at least in bulk terms.

(g) *Radioactive and fluorescent tracer dispersion*

This is an avenue of research with a high potential which has so far been too little used in tidal seas. However, there are already sufficient data to show that this method can indicate net sand transport directions and so lend valuable support to those directions of transport shown by other methods. An early example of such work (Reid, 1958) demonstrated substantial net transport of sand during a period of three months at a locality near the East Anglian coast. Associated near-bottom current measurements (1 m above the bed) showed an unusually large peak ebb-flood speed asymmetry of 60 cm/s (1.1 knot) during a spring tide, with the

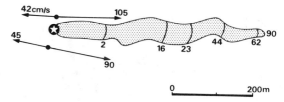

Fig. 4.9 The spread of radioactive tracer sand along the bed from the launch site (star) into a plume during six periods up to the maximum of 90 days. The plume is elongated in the general direction of the stronger (flood) near-bottom tidal current (after Reid, 1958).

stronger tidal current heading in the same southerly direction as the plume of radioactive tracer (Fig. 4.9). Later radioactive tracer studies have confirmed sand transport paths at other locations with much smaller peak speed asymmetry, for example the northerly flow past southern Holland (Morra, Oudshoorn, Svasek and Voss, 1961), easterly flow on the French side of the Eastern English Channel (Anguenot, Gourlez and Migniot, 1972), as well as in Swansea Bay (Heathershaw and Carr, 1977), and near Lowestoft (Jolliffe, 1963). Tracer plumes are also of value because they show the degree of dispersion to either side of the net sand transport direction, and can give approximate magnitudes of integrated sand transport over the survey period if cores are taken to give the depth of mobile sand.

(h) *Mineral tracer dispersion*

Spatial variation in the relative abundance of naturally occurring mineral grains or waste materials introduced by man can be used to identify net transport directions, after allowance has been made for progressive sorting by size, shape or density (Section 4.2.4). This method has confirmed the long-term existence of the Southern Bight Bed-Load Parting in the North Sea and northerly transport from it (Stride, 1970). Off north-east England a preferred southerly direction of sand transport is implied by the observed spread of hydraulically equivalent coal particles originating from dumps of colliery waste (Eagle, Hardiman, Norton, Nunny & Rolfe, 1979).

(i) *Biogenic tracer dispersion*

Fossil bryozoa indicate northerly sand transport in a

zone close to the Dutch coast (Lagaaij, 1968). Another useful example is referred to in Section 4.4.3 for ground to the north-west of France. It shows that there is some transport oblique to the main westerly sand transport direction. There is considerable potential in the future use of a wide range of faunal and floral tracers.

(j) *Magnetic fabric polarity*

The magnetic fabric of sedimentary grains can reveal the depositional fabric of a deposit and thus serve as an additional line of evidence for indicating transport direction at the moment of deposition. This method was attempted for some Irish Sea silts (D. Frederick, pers. comm.) and has now been exploited more fully in the Bristol Channel and North Sea (E. A. Hailwood, pers. comm.).

(k) *Sand trap data*

Direct evidence of sand transport in a tidal sea was provided for Dover Strait (van Veen, 1936), for example. His sand trap measurements give some indication of the quantities of sand in motion during both near-maximum flood and near-maximum ebb currents, from which the net transport direction can be derived.

4.4.3 REGIONAL NET SAND TRANSPORT DIRECTIONS AROUND THE BRITISH ISLES

A compilation has been made of all the available lines of sea bed evidence about the directions of net sand transport by tidal currents around the British Isles and west of France (Fig. 4.10). Such a composite map can be produced because there is such a striking measure of agreement about the main features, and because the directions correspond with the direction of the stronger tidal flow as given on charts and tidal atlases. In some regions the net sand transport occurs on paths that extend parallel with coasts, while in others it may be at right angles or at some intermediate direction, depending on the tidal current regime. Individual sand transport paths in this region are up to 550 km (300 nautical miles) long and up to 170 km (90 nautical miles) wide.

The main paths originate at five bed-load partings. These features can determine, for example, that sand moves towards the open ocean from one side of a parting, while from the other side sand may move towards an inner sea, as in the English Channel or Irish Sea. The paths may terminate in bed-load convergences, such as at Dover Strait (although it is surmised that there may be some sand escaping along the southern edge of that Strait and so passing into the North Sea). The origin of this compartmenting of the sand transport paths by partings and convergences is discussed in Sections 4.4.5 and 4.4.6. These features are characteristic of the pattern of offshore tidal current transport and should be recognizable in the assemblages of transported shelly faunas. However, it is not easy to demonstrate this pattern for fragments smaller than 2 mm, as they cannot usually be identified.

A second characteristic aspect of the tidal current determined pattern is the transport of sand from areas of floor in the open sea inwards towards many estuaries and embayments. These include the Solway Firth, Liverpool Bay, the innermost part of the Bristol Channel, the Thames Estuary and The Wash, around the British Isles, together with the Gulf of St Malo off France, the estuaries of the Dutch coast and estuaries along the edge of the German Bight. Explanations for this up-estuary transport are (a) the density stratification which gives up-estuary transport on the bed (and down-estuary transport near the surface; (b) the tide in an inner estuary tends to develop a 'bore' form, with a steep front and correspondingly stronger up-estuary than down-estuary peak current speed; and (c) the tidal wave is largely dissipated by friction while travelling up an estuary and being reflected at its head, so that the reflected down-estuary wave is weak.

A third characteristic feature of the pattern is that there are local sand transport paths which may often trend counter to the prevailing, regional ones, as in the western part of the English Channel. Near coasts an important cause of such effects are the eddies associated with flow around headlands. The numerous longitudinal sand banks of estuaries and further offshore also have their own local sand transport systems, the crest of a sand bank being, in effect, a local convergence. For most sand banks in the

76 Offshore Tidal Sands

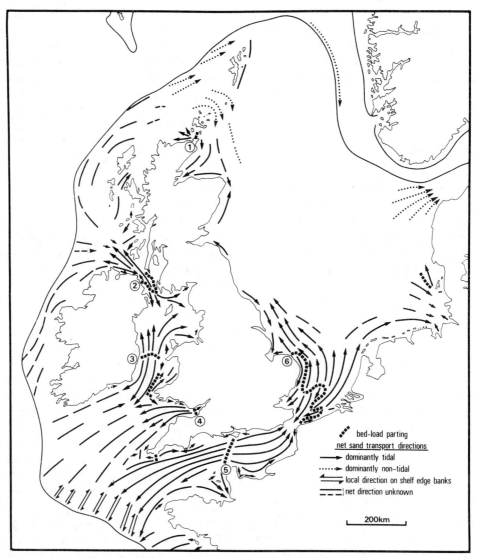

Fig. 4.10 The net sand transport directions on the continental shelf around the British Isles, based on IOS bedform data, tidal current data, published bedform data and other lines of evidence listed in the text. Some of the most useful published bedform data includes Auffret, Alduc, Larsonneur and Smith, 1980; Derna, 1974; Dobson, Evans and James, 1971; Dyer, 1970b; Houbolt, 1968; Neumann and Meier, 1964; Pendlebury and Dobson, 1976; Samu, 1968; Sly, 1966. (1) Pentland Firth Bed-Load Parting: (2) North Channel BLP: (3) St George's Channel BLP: (4) Bristol Channel BLP: (5) English Channel BLP: (6) Southern Bight BLP.

northern hemisphere that are not associated with headlands, the sand waves turn in a clockwise direction as they approach the crest of the bank from either side (Section 4.7).

Some sand is being lost from the continental shelf on to the continental slope to the west of the Celtic Sea, as shown by the presence of sand in canyon axes. Available data suggest that it passes outwards from the continental shelf on the spurs between the submarine canyon heads, while sand transport in line with the canyon heads is towards the inner part of the continental shelf.

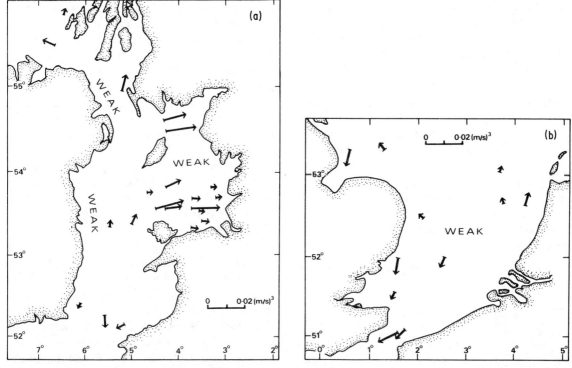

Fig. 4.11 Net sand transport directions and relative magnitudes shown by arrow lengths (at the location of arrow tips) for the Irish Sea (a), and the Southern North Sea (b), computed from the M_2 current and its first harmonic M_4 tidal current observations for approximate mid-depth. The relative magnitudes will be approximately the same as for mean spring tides. The net sand transport vector over a tidal cycle is computed as the integral, over ebb and flood currents, of the vector in the current direction with magnitude equal to the cube of the excess of current speed above the threshold speed (analysis by M. J. Howarth).

In drawing Fig. 4.10 a few choices have had to be made. In general, the data from bedforms and tracer observations are preferred, so long as the net sand transport direction is clearly determined by the tidal currents. For example, the westerly directed sand waves in the western half of the English Channel are preferred to three current observation stations hereabouts which show a dominant near-surface tidal current flowing east (Hydrographic Department, Admiralty, 1948). This is a reasonable choice because the sand waves are widespread and their steeper slopes have been observed to face west on numerous occasions during the past 20 years. In contrast, the three aberrant tidal current observations were obtained during one or two days by a well established but primitive technique and it is now known from a current-meter recording over several weeks at a nearby station that the westerly flow is the stronger (Channon and Hamilton, 1976). Nevertheless at other localities where bedforms provide little guidance, use is made of the available current data, as for example in some parts of near-coastal zone off northern France and in the innermost part of the Bristol Channel. (In most areas current measurements are available at sufficient positions for aberrant data from one position to be detected and not reach the tidal current charts.) The conflict between the two lines of evidence at the southeastern corner of the German Bight, where the strongest tidal currents imply an easterly direction of net sand transport, while the underlying floor is muddy, is resolved by concluding that sand is only being transported eastwards beyond the south side of the muddy floor.

The composite map of net sand transport directions (Fig. 4.10) based on all the sea bed and tidal current chart data listed above, can now be compared with computer simulations (alternatively called numerical models). The computed net sand transport vectors from M_2 and M_4 current constituents (M. J. Howarth, pers. comm.: derived from his extensive current recordings in the Irish Sea and Southern North Sea) are shown in Fig. 4.11. Other studies have used one of the methods for estimating currents due to the main constituents by numerical modelling. One such simulation integrated sand transport rate over the modelled twice-daily and spring–neap cycles of tidal currents, for the North Sea (Sündermann and Krohn, 1977). It reproduced the expected northwards net sand transport over the sand sheet to the west of Holland, with net erosion in the south and net deposition in the north (Fig. 4.12).

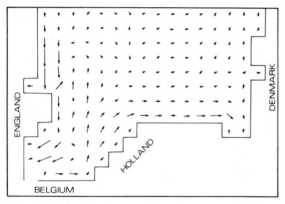

Fig. 4.12 Computed net sand transport vectors for the Southern North Sea (after Sündermann and Krohn, 1977). This figure applies to a spring–neap tidal sequence and is based on those authors' numerically modelled twice-daily tidal currents.

The whole area of interest around the British Isles is covered by a numerical model (Pingree and Griffiths, 1979) of maximum (our 'peak') bottom stress (Fig. 4.13) over the twice-daily tidal cycle due to modelled M_2 and M_4 current constituents. The maximum bottom stress is computed as peak current speed squared, times a drag coefficient as discussed in Section 4.2.2. The directions of computed maximum bottom stress, and the net sand transport direction which would result from calculating the integral, over the M_2 tidal cycle, of the cube of current speed minus threshold, using the same computed currents, would be the same.

The numerical model presentations show remarkable basic similarity with Fig. 4.10 for regions with high rates of sand transport. The main differences are that the computer simulation does not show the exact locations of the bed-load partings as known from the bedforms, and, although a small grid size was used in relation to the large area covered by the whole simulation it is not able to resolve fully the known small-scale convolutions of the North Channel Bed-Load Parting in the Irish Sea or the Southern Bight Bed-Load Parting in the North Sea. In the western part of the English Channel and also to the west of France the asymmetry of numerous sand waves shows conclusively that there is westerly net sand transport without any indications of weak easterly transport in the regions suggested by the simulation. This local failure of the simulation to give significant directions for the low values of bottom stress in these regions may result from the uncertainty concerning the M_2 water level variations which have to be specified for the shelf edge on the western boundary of the model; also zero M_4 was specified there. However, in other areas the simulation does show the narrowness of the bed-load partings and the progressive decrease in maximum bottom stress (and thus the progressive decrease in sand transport rate) along the sand transport paths where deposition is occurring. It also provides data for the relatively small areas for which bedform data are presently lacking. However, the real importance of the overall similarity between the two approaches is that it provides the confidence in the value of a numerical model that will encourage such modelling for regions of continental shelf with strong tidal currents, for which there are scant bedform or tidal current observations.

The numerous lines of evidence given above show conclusively that at most offshore locations in these seas the net sand transport direction is mainly the result of distortion of the twice-daily tidal current speed curve by its first harmonic (Fig. 2.8). This distortion gives the peak transport in the direction of the stronger of the ebb or flood peak tidal current (Fig. 4.1). Thus, Allen's (1980a and b) treatment of net transport of sand as determined by a net flow of water superimposed on the undistorted twice-daily tidal current speed curve is not seen as the general

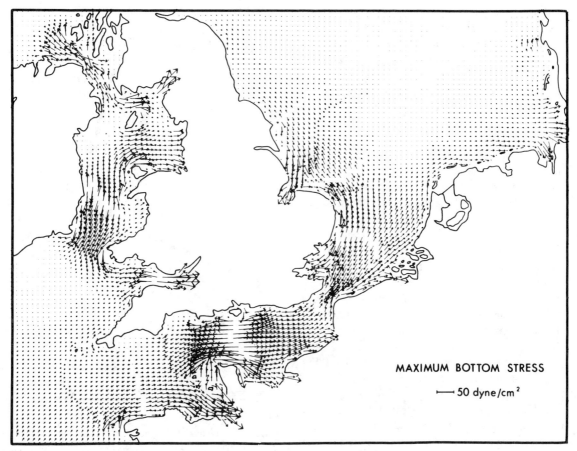

Fig. 4.13 Numerical simulation of maximum bottom stress vector during the twice-daily tidal cycle due to M_2 and M_4 constituents (after Pingree and Griffiths, 1979). The majority of these directions correspond with the net sand transport directions of Fig. 4.10.

case. However, near sand banks and headlands the tidal residual currents, particularly in eddies, can be important. The local importance of non-tidal flow of water is discussed in Section 4.5.3.

The summary of evidence about the net sand transport directions in these seas is now so overwhelming that earlier objections to the northerly transport past Holland (Oele, 1969, 1971) can be sustained no longer.

Sand transport oblique to main tidal flow

In addition to the main net sand transport directions, described above, there is evidence for net transport of some sand at an oblique angle to the main flow during the tidal cycle. The reasons for this are several: (a) tidal current vectors when plotted for a common zero describe an envelope so that sand transport will be spread out over an arc, increasing in width as the envelope becomes progressively more circular; (b) this angular spread will also be increased by the usual turbulent fluctuations in current direction; (c) it will be shifted to one side of the strongest flow direction by the lag effect (Section 4.2.3) in picking up compared to putting down of sand after a change in current speed, especially for suspension transport of sand; and (d) the sideways transport will also be enhanced when there is a lateral velocity gradient of current speed (i.e. such as normal to the peak tidal flow direction). Although much of the sand will more or less follow the strongest current direction, a part of it will be moved by the above mechanism into the adjacent region of slightly weaker current. By successive zig-zag paths it will move progressively across current. A similar result can be produced by peak ebb and peak flood current speeds equal but not quite opposite in direction, as often happens near to curved coastlines.

The spread of a tracer into a plume shows this dispersion of sand across the main net sand transport direction. A gross example of such lateral dispersion may be indicated by the preferential presence of sand in the northern half of the Western English Channel and its virtual absence in the southern half. Even more significant may be the asymmetric development of the barchan style sand waves thereabouts, where the more northerly of the two westward pointing arms is the longer (Section 3.3.3), suggesting some net north-westward transport of sand (by analogy with similar bedforms in deserts). In addition, there is the presence north-west of France (Bouysse, Le Lann and Scolari, 1979) of bryozoa that grow only in shallower water nearer to that land (Fig. 6.9). These must imply north-westward transport of material obliquely across the path of strongest tidal flow. Unfortunately, bottom sampling was not continued far enough to the west and south by these authors. This might have shown the expected even greater relative importance of the main net transport direction in moving the two useful biological tracer species studied by them.

4.4.4 NET SAND TRANSPORT PATHS ON OTHER CONTINENTAL SHELVES

Few tidal current dominated sand transport paths have so far been described for ground beyond the British Isles, although there is scope for their recognition in the extensive tidal seas off northern Australia, off northern Canada and south-western Alaska (Bouma, Hampton and Orlando, 1978; Bouma, Hampton, Rappeport *et al.*, 1978), west of Korea, in the Malacca Strait, and off south-eastern Argentina, for instance. However, net sand transport paths have been suggested for the entrance to the White Sea, USSR (Fig. 4.14) and between Cape Cod and Georges Bank, USA (Fig. 4.15) on the basis of

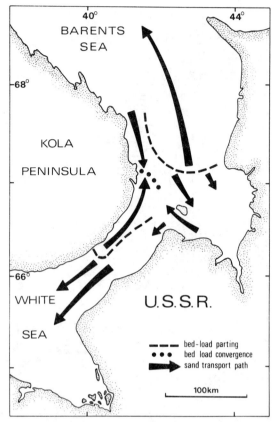

Fig. 4.14 Net sand transport directions proposed for the channel between the Barents Sea and the White Sea, USSR as inferred from the stronger of the ebb or flood tidal currents (after Belderson, Johnson and Stride, 1978), and by sand bank asymmetry.

published tidal current and bedform data. The earlier interpretations of these areas (Belderson, Johnson and Stride, 1978) were clarified by making use of sand bank morphology. The proposed existence of bed-load partings and convergences in these regions serves to provide the first rational explanation of the regional location of the grain size pattern of the deposits. Additional sand transport paths have now also been suggested for the vicinity of Seoul and for Korea Bay (Kenyon *et al.*, 1981).

4.4.5 BED-LOAD PARTINGS AND BED-LOAD CONVERGENCES

There seems to be a tendency for bed-load partings to be located in narrows between seas where tidal

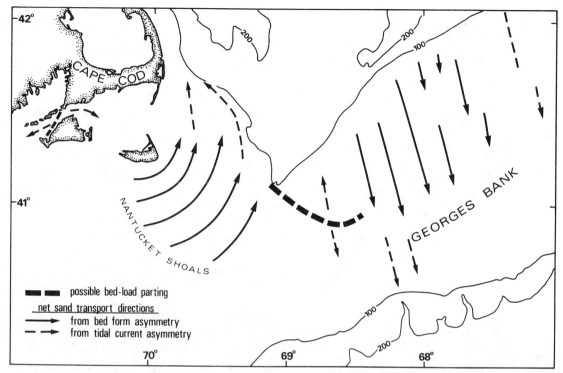

Fig. 4.15 Net sand transport directions for the sea floor between Georges Bank and Cape Cod, USA, as inferred from sand bank asymmetry and for locations far removed from the sand banks by sand wave and tidal current asymmetry (partly after Belderson, Johnson and Stride, 1978).

currents are particularly strong (Fig. 4.10). Notable exceptions are at Dover Strait (where there is a bed-load convergence), and in the narrows between the Orkney and Shetland Isles where sand waves indicate easterly sand transport due to the superimposition of a relatively weak south-easterly current upon the strong tidal currents (Section 4.5.3).

The bed-load partings tend to be narrow, with a plan that is 'S' or bow-shaped across the main flow, with extensions parallel with the main tidal flow, so that net sand transport although in the opposite sense, can be almost parallel on the two sides of each limb. Although both bed-load partings and bed-load convergences may be thought of theoretically (for a flat floor) as lines, in practice they generally cannot be defined more closely than as more or less narrow zones. Their locations are defined most certainly by sand wave asymmetry (Section 3.3.3). In the North Channel of the Irish Sea (Fig. 4.16), in the entrance to the White Sea (Fig. 4.14), and in the Southern Bight of the North Sea (Fig. 4.8) the basic bow or 'S' shape of the bed-load parting is complicated by invaginations which are associated with local bed-load convergences. The general location and form of the bed-load partings are also shown by computer simulation of bottom stress (Fig. 4.13). However, the hints of such partings suggested by low shear stress values in the Western English Channel are not supported by sand wave asymmetry and so probably do not exist. The approximate locations of the St George's Channel Bed-Load Parting and the Southern Bight Bed-Load Parting are also shown by the available computations of sand transport directions for these areas (Figs 4.11 and 4.12).

4.4.6 ORIGIN OF BED-LOAD PARTINGS AND CONVERGENCES

The sense of tidal current peak ebb–flood asymmetry and the direction of net sand transport are

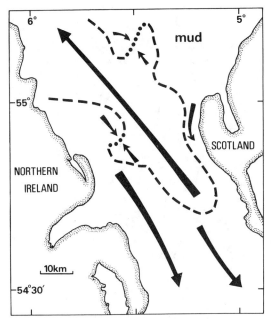

Fig. 4.16 Net sand transport directions in the North Channel of the Irish Sea, British Isles, based on abundant bedform data and some tidal current observations (G. F. Caston, 1976). The North Channel Bed-Load Parting (pecked line) is sinuous and is interrupted by two local bed-load convergences (dotted lines).

often found to be uniform over large offshore areas. Thus it may be inferred that the sum of M_2 and M_4 constituents (Pingree and Griffiths, 1979) largely determines the positions of the partings and convergences in offshore areas.

Marked variations of depth or indented coastlines, as through a strait, may often cause bed-load partings or convergences to be present in the neighbourhood. In such areas of marked topography the near-bed tidal residual current is likely to be more important than the M_4 constituent (Pingree and Maddock, 1977; Pingree and Griffiths, 1979). Analytical models of near-bed tidal residual current distribution in idealized estuaries (Iannello, 1979) and numerical models of its near-bed distribution in a particular estuary (e.g. Johns, 1967) have predicted spatial reversals of tidal residual current direction. A major factor is that eddies will develop with respect to headlands, with different eddies for ebb and flood currents. These are shown in large-scale tidal-current charts and mentioned in navigational 'Pilots'.

Whether a given point lies on one side of the resulting parting will largely depend on its position relative to the nearest flood-current and ebb-current eddies. This should help to explain the curvature of some bed-load partings referred to above.

In the case of large continental shelf seas without rapid offshore depth variations, the approximate offshore positions of bed-load partings and convergences can be found (in principle) for any particular sea by numerical modelling of M_2 and M_4 currents, such as that by Pingree and Griffiths (1979). For some areas the required M_2 range and phase along the open sea boundary of the area modelled may be uncertain. If there are no observations, predictions from syntheses or numerical models of the M_2 tide over the ocean concerned may have to be used. When predictions from models of M_2 and M_4 for other areas than the British continental shelf modelled by Pingree and Griffiths become available, empirical generalizations about the spacing and patterns of bed-load partings and convergences may be possible. Meanwhile, it is noticeable in Fig. 4.13 that no partings or convergences are indicated well offshore in the Celtic Sea or the North Sea, and this is probably due to the progressive character of the M_2 oscillation there. Support for this comes from Heath's (1980) finding that, if the M_4 harmonic is mainly generated by the bottom friction rather than by other non-linear effects, as concluded for offshore parts of the English Channel by Pingree and Maddock (1977), the phase relationship between M_2 and M_4 in a progressive M_2 wave is in general such that the flood peak current speed exceeds the ebb peak speed. This should apply throughout the area covered by the progressive M_2 wave and would correspond to no reversals of peak current direction and no bed-load partings or convergences would be present.

Fig. 4.13 shows that bed-load partings and convergences can occur in continental shelf seas where the M_2 oscillation is largely of standing wave character, such as the Irish Sea and English Channel. However, these are also located near (large-scale) constrictions, so that the latter could be the cause. There is no simple generalization possible about the relative phases of M_2 and M_4 (Heath, 1980), other than in the particular progressive M_2 wave case just mentioned.

For a very long and wide continental shelf sea or gulf at least one M_2 tidal wavelength long (such as those discussed in Chapter 7), so that a standing wave pattern could be well developed, it can be argued from the form of the equations that the M_4 harmonic will have half the (space) wavelength in a similar way to having half the (time) period. Results of modelling of the M_4 constituent show a more complex pattern than that of the M_2 constituent for the same area, but Pingree and Griffiths (1979) note that the spacing of amphidromic points is approximately halved compared to M_2. Although amphidromic points concern tidal range (i.e. where it is zero) and have no direct relationship to the distortion from sinusoidal of the tidal current speed–time relationship, reversals of the sense of distortion may be separated by a similar amount or there may be two reversals in an M_4 wavelength. In such a case bed-load partings and convergences will be separated by a quarter to half an M_2 wavelength. A quarter wavelength of M_2 is about 190 km and 350 km for 30 and 100 m depth respectively.

4.4.7 BED-LOAD PARTINGS AND CONVERGENCES WITH NON-TIDAL CURRENTS AND IN DESERTS

Where a strong ocean current impinges on a continental shelf and coast at a high angle (such as the South Equatorial Current does off the east coast of South America) the current may split into oppositely flowing components with an intervening bed-load parting (Fig. 4.17). A similar effect could be produced by a strong ocean current flowing broadly parallel to a coast but over some stretches being separated from it by a counter-current, or eddies (Fig. 4.18). An example of this occurs off the east coast of South Africa (B. W. Flemming, 1980). There may be seasonal shifts in position for both of these types of parting.

Somewhat analogous features to the bed-load partings and convergences of tidal seas can also occur in some deserts. Here their positions and spacings may be expected to depend mainly on the geographical variations of the prevailing peak wind speeds and directions over periods of hundreds or thousands of years. Fig. 4.19 shows a 'sandflow divide' separating

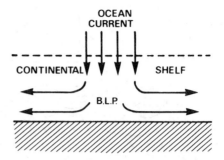

Fig. 4.17 An ocean current flowing towards a continent can split into two portions that flow along the continental shelf in opposite directions. Cross-bedding in any resulting sand deposits could show opposite directions of dip. Analogous deposits in the stratigraphic record might, perhaps, be mistakenly interpreted as indicating a bed-load parting due to tidal currents.

broadly opposed regions of sand flow proposed for the Saharan region (I. G. Wilson, 1971).

4.5 Temporal variations of sand transport rate and direction in a tidal sea

4.5.1 VARIATIONS DUE TO THE TIDAL CYCLES

In areas of predominantly twice-daily or predominantly daily tidal currents sand transport rates will increase progressively from the smallest neap tides (for which they are commonly zero) to the largest spring tides. Furthermore, the tide-raising forces will vary in their peak value not only through the year, but from year to year, century to century and millennium to millennium (Section 2.2). A significant long-term variation arises from the 18.61 year cycle of revolution of the Moon's nodes (Darwin, 1898). The main twice-daily constituent M_2 was or will be about 3.8% greater, and its first

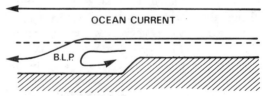

Fig. 4.18 An ocean current flowing approximately parallel with a coast can generate a local counter current in the vicinity of a headland, so causing a bed-load parting to develop on the continental shelf between them (after Flemming, B. W., 1980).

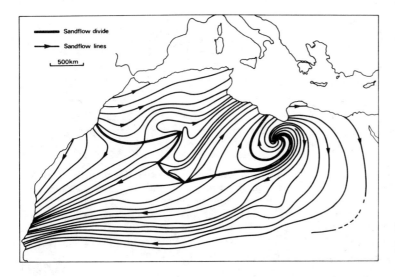

Fig. 4.19 A wind induced sandflow divide in the Sahara desert (after Wilson, I. G., 1971).

harmonic M_4 about 7.6% greater, in 1904, 1922, 1941, 1959, 1978, 1997, etc., and the same amounts less in 1913, 1931, 1950, 1969, 1988, etc. However, the solar constituents like S_2 are unaffected and the two largest once-daily constituents K_1 and O_1 vary in the opposite sense. In 1904 etc. these latter constituents have respective values of 0.88 and 0.81 of their long-term mean values, and other constituents have various other values (Table 14 of Schureman, 1940). As the peak tidal range and current speed in any year will largely depend on the relative phases of the various constituents, they have to be found by computation.

Confining attention now to the areas of the British continental shelf with predominantly twice-daily tides, the sequence of spring and autumn maximum values of the French Tidal Tables coefficients (two values per 25 hours; proportional to the predicted tidal range at Brest), for the years 1960 to 1981 for which we have seen the predictions, shows no clear 18 or 19 year variation. Their range was 118 to 107, although the probability distribution in any year of the twice-daily values above say the mean springs value (95) would also be important. The maximum value for the present century is likely to have occurred in September 1922 (Section 2.2.3) and the observed maximum tidal range at Brest (Cartwright, 1974) would correspond to a coefficient of 121. (The predicted maximum tidal range at Brest for 1922 – from the British Tide Tables for that year – corresponds only to a coefficient of 118, but the tide gauge records and analysis techniques then available were less complete.)

Corresponding coefficients could be defined for other Tide Tables 'standard ports' on the British and NW European coasts. If mean springs range were taken as 95 the coefficients corresponding to extreme neaps, mean neaps, mean tides and extreme springs would differ from their values 20, 45, 70 and 120 at Brest. The most important for sand transport is the 'extreme springs' value and for 'standard ports' around the British coast the corresponding coefficient would vary between 116 (Shoreham) and 125 (Aberdeen and Dublin). Locations well up estuaries have been excluded, also Harwich and Lowestoft which are near amphidromic points of the twice-daily tides, so that their values would not be representative of tidal currents near them. We have not seen any estimated extreme springs values from tidally-analysed current meter recordings in this region.

The importance of tidal variations, such as the above, to sand transport will be discussed in Section 4.6. It is noted now that offshore tidal current speeds will rarely, if ever, be high enough on their own to give sand bed states in the upper flow regime (high-transport plane bed, or antidunes), except possibly on the tops of sand banks.

4.5.2 VARIATIONS DUE TO SEA SURFACE WAVES

Wave effects in the nearshore areas are largely outside the scope of this book and so will only be

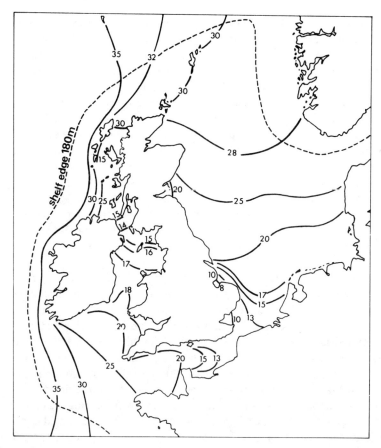

Fig. 4.20 The 50-year maximum wave heights (in metres) around the British Isles, based on instrumental wave measurements and wave forecasts from wind data (after Draper, 1973).

mentioned in passing. During storms the zone of breaking waves could extend out from the coast to water depths of 15–20 m. This zone would also include the upper portions of many of the longitudinal sand banks, on which the wave activity may be locally enhanced by refraction. Probably sand waves there will be smoothed out (but reform when less stormy weather returns). Outside the nearshore zone, near-bed oscillatory speeds due to waves can be calculated from linear wave theory. Such theory is found from various comparisons with offshore measurements to give speeds to about ±10% accuracy when applied to surface wave measurements made by one of the standard forms of wave recorder. Visual wave observations are not reliable here because small errors in estimating significant height and, particularly, significant period cause greater errors in the estimated near-bed oscillatory speeds. Surface wave prediction from wind data is not yet reliable, particularly in the shallower shelf depths. Also, second-order wave theories would give no improvement in accuracy over linear theory (Grace, 1973).

Around the British Isles the calculated near-bed oscillatory speeds (Hadley, 1964; Draper, 1967; Ewing, 1973) are on occasions several times greater than probable sand-movement threshold values for near-bed oscillatory speeds (such as those arrived at in Komar and Miller's 1975 review for fine and medium sand). These speeds would be associated with appreciable rates of sediment erosion and would make symmetrical ripples in the absence of any current. The waves of the '50-year storm' (Fig. 4.20) should make ripple marks on all parts of the continental shelf around the British Isles where the floor is sandy. Furthermore, the wave oscillatory movements support the sediment off the bed while it is transported in the direction of the current prevailing at the time (Bagnold, 1963). The waves have a slight direct effect by imposing a small bias towards their travel direction because of their mass-transport contributions to total current velocity (Longuet-Higgins, 1953).

The presence of sand ripples can cause a complication in sediment transport calculations which is not considered in Bagnold's theory (e.g. Inman and Bowen, 1963; Komar, 1976). Strong wave oscillatory movements associated with weak currents of a few cm/s flowing over rippled sand beds can give a net sand transport rate opposite to the current direction. This arises from the asymmetry of the sand ripples and hence in the strengths of the eddies released from between the ripples when the water velocity reverses. Most of the sand transport in such conditions results from suspension in these eddies. However, when these same wave oscillatory movements are associated with stronger currents the net sand transport rate is greater and the sand moves in the direction of the current, and the integrated sand transport rate over a tidal cycle would be in that direction also.

Some preliminary estimates (Johnson and Stride, 1969) of typical orders of magnitude show that in the absence of sea waves the rate of sand transport at neap tides for a site in the Southern Bight of the North Sea is typically one tenth or less of that at spring tides. However, with sea waves present the neap sand transport rate can equal the rate for mean spring tides that are not aided by sea surface waves. Furthermore, when mean spring tidal currents are themselves aided by storm waves the transport rate is estimated as up to ten times greater than when these tides occur during calm periods (on the basis of Bagnold's 1963 approach). Indeed, as much sand can be moved during one day as during the rest of the previous month (Johnson and Stride, 1969).

Sea waves are also responsible for lowering the height of sand waves in fairly shallow water (Section 4.6).

4.5.3 VARIATIONS DUE TO NON-TIDAL CURRENTS

Considerable areas of the World's continental shelves have weak tidal currents not exceeding 25 cm/s, say.

In some of these regions, as well as in other regions where tidal currents are strong, the non-tidal currents can be dominant over tidal currents on occasions, with the direction of net water transport varying with local conditions (see also Section 5.6). This situation has been explored by Allen (1980a) with regard to sand waves, although he only considered tidal currents that were simple undistorted sine curves. Any tidal residual current may not be distinguishable in observations from non-tidal currents. Seasonal winds in consistent directions cause relatively long term strong unidirectional currents, which can determine the net sand transport direction on the continental shelf during a particular season, even if strong tidal currents are also present. This effect becomes more important for strong winds, shallow water and where there is only a small ebb–flood tidal current asymmetry. An interesting example of this net current effect occurs in Torres Strait, between Australia and New Guinea (Hydrographic Department, Admiralty, 1945). During the NW monsoon the west-going mean spring tidal current reaches about 200 cm/s (3.8 knots) and the east-going mean spring tidal plus the non-tidal current reaches 270 cm/s (5.2 knots), while during the period of the SE trades the west-going mean spring tidal and non-tidal currents reach a sum of 300 cm/s (5.8 knots) and the east-going tidal current reaches about 200 cm/s (3.8 knots). Thus, the direction of non-tidal current reverses between the two seasons, with the currents in each direction varying in strength by as much as two knots from season to season. However, the west-going current reaches an annual mean spring peak speed which is 31 cm/s (0.6 knot) greater than that of the east-going current, suggesting a net westerly transport of sand. This is substantiated by the numerous sand banks extending westwards in the lee of islands and reefs as seen on aerial photographs (Plate 3.20) as well as on navigational charts. A full account of such effects is outside the scope of the present book. However, mention must be made of the effects of non-tidal currents around the British Isles. Their usual pattern of flow is shown in Fig. 4.21. In this region the coast lines ensure that these currents run largely parallel to the main tidal flow, although locally they may flow even at right angles to one another (Fig. 5.18). Where they are caused by

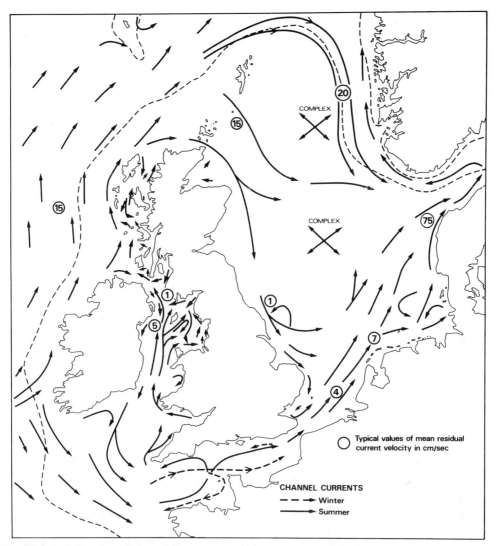

Fig. 4.21 The usual directions of flow of the mean non-tidal currents around the British Isles with some typical values of near-surface speed in cm/s (after Lee and Ramster, 1979). Some extreme values of speed are given in the text.

wind stress on the sea surface, their speed should, in theory, decrease more rapidly beneath the surface than does the tidal current speed. Examples are given of where these currents are weak, dominant or where they are relatively weak yet nevertheless play a significant role in sand transport.

The weak easterly drift of water along the eastern half of the English Channel due to the prevailing westerly wind may, if it reaches the sea floor, slightly increase the effectiveness of the east-going tidal flow of about 130 cm/s near surface and slightly decrease the effectiveness of the weaker west-going tidal current of 125 cm/s at the same station (the east-going tidal current is also stronger at the other stations in the area). This would enhance easterly sand transport above what could be achieved by the tidal currents on their own.

The profiles of sand waves show that the weak northerly net water flow through the Irish Sea is not strong enough to halt the net southwards sand transport at the southern end of St George's Channel. Its strength of order 5 cm/s may, however, enhance the

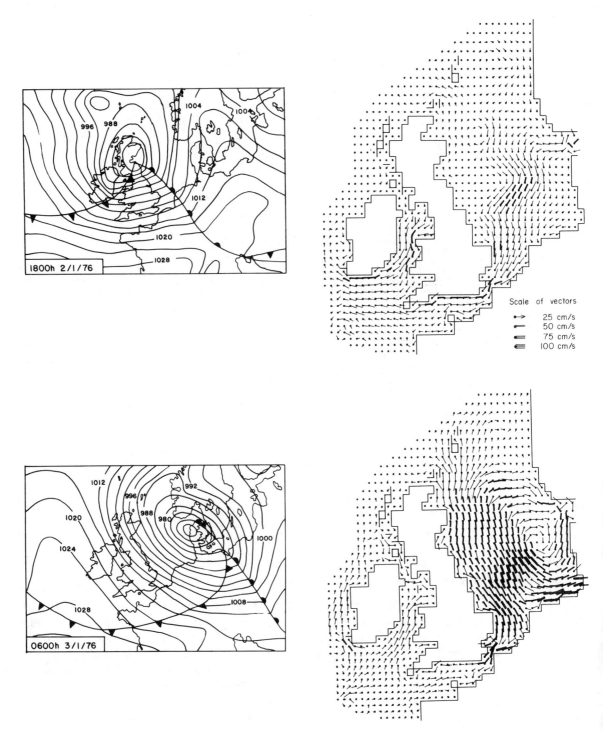

Fig. 4.22 Numerically modelled depth-mean currents (right) resulting from the pressure distributions (left), shown at two stages during the passage of a storm over the British Isles, on a path which often causes 'storm surge' sea level changes at the coast. The currents are short lived and vary widely in direction (after Flather and Davies, 1978).

northerly transport on the north side of the St George's Channel Bed-Load Parting (Fig. 4.10).

The non-tidal Fair Isle current (Dooley and McKay, 1975) flowing round the north side of the Orkney Isles and then south-eastwards on the south side of the channel between the Orkney and Shetland Isles (northern Scotland) reaches usual near-surface speeds of 50 cm/s and near-bottom strength of 20 cm/s at times, whereas the associated tidal currents reach near-surface values of about 110 cm/s towards the west and 100 cm/s towards the east, with smaller near-bottom values. The Fair Isle current, thus, may exert an overriding influence on whether sand transport within it is in an easterly or westerly direction (Fig. 4.21). This easterly transport will occur when its near-bottom strength is greater than the difference between the peak ebb and flood tidal current speeds. However, as shown by sand wave orientations the sand moves virtually along the path of the tidal flow, rather than in the direction of weaker net flow of water. Sand wave profiles suggest that this is true at least on the 3 occasions on which echo-sounder records have been obtained between 1974 and 1981.

The non-tidal currents of the North Sea, known from observations and numerical models, provide valuable indications of occasional sand transport directions. Of particular interest will be the exceptionally strong flows. For example, the usual northerly net flow of water through Dover Strait (Fig. 4.21) is reduced to zero by a steady wind of only about 8 m/s (15 knots) from the north-east (Prandle, 1978). However, much stronger northerly or north-easterly winds cause a substantial water flow to the south. For example, in 1953, such gales caused an exceptional storm surge in the Southern North Sea and an observed out-flow from the North Sea towards the south-west for 6 days (Lawford, 1954). Davies (1976) estimated an outflow of 100 cm/s for another surge. In the Southern Bight of the North Sea these short-lived flows (Fig. 4.22) are strong enough in theory to cause widespread southerly sand transport, thus eliminating the Southern Bight Bed-Load Parting for a day or two and moving some sand from it to the south. For the northern side of that parting, however, the occasional conditions of most interest are those (Fig. 4.22) causing enhanced sand transport to the north.

Wind induced currents are more extreme in the North Sea off north-western Denmark where a north-easterly-going current occasionally reaches 200 cm/s near the sea surface (Stride and Chesterman, 1973). Strong north-easterly currents have been computed here for storm surges (Fig. 4.22) and for steady strong south-westerly winds (Pingree and Griffiths, 1980). Such temporary flows are fully able to determine the local sand transport direction well out to sea as is indicated by the presence of sand waves, where the associated tidal currents only reach about 25 cm/s at mean springs (Stride and Chesterman, 1973). Non-tidal currents associated with storms have also been postulated as causing a net transport of sand from the sand banks along the edge of the German Bight and depositing it in deeper water offshore (Section 5.2.5).

Where the non-tidal current is flowing oblique to the peak tidal current the net sand transport will also be oblique to the peak tidal current, but by a much smaller angle (approximately that of the vector sum of the near-bed mean unidirectional current and the greater of peak flood and peak ebb tidal current). This situation is shown for equal ebb and flood tidal currents in Fig. 4.23. In the case of the anticlockwise net water circulation around the Southern North Sea this will enhance the sand transport eastwards from East Anglia and help to cause an easterly shift of the large Norfolk sand banks. However, this occasional unidirectional current is not the sole cause of their asymmetry. (The more general explanation for these and the other asymmetrical sand banks is given in Section 3.4.5.)

4.6 Growth, migration and decay of sand waves in the Southern Bight of the North Sea by total water movements

It was convenient, in Section 4.3.2, to consider the bedform zones as if they were dependent solely on the peak tidal current strength occurring at mean springs. In practice the bedforms are an integrated response to all water movements that have been operative during quite a long period. This makes it more difficult to attribute cause and effect correctly. Large sand waves will adapt much more slowly than small sand waves to the changes in current strength.

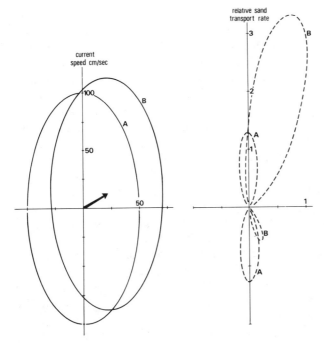

Fig. 4.23 Equal rates of sand transport by ebb and flood (loops A and A with pecked lines) associated with a symmetrical tidal current ellipse (A) are contrasted with the net sand transport shown by the markedly unequal ebb and flood sand transport rate envelopes (pecked loops B and B) that result from the presence of a small superimposed steady current (arrow) displacing the tidal current envelope to B. The obliquity of the steady current adds a much smaller obliquity to those ebb and flood sand transport envelopes.

They therefore give a particularly good indication of net transport direction over a longer period, while the small sand waves can indicate short term local current flow events. They can also reveal the effects of storm waves (Plate 3.1).

The known current speed required to make small sand waves from a flat sand bed in a flume (Fig. 3.4) provides a way of understanding some of the effects of the known changes in total current strength that occur in the sea (Fig. 4.24). The region chosen for review is the sand sheet in the eastern part of the Southern Bight of the North Sea. Here there is a gentle northwards longitudinal decrease in tidal current strength and northwards decrease of grain size down the net sand transport-deposition path. Allowance will be made for sea temperature (as this substantially affects water viscosity, Section 3.2.1) and also for the grain size of the sands as they occur now (Johnson, Stride, Belderson and Kenyon, 1981). This uses McCave's (1971b) diagram of median diameter which is based on Netherlands Institute of Sea Research data. Smaller scale variations probably occur, such as are shown on Jarke's (1956) chart of sediment grade. Fig. 4.24 shows the predicted outer limit of generation of sand waves up to about one metre high during mean spring tides (French Tide Tables coefficient 95) on their own and also for the tides of extreme springs on their own (French Tide Tables coefficient 120). The position of the boundary of small sand waves during mean springs with summer water temperatures is about 40 km seaward of the boundary for winter water temperatures (ignoring the effects of any other water movements) while the extreme tidal currents enclose a band of floor up to 40 km further seaward again.

The narrow northernmost zone of known small sand waves may indicate that a slightly lower current speed will serve to move them once they have been formed. However, it is more likely that the sand waves in the outermost band of floor are only made by the tidal currents when they are aided by some non-tidal water movements. Furthermore, the growth and migration of sand waves in the whole of the Southern Bight will be facilitated by these non-tidal water movements. Thus the periods of activity need not be restricted to occasions when the tidal currents on their own are adequate.

When the combined currents are too weak to form sand waves they will tend slowly to destroy them by reducing their slope angles. Such has been observed on drying banks, between spring and neap tides (Cornish, 1914), although some may have been

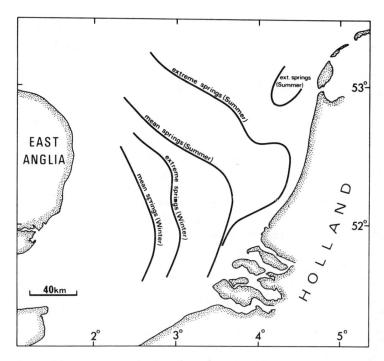

Fig. 4.24 The predicted outer limits of generation of small sand waves in the Southern Bight of the North Sea for mean spring (French tidal coefficient 95) and for extreme spring tidal currents (coefficient 120), but excluding waves and non-tidal currents, for both summer maximum and winter minimum water temperatures. Median grain size contours have been taken from McCave (1971b).

affected by different currents just before drying out.

Storm waves, even aided by the associated weak mass transport current, are not able to make sand waves but only symmetrical sand ripples (Section 3.3.2). What current strength would be adequate to give sand waves is not known but it should be sufficient when the storm waves travel in the direction of a tidal current, whose peak speed near the bed is not less than twice the 'significant' wave oscillatory speed (i.e. the speed exceeded in one third of the oscillations). These waves would have the effect of increasing the sand transport rate in the peak tidal current direction and would only reverse it momentarily, which should only have minimal destructional effect on the sand waves. Perhaps fairly big storm waves, such as would occur for about 10 days in an average winter around the British Isles would accompany tidal currents stronger than mean spring tide values at least once.

The dominant effect of tidal currents in controlling sand wave movement directions in the Southern Bight of the North Sea is shown by the agreement between the asymmetry of the large sand waves and the ebb–flood tidal current asymmetry (Section 4.4.2). Accordingly, all the other water movements may be considered approximately as providing a random addition of strength and direction, so temporarily increasing the effectiveness of the tidal currents to transport sand in the direction determined by their own asymmetry. The effectiveness of the tidal currents must therefore decrease northwards up the Southern Bight, both in terms of their strength and in the number of days during which they can cause appreciable sand transport each year. Inevitably there will be more sand transported on the southern part of that path than on its northern part and therefore more opportunity for making sand waves in the south. This is probably the main explanation of the obvious northwards decrease in the height of the sand waves (shown in Stride, 1970; McCave, 1971b). That the highest sand waves of all in this region are located not far from the Hook of Holland (52°N) is probably due to the steeper longitudinal velocity gradient of tidal currents hereabouts, so locally giving a more rapid decrease in the northwards sand transport rate.

Gradual smoothing of sand waves will commonly occur when the currents are below the value necessary for making sand waves so that they cannot transport sand up their slopes. Smoothing due to very strong currents is less likely as discussed later in this section. The situation for the tidal currents on their own is shown in Fig. 4.24. Sand wave smoothing or even destruction will be increased by extreme storm wave disturbance.

A good correlation between partial smoothing of sand waves from one survey to the next, and occurrences between the surveys of sea waves or swell with mean height above 1.5 m, has been reported by Ludwick (1972) from the eastern seaboard of the USA. Those sand waves had crests located mostly 5 m to 8 m below mean low water.

Repeated echo-sounder surveys made with good navigation in the Southern Bight of the North Sea suggest fairly certainly that storms can affect sand wave heights in the considerably greater depths of 18 to 23 m of water (Terwindt, 1971b). Here the water waves would have much greater heights than Ludwick's 1.5 m. In general it was concluded that height reduction took place when the calculated near-bed oscillatory wave speeds exceeded 56 cm/s (for 10% of the sea waves) whereas up-building of the sand waves took place particularly in summer. A survey with more widely spaced lines, less repetition of them and poorer navigational control led to the belief that there was a band of floor around the northern and eastern edges of the Southern Bight, where substantial height changes or complete smoothing took place (McCave, 1971b). However, some at least of these supposed changes may be due to stated possible navigational errors of up to 2 miles (3.7 km), as the northernmost sand waves (at least) are known from surveys with good navigation to be patchy in distribution (Caston and Stride, 1973). Furthermore, it is unlikely that large enough sea waves were generated in this area during the period of the surveys (Johnson et al., 1981).

Some evidence on smoothing of sand waves by strong currents comes from rivers. A well established example occurs in the River Missouri near Omaha, USA. Over 90% of its bed is covered with sand waves during summer and up to at least early September, whereas by late November less than 25% of the bed has sand waves on it, and these are usually of smaller height. The river water transport rate is held almost constant by upstream reservoirs during September to November, but during this period the current speed increases and water depth decreases corresponding to the reduced bed drag coefficient. A reasonable explanation of this change seems to be provided by the theoretical treatment made by Fredsøe and Engelund (1975). They used the reduced grain settling speed resulting from increased water viscosity that is due to the lower winter water temperature. This increases the ratio of suspended-load to bed-load transport rate in winter. The Froude number (around 0.3 in winter and 0.24 in summer) in the River Missouri would correspond to a stationary water-surface wavelength of only about 1.8 times the water depth, whereas the spacing of sand waves is observed to be 9 to 28 times the water depth. Therefore, interference from stationary water-surface waves and influence of Froude number are presumed to be negligible. Current speed will therefore be taken to be the main factor in causing smoothing, since it determines the ratio of suspension-load to bed-load transport rate. The current speeds were about 1.6 and 1.3 m/s in winter and summer respectively. However, higher speeds would be needed to smooth sand waves in coarser bed sediment than in the Missouri (median diameter 0.21 mm), and probably slightly higher speeds in greater water depths. This last suggestion is supported by the persistence at current speeds exceeding 200 cm/s in the Brahmaputra River (Coleman, 1969) of large sand waves, of heights up to 7 m (not the even larger 'transitional flow sand waves' of the Brahmaputra River discussed in Section 3.2.2). The largest median diameter in 72 bed-load samples was 0.34 mm and probably most were in the range 0.20–0.25 mm. At flood times in the Mississippi River at St Louis, USA (Jordan, 1965) some sand waves persist, though most of the bed becomes fairly flat. The cross-sectional mean flow speed was 240 cm/s and the mean depth 14 m, with median diameter of the sand 0.54 mm.

Current speeds so high as these occur very rarely, if ever, near the floor of the Southern Bight of the North Sea (or even near the water surface, as given by the Dutch light vessel measurements over many years, Otto, 1971). Thus, complete smoothing of sand waves in this region at depths over 10 m, say, is likely to be far less frequent than partial smoothing and to require either a major storm with large wave oscillatory speeds or for the sand waves to occur in an area of very strong tidal currents. Any ripples on top of sand waves would, of course, be smoothed out by much weaker water movements.

4.7 Local sand transport on modern sand banks

The rate of sand transport on sand banks will be dependent on the same forces that affect the adjacent sheets of sand (Section 4.5). The low gradients of the sand banks will not reduce upslope rates much. In contrast, though, the considerable height of a sand bank and its usual oblique alignment (of up to 20 degrees) to left or right with respect to the strongest tidal flow (Fig. 4.25) means that one side is preferentially swept by the flood tidal current, and the other side is exposed preferentially to the ebb tidal current. A synoptic map of the resulting sand transport directions can be derived from sand wave morphology. The asymmetry of the sand waves in profile confirms expectations of flood dominance on one side and ebb dominance on the other side. In addition, it is seen that on a traverse towards the top of a sand bank the sand wave crest orientation in plan view turns progressively towards parallelism with the crest line of the bank (Fig. 4.25). This pattern is found even for the head and tail of a sand bank, so that the morphology provides no evidence of a simple pattern of sand circulating around the bank (Fig. 5.17) (G. F. Caston, 1981) implicit in earlier interpretations based on much less data (see also Section 5.3.3). The theoretical tidal residual circulation of water round the bank (Huthnance, 1973) does not imply a sand circulation.

At most localities the tidal flow in one direction is the more competent to move sand so the sand bank will have an asymmetrical cross section, with the steeper slope facing obliquely in the regional net sand transport direction (as used in Section 4.4.2(b)). The implication is that from time to time sand will be removed from the broad gentle slope of the sand bank and pass onto its relatively steep lee slope. The internal structure (Section 5.3.3) shows, in general, that this material spreads over the whole face, so causing the lateral migration of the bank. The

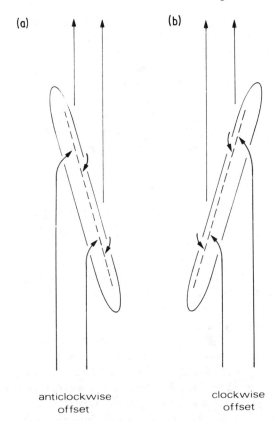

Fig. 4.25 The sand banks of tidal seas are generally aligned at a small angle to the main flow directions, mostly in an anticlockwise sense (a). The long arrows show the regional net sand transport direction. The short arrows show the local, opposite, transport direction on the steeper slope of each sand bank due to the regionally less competent, but locally more competent, tidal current on that side.

spreading process will be greatly facilitated when much sand travels in suspension. This will occur as the sand is of fine grain size, the sand banks have shallow water over them and they are exposed to considerable storm wave activity (even in a relatively stormless winter in the Southern Bight of the North Sea; for example Draper, 1967). Wave action is locally further enhanced by wave refraction. The across-bank flow is strong and on the far side of the crest during calm seas a band of strong turbulence is visible at the sea surface with a band of surface waves beyond. These bands of rough water are recognizable on side-scan radar images of the sea surface in the Southern Bight of the North Sea (Kenyon, 1980).

The disturbances associated with across-bank flow are particularly notable where there is a wide tidal ellipse, even away from the sand bank, as has been figured for Georges Shoal, USA (Stewart and Jordan, 1964). Divers obvserved a dense layer of suspended sand more than a metre thick flowing over the top of that bank, even without the presence of storm waves. During storms the sand waves on the tops of sand banks can be flattened out (Section 4.5).

4.8 Main conclusions

1. Net sand transport directions caused by tidal currents on the continental shelf are shown primarily by the shape and movement of bedforms, by tracer dispersion and by ebb-flood peak tidal current asymmetry. This empirical approach has encouraged workers to make numerical models of bottom shear stress and of net sand transport due to mean spring tides.

2. The rate of sand transport has an approximate cube law relationship with the excess of current speed over a threshold value. Therefore, the stronger of the ebb or flood tidal currents at any point, generally moves more sand than the tidal flow in the opposite direction even when the stronger current lasts a shorter time. This causes a net transport of sand in the direction of flow of the stronger current, not necessarily the direction of flow of the residual current.

3. The net sand transport direction tends to be the same for a sizeable area of offshore sea floor and opposite to the net direction in adjacent areas of it. The transport paths can diverge at bed-load partings and meet at bed-load convergences, corresponding to zones of erosion and deposition, respectively.

4. Sand is being transported by tidal currents towards many estuaries so that sand banks occurring there will increase in volume and eventually amalgamate as the estuary is filled in.

5. Local directions of net sand transport apply to each linear sand bank. These directions are generally opposite on its two sides and converge at its crest; they do not necessarily imply circulation of sand around the bank, as has been claimed by some workers.

6. In a tidal sea the sand transport rate at any point will vary in time, in keeping with the various tidal rhythms. The rate will be increased by any unidirectional currents flowing in the same direction as the tidal current and will also be increased by wave-induced fluctuations in current strength. The net rate can be temporarily reversed by storm-induced currents.

7. The occasions when both the generation and movement of sand waves occur in regions of relatively weak tidal currents may be separated by years during which they are static or are being destroyed.

8. Sand waves associated with relatively weak tidal currents will in general move less often and more slowly than ones of about the same grain size associated with stronger tidal currents (ignoring other water movements that may be present). They will have relatively much less opportunity to grow in size.

9. Net sand transport direction at a given point may diverge from net mud transport direction at the same point by as much as 180°, as the sand is driven by the stronger of the peak ebb or flood tidal flow and the mud by the weak net (residual) transport of water.

Chapter 5

Offshore tidal deposits: sand sheet and sand bank facies

5.1 Introduction

Sand deposits are the most significant products of strong modern offshore tidal current activity. The high porosity of these modern sands implies that their fossil analogues could be economically significant as reservoir rocks. Around the British Isles the associated tidal current gravels are usually only a few centimetres thick but as yet have been little studied. The associated muds are up to about 30 m thick and are potentially much more extensive than sands or gravels because of the large areas of continental shelf with relatively weak tidal currents. However, as shown below, mud deposits are much less of an indicator of tidal current sedimentation than the sands and gravels, and will not be treated in detail.

The offshore tidal sands are represented by two main facies which are different in geographical context, geometry, internal structure and in their expected future development. The first main facies includes the extensive sand sheets whose grain size is related to mean spring peak tidal current speed (Section 5.2.1). The known examples of these sheets around the British Isles are at present relatively thin (up to about 12 m seaward of Holland) but are expected to thicken up in the future as deposition continues (Section 5.5). The second main facies consists of the relatively thick (up to 55 m in the Celtic Sea, for example), isolated sand banks. These are generated where sand has been trapped in an area with tidal currents that are too strong for deposition of the sand sheet facies.

The shelly faunas (Chapter 6) in the tidal current deposits of the continental shelf around the British Isles show marked geographical differences in both the number of species present and in their relative abundances. This variation is closely tied to the bedform zones described in the present chapter and should be read in conjunction with each of them. The tie up between bedforms and faunas is also highly relevant for palaeogeographic reconstructions.

The offshore tidal current deposits need not be related to the regional slope of the continental shelf. This conclusion follows from the observed distribution of tidal current strength (Fig. 2.6) with respect to the depth of the sea (Fig. 1.2).

(a) *Age of the deposits*

The deposits here described seem to span the whole of the Holocene. During approximately the first 5000 years of this period the sea level was rising rapidly, as a result of the melting of the Pleistocene ice sheets. Large areas of land surface were being submerged and the resulting shelf sea conditions at any point were changing (Section 1.3). This early part of the Holocene was followed by a late Holocene period of about 5000 years when sea level was effectively static. Some at least, of the early transgressive deposits have been preserved. They must be considered separately from those laid out during the late Holocene, when conditions were truly modern. These modern deposits must also be considered separately from those materials that are still being moved along transport paths (Fig. 4.10) into regions of deposition.

Five methods of dating the deposits have been utilized. These make use of fauna, flora, radiocarbon, sedimentology and stratigraphy. All together they provide a reasonable spread in time and space. The most obvious method is the radiocarbon dating of the molluscan and other biogenic debris in the shell gravels (Section 6.17). Pollen has been useful for dating some nearshore muddy deposits in the North Sea (Zagwijn and Veenstra, 1966). The faunas of the sands have provided a useful indicator of age (e.g. Clarke, 1970; Oele, 1971; Pantin, 1978; Chapter 6), although some sands are almost barren. Stratigraphic relationships are also particularly useful, providing relative ages for some portions of both the sand sheet and sand bank facies.

The intimate relationships between current strength, grain size and bedform types in modern seas provide a scheme for searching for and recognizing modern sheet deposits, together with the extensive associated areas of sediment erosion and transport. This knowledge facilitates recognition of analogous early Holocene deposits, as well.

(b) *The nature of the sedimentary record*

No single section of the offshore deposits of a tidal sea is likely to contain a complete record of its geological evolution. The events can vary in intensity from place to place but may be recorded only locally. Yet if all the data are taken together they can even allow recognition of details such as the deposits of a single ebb or flood tide, a single storm or the effects of seasonal changes in water temperature.

The effects of strong currents are shown particularly well by the presence of cross-stratified sands. When the currents are dominantly tidal this is revealed in the sands in a number of different ways. These are, (a) the presence of two main offshore facies (sand sheet facies of Section 5.2 and the sand bank facies of Section 5.3), that can have a vertical or lateral association with the tidal flat facies of the shoreline; (b) unipolar and bipolar cross-stratification and reactivation surfaces in the sands; (c) cross-strata where the thickness of sand bundles varies with a tidal rhythm, allowing recognition of ebb and flood, as well as spring and neap cycles. This vital part of the record was recognized unambiguously for the first time in sand wave deposits buried by laterally migrating sand banks at the bottom of a Dutch estuarine channel up to 15 m deep (Visser, 1980). The most perfect examples of this type of depositional pattern are illustrated in Fig. 5.1. They are interpretable as tidal in origin largely because of the presence of intervening mud layers laid down during slack water. The dominant tidal current (the ebb in this case) causes deposition of a bundle of sand laminae on the lee slope of a sand wave. This is then covered by a thin mud layer at slack water. When the tide turns the shallowest part of this mud layer and of the underlying sand bundle is eroded by the weaker (flood) tidal flow giving rise to a reactivation surface and a thin layer of sand on the lee slope. At the next slack water a thinner mud layer is laid down. The cycle is now ready to be repeated. The progressive change in peak tidal current strength throughout the month is indicated by a progressive change in the thickness of the sand bundles (Fig. 5.1) corresponding to each (twice-daily) ebb tide. Thus neap and spring tides can be recognized as well as the number of days in a month (cf. Section 2.7). Of course, this cyclic pattern can be complicated by non-tidal events at times. For example, waves could prevent mud deposition at times, but probably not often in this estuarine channel. Should such sequences prove to be recognizable in ancient rocks laid down in shallow water they would offer a further avenue of research into past changes in the orbit of the Moon relative to the Earth (Section 2.7).

Another important aspect of Visser's observations is that the mud layers of both ebb and flood tidal currents are found in the bottom sets. They offer a criterion for distinguishing between sub-tidal and

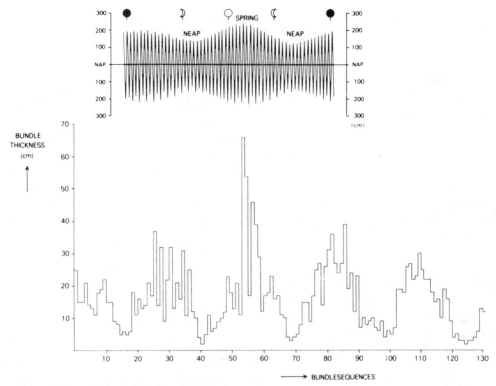

Fig. 5.1 Diagram to show tidal rhythm recognized in the cross-bedding of sand waves in the Oosterschelde, Holland. The bundles of cross-laminae increase in thickness towards spring tides and decrease in thickness towards neap tides. Each slack water period is shown by a mud layer. The mean number of sand bundles in each cycle corresponds with the number of days in a month (Visser, 1980).

intertidal facies because in the latter a mud layer can only be deposited during the slack water at high tide.

Non-tidal currents due to an isolated storm superimposed on tidal currents can cause deposition of thin but extensive layers of sand in an offshore region with a high mud deposition rate (Section 5.2.5).

Deposits showing seasonal changes have been recognized in a deep excavation made in a Dutch estuary. These show progressive upward changes in the number of juveniles of certain species and in the relative amounts of bioturbation. These are associated with differences in lithology and deposition rate (van de Berg, 1979).

(c) *Mud deposition*

The relation of erosion, transport and deposition rates of mud (clay, silt or flocs of these) to current speed is quite different from that for sand. The mud is generally in motion down to lower current speeds and so travels for longer periods during the tidal cycle. It travels in suspension. Thus its displacement will frequently be quite different from that of sand and will not depend so much on the peak speeds of ebb and flood currents. An extreme case is when mud has been carried in suspension continuously for several tidal cycles or more, with current speeds continuously above 10 cm/s or so, before deposition. It will then have responded during transport to net (residual) flow of water, including tidal and non-tidal currents.

Mud deposition can take place around slack water. It can also continue up to quite high current speeds if the mud concentration in suspension is particularly high. In some instances, for example in the inner part of the Bristol Channel (Kirby and Parker, 1974) a large quantity of mud may be trapped in a high energy environment such that a temporary deposit of 'fluid mud' up to 3 m thick is found in sea floor hollows during slack tide, only to be swept away again when the tide has turned. McCave (1970) and Nihoul (1977) developed a 'quasi-continuous deposition' model in which suspended mud that comes near to the bed is deposited, through a partly laminar bed boundary layer, for bottom stress smaller than would correspond to near-surface current speeds of up to 30 to 40 cm/s. Deposition becomes more marked when enough mud has been deposited to make the bed hydraulically smoother. This model is used to explain thicknesses which are greater than those which would result from slack tide deposition of the muddy layers or laminae between sand layers in sand waves and in sand banks. Wave activity will tend to inhibit mud deposition at the time (but might cause more mud deposition after the waves have died down if the waves erode large quantities of mud from a tidal flat, say, and currents carry it in suspension to the location of interest).

The break-up of muddy layers may account for the presence of 'silt-balls' or mud-clasts among sandy deposits (Section 5.2.8).

(d) *Mineral composition of the sands*

The sheets of sand show considerable variation in composition. They can consist almost entirely of quartz, as in the Southern Bight of the North Sea (Jarke, 1956), or almost entirely of calcareous debris as in parts of the English Channel, west of Scotland and on Rockall Bank, and in many tropical areas (Section 6.5). The expected temporal increase in carbonate abundance in some regions is referred to in Section 5.5. Heavy minerals also vary in abundance. They are known to be unimportant in the Southern Bight of the North Sea (Baak, 1936) for example, and to vary in abundance with distance from source, because of selective transport and deposition (Stride, 1970). Mineral studies (considered in terms of metal concentrations) for part of Cardigan Bay (Moore, 1968) and in the north-eastern part of the Irish Sea (Cronan, 1970), for example, show that titanium and manganese are the most abundant elements there, but even these are less than 0.3% by weight in sand samples from the north-eastern Irish Sea and less than 0.5% in Cardigan Bay. Glauconite is referred to in Section 5.4.

(e) *Porosity of sands and shell gravels*

A notable feature of the sheet facies is the excellent sorting of the sand (Section 5.2.3). It is presumably associated with high porosity. The only available value of porosity for offshore tidal sands is stated as 30%, for the sand bank facies of Well Bank, off East Anglia (Houbolt, 1968). That author also pointed out that for a bank volume of about 2.4×10^9 m^3 and a recovery of 20%, this bank would yield a potential volume of about 140×10^6 m^3 of oil, say. The shelly sands composing some of the sand patches and some of the sand banks, such as the Outer Gabbard (Houbolt, 1968) as well as the shell gravels, are also expected to be relatively porous. The high porosity of analogous material in the stratigraphic record should be readily detectable in electric logs of boreholes.

5.2 Late Holocene sand and gravel sheet facies

General

Modern sheets of sediment laid out by tidal currents can range from gravel into sand and into mud. They are known from large and small areas on the continental shelf around the British Isles, as well as on Georges Bank, USA, for example. Similar sheets will be found in other seas with strong tidal currents. At present the known sand sheets are generally much more extensive but much thinner than many of the sands of the sand bank facies, although they will gradually thicken so long as sand is supplied and present day tidal current speeds are approximately maintained. The sheets represent an integration of the deposition rate since the end of the Holocene transgression.

The present-day deposition rate of a well sorted sand has a spatial distribution determined by that of the sand transport rate. If sorting is poor the fractions may have to be considered separately. Quantitatively, the 'equation of continuity' states that the local deposition rate is proportional to the sum of (1) the rate of decrease of net sand transport rate in the longitudinal direction and (2) the rate of decrease of net sand transport rate in the transverse direction. Three particular cases are considered.

Offshore tidal deposits

strong enough to move appreciable amounts of sand. Where these transverse currents decrease in strength in their direction of travel, deposition will take place. In the extreme case the deposits might be thickest in a band perpendicular to them. Thus, these deposits would be elongated parallel to the longitudinal direction of transport and at right angles to the deposits of the first case, mentioned above.

The third case for consideration is for tidal envelopes that are fairly narrow. Even here there will be some transverse net sand transport because of turbulence, lag effects and zigzag movement of grains (as discussed in Section 4.4.3). This sand transport will give net deposition wherever the sand is moved laterally into regions of progressively lower current speeds.

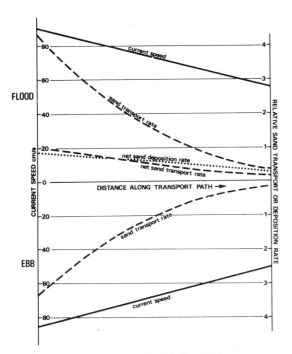

Fig. 5.2 Diagram to show the relationship between net deposition rate and current speed for a uniform speed gradient.

First, if almost all the sand transport is longitudinal and unidirectional (to within a few degrees) the deposits should be thickest where the rate of decrease of sand transport rate is greatest. The transport rate is taken as proportional to the cube of current speed minus threshold speed. Thus, this rate of decrease is proportional to the square of the current speed minus the threshold speed, times the longitudinal rate of decrease of current speed. For example, take a threshold speed of 30 cm/s. Then a decrease of current speed from 90 to 85 cm/s in a certain distance would give over 4 times the deposition rate compared with that for a decrease from 60 to 55 cm/s in the same distance. Application of these unidirectional current results to a flood tidal current, which is taken as 5 cm/s less than the ebb tidal current at each position, gives net deposition rates of about 2:1 (Fig. 5.2).

The second case is for the situation where the longitudinal current speed only gradually decreases along the sand transport path while the tidal envelopes are well rounded, so that transverse tidal flows are

5.2.1 GRAIN SIZE AND CURRENT SPEED

Widespread regional comparisons for the sea floor around the British Isles show empirically that the median grain size of a modern sand and gravel sheet decreases with decreasing mean spring peak tidal current speed (as anticipated in Fig. 1.1), and so is approaching equilibrium with it. The anomalies in Pratje's earlier (1950) geographical correlation are now understood in terms of (a) the local deeps and shoals of rock or Pleistocene materials that are associated with local areas of anomalously fine or coarse grades of modern material, respectively, and (b) sand that remains to be removed from a region of gravel floor and is recognizable as sand ribbons and asymmetrical sand waves. The observed boundary between gravel and sand lies at about the position of the 100 cm/s (2 knots) line of near-surface mean spring peak tidal current strength (Fig. 2.6). In reality this empirical correlation needs some qualification. This is because of the uncertainties about the location of the 2 mm median grain size boundary, as well as about the movement of different grain sizes in a mixture of sand and gravel, and the regionally varying effectiveness of non-tidal water movements.

The total water movements concerned would call for the presence of a near-surface mean spring peak speed tidal current of only about 2/3 of what would be needed without non-tidal currents and waves. On balance it seems likely that the near-surface mean

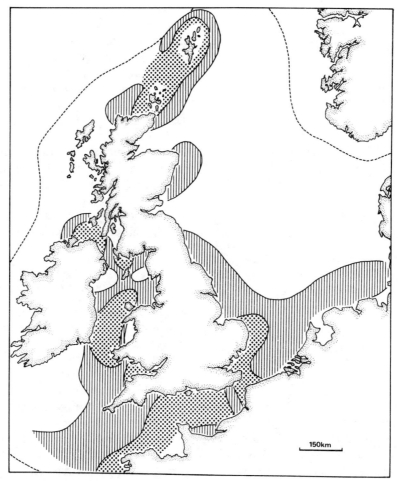

Fig. 5.3 Predicted location of the eventual sheet facies of gravel (dotted) and sand (lines) around the British Isles, within the 200 m depth contour (pecked line). Grain size is largely related to mean spring peak tidal current speed. Mud floors are left blank.

spring peak tidal speed that should be chosen is more likely to be less than about 100 cm/s.

The outer limit of sand is difficult to define in terms of tidal current strength. In this connection, the boundary between clean sand and muddy sand should exclude mud that is deposited at higher peak current speeds where there are abnormally high clay and silt concentrations in suspension (the 'quasi-continuous' settling referred to above). In the central North Sea the boundary is indicated by a near-surface mean spring peak tidal current speed of about 40 cm/s (Fig. 2.6). This has 'theoretical' support in that it is a little less than the speed predicted as being required for the formation of asymmetrical sand ripples (on the basis of the bottom stress found to be required in open channels (Fig. 3.4)). Such a difference is to be expected because of the occasional enhancement of water movements by non-tidal currents and wave action. For convenience, in constructing Fig. 5.3 the boundary between gravel and sand is nevertheless put at the position of the 100 cm/s line and the outer limit of sand at the 50 cm/s line, using the limits shown on Fig. 2.6. The predicted areas of sand and gravel are extensive. Those of the English Channel and Irish Sea are largely shown as they occur now. However, the small

Offshore tidal deposits 101

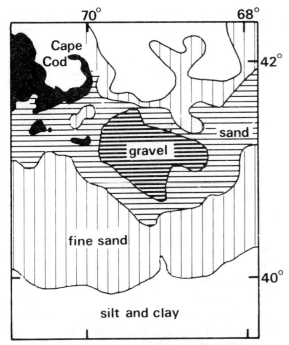

Fig. 5.4 Location of the sheet facies of gravel and sand to north and south of the proposed bed-load parting in the saddle between Georges Bank and Cape Cod, USA. Grain size is related largely to mean spring peak tidal current speed. Much of the ground at either end is occupied by sand banks. Land is shown in black. The silt and clay occur on the continental slope.

area of mud floor now occurring in a local deep to the west of the Bristol Channel is assumed to be covered in sand, once the hole has been filled in. Similarly, with continued infilling part of the present mud floor in the North Sea near to the south side of the German Bight is assumed to be covered in sand. The future status of the patches of sandy floor on the shallow Dogger Bank in the North Sea and to the west of Ireland and Scotland is not clear. Both of these areas are examples of sea floor on which storm-induced water movements are equal or greater in their effect than the relatively weak tidal currents. However, since muddy sands are already present in the north-central Celtic Sea, it is assumed that continued mud deposition here will result in the more extensive mud floor shown in Fig. 5.3. An absence of a sand sheet to the west of the Fair Isle Channel, between the Orkney and Shetland Isles, should result from the continued removal of sand from west

to east. Overall these offshore tidal sheet deposits do not necessarily show a decrease in grain size outwards from the coast.

The location of sand and gravel sheets in the vicinity of Georges Bank, USA (Fig. 5.4) is consistent with available data on tidal current strength (Haight, 1942) and the proposed sand transport paths (Fig. 4.15).

5.2.2 GRAVEL SHEET FORM, COMPOSITION AND STRUCTURE

Modern gravel sheets around the British Isles have been little studied but seem to be extensive in area (Table 5.1), although generally very thin. The gravels may include pebbles, cobbles and even boulders resulting from the early Holocene marine reworking of deposits of the Pleistocene land surface or produced by subsequent tidal current and faunal erosion of such materials and even exposed bed-rock. They can also be rich in Holocene shell debris (Chapter 6). The location of gravel deposits, gravelly-sand and sandy-gravel have also been outlined for Georges Bank (Schlee, 1973) and neighbouring ground off eastern USA (Fig. 5.4).

Table 5.1 The maximum known dimensions of modern equilibrium grade sheet facies laid out by the dominant tidal currents of the seas of north-west Europe.

General sediment grade	Maximum sheet dimensions			Corresponding mean springs near-surface peak current speed	
	Length (km)	Breadth (km)	Thickness (m)	knots	cm/s
Gravel	1000	100	a few cms?	>2	>100
Sand	400	50	12	1–2	50–100
Sand patch ground	700	100	4	<1	<50
Mud	900	400	30	<0.5	<25

The internal structure of the gravel sheets has not been determined. However, the presence of gravel waves with a wavelength of about 10 m and heights of up to about 1 m, formed by tidal currents (for example, Belderson et al., 1972, Fig. 44) suggests that some gravel sheets could be cross-stratified. The dip of these layers of cross-bedding may be indicated by the gravel waves in the West Solent, England, which have lee-slope angles of 11–17° (Dyer, 1971). The tidal-current formed gravel deposits contrast with the lag gravels of weak current regions such as off Plymouth (Flemming and Stride, 1967). Here there are large symmetrical gravel ripples with wavelengths of up to 125 cm and heights of 20–25 cm, formed by storm waves (Plate 3.1). Such large gravel ripples are expected to be widespread in regions swept by large storm waves, but where the tidal currents are too weak to move gravel.

5.2.3 SAND SHEET FORM AND TEXTURE

Sheets of sand laid out by modern tidal currents occupy large areas of sea floor around the British Isles (Table 5.1) and off eastern USA (Fig. 5.4) and must be extensive in many other regions. Some sheets are longer than wide (Fig. 5.3), with known thicknesses of up to about 12 m in the Southern Bight of the North Sea (Oele, 1971), for example.

The sheets can include the full range of sand sizes. However, the actual range of grain sizes present in any particular sand sheet will be dependent on the sizes that were available for transport, together with the range of peak tidal current speeds (and to some extent plus the other water movements) with which they are in equilibrium (Section 5.2.1). The full range of sand sizes could be present in the deposits of the western part of the Bristol Channel, for example, as that transport path covers a wide range of current speeds. The coarsest sand occurring west of Holland is 0.45 mm median diameter because of the relatively weaker currents occurring there.

The sorting of the sand, where known, is good to very good, being best, as would be expected, for the 0.2 mm sand (e.g. Irish Sea, Cronan, 1969) and equivalent to that found for beach sands. This point has also been made for the sands of the outer part of Georges Bank, USA (Schlee, 1973). Some samples of sands off north-west Holland have 95% in the same fine sand grade of 0.125 to 0.25 mm (Völpel, 1959) and many nearby, but somewhat coarser, samples of sand are almost as well sorted.

5.2.4 STRUCTURE OF A SAND SHEET IN THE SOUTHERN NORTH SEA

The best known sand sheet near to the British Isles is located in the north-eastern part of the Southern Bight of the North Sea. The deposits can also be followed as a narrow band along the southern edge of the German Bight, as discussed later. Study of the Southern Bight portion is particularly valuable because sand is abundant, water depth averaged over the sand waves varies relatively little and because there is a very gentle northwards decrease in peak tidal current strength. The associated sand transport-deposition path is about 130 km long. The net sand transport direction is northwards. However, as shown in Fig. 4.10 the path commences at the Southern Bight Bed-Load Parting where there are symmetrical or nearly symmetrical sand waves, so that this sand sheet will show complexities in sand wave shape and resulting internal structure that will not be seen on some of the other sand transport-deposition paths. For example, the English Channel and Bristol Channel Bed-Load Partings are devoid of sand waves, because of the greater strength of the tidal currents. The sand sheets associated with their transport paths should thus have a less complex structure.

In the Southern Bight of the North Sea the gentle northwards decrease in peak tidal current strength and median grain size allows recognition of significant differences in the form of sand waves which are present in all but its northernmost part. This variation is well shown by a profile and sonographs extending up the Southern Bight that were obtained by RRS Discovery in 1972 (Fig. 3.5). The morphological information is combined with the meagre data obtained from box cores, gravity cores and vibrocores taken from these bedforms to illustrate the internal structure of a sand sheet. The last two coring methods unfortunately cause some distortion of the structures and there can also be some doubts about the inclination of the corer with respect to the vertical and, thus, of the true dip of any stratification.

A corresponding sand sheet is not present on the southern side of the Southern Bight Bed-Load Parting as the tidal currents increase in strength southwards and cannot deposit the sand until the end of the path is reached (at the Dover Strait Bed-Load Convergence). The large sand waves on that path have relatively high lee slope angles of about 12 to 31°, which are consistent with non-deposition.

(a) Zone of large sand waves

At the Southern Bight Bed-Load Parting the floor has a broad-scale irregular relief, with superimposed smaller-scale but clearly defined sand waves (Fig. 3.5, profile (a)). This is presumably because the bed-load parting is essentially an area of erosion.

The large sand waves throughout the zone have small sand waves on them, although the smallest sand waves may be visible only on the sonographs and not the echo-sounder profiles. Available data show that at the bed-load parting there are large sand waves with symmetrical profiles, with both slopes of 10° or even up to 19° (Fig. 3.5, profiles (a) and (b)). In addition, there are many other sand waves with almost symmetrical profiles such that the two slopes of a sand wave can be 7° and 10° or 14° and 18°, for example. Some groups of these have their steep slopes facing south. These are mixed up with other groups with steep slopes facing north such that the bed-load parting is a complex area of diverging sand streams, whose details remain to be worked out.

The internal structure of these symmetrical sand waves has not yet been determined. Nor has it been revealed by illustrations of large symmetrical sand waves of the St Lawrence Estuary (D'Anglejan, 1971) as their seeming internal structures are now attributed to side echoes from small sand waves on their surface (D'Anglejan, 1979, pers. comm.). The internal structure of symmetrical sand waves is generally expected to include equal contributions from both sand transport directions. The arrangement of these sets may well be as complex as has been shown for sand waves with some asymmetry in the Jade Estuary (Fig. 5.5(a)) as schematized in Fig. 5.5(b). However, the slope will not be about 3° as Allen (1980a) surmised, but is between 9° and 19° as stated above. Moreover, it seems likely that the steep slopes of such a sand wave will encourage movement of small sand waves up each side, so that the cross-bedding will dip in opposite directions on the two sides, rather than as shown in Fig. 5.5(b). It must be appreciated that the currents associated with these examples are much stronger so that the contribution by each ebb and flood tidal flow will be larger than in the case of the weaker currents at the Southern Bight Bed-Load Parting.

North of the symmetrical sand waves there is a larger area of sand waves with asymmetrical profiles. These large sand waves decrease in height northwards (Stride, 1970; McCave, 1971b) down to a minimum value of about 2 m (Fig. 5.6). The westward decrease in their height is due to lack of sand. Successive crests of these sand waves show a northwards decrease in the angle of their more gentle southern slope, down to values of less than one degree, although there is some scatter. The steeper northern slope decreases progressively from about 12° to about 3° along the same path, again with scatter and peak values up to about 17°. This is the lee slope in terms of net sand transport.

The more southerly style of asymmetrical cross section is described as 'catback' (van Veen, 1935) because of their characteristic humped profile. Although no data are yet available about their internal structure it is inferred that, as both ebb and flood tidal currents will be strong enough to move appreciable sand and make sand waves, then cross-bedding due to sand transport in both directions could be present.

The catback form gives way northwards to the much more usual asymmetrical profile with a flatter gentle southern slope. The more southerly of these sand waves has a lee slope with tangential base (Fig. 3.5(c)). The lower part of this slope can have some small sand waves on it, with their own lee slopes indicating movement up towards the crest (Terwindt, 1971b, Fig. 5). This may imply an internal structure such as that shown in Fig. 5.5(c). However, it is not clear from Terwindt's data that there really is a significant grain size decrease from top to bottom of a sand wave, as has been argued by McCave (1971b) from five samples taken by Houbolt (1968), nor that this supposed difference is due to fall out from suspension. Indeed, the gentle lee slopes of the large sand waves are unlikely to give rise to flow separation originating at their crests, except perhaps for a small one behind any near-crest subsidiary steep slope due to the presence of a small sand wave. The suggestion (McCave, 1971b) that large sand waves are due primarily to suspension transport whereas small sand waves (his 'megaripples') are due primarily to bed-load transport is not accepted as valid generally. On the contrary the southward

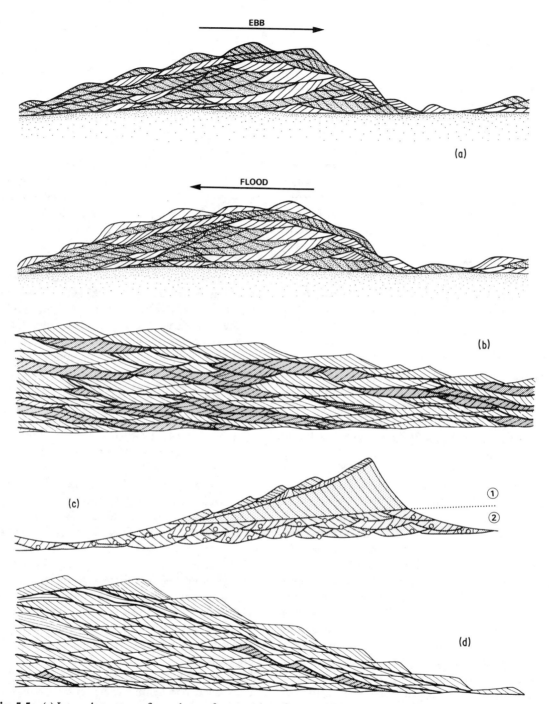

Fig. 5.5 (a) Internal structure of a sand wave from a region of strong tidal currents capable of moving sand on both ebb and flood, amongst sand banks of the southern edge of the German Bight (after Reineck, 1963).
(b) One side of a sand wave with symmetrical cross-section to show supposed internal structure, as visualized by Allen (1980a).
(c) Hypothetical cross-section to show presumed internal structure of a large sand wave with tangential toeset and small sand waves climbing up both slopes (after McCave, 1971b).
(d) Supposed internal structure of an asymmetrical sand wave as visualized by Allen (1980a).

Offshore tidal deposits 105

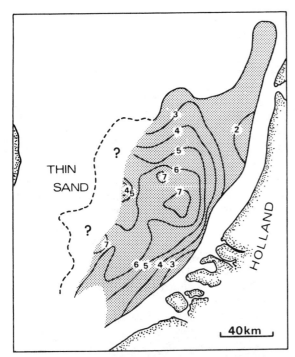

Fig. 5.6 The mean height of the highest one third of the large sand waves on the sand sheet located west of Holland, from surveys during the period August 1968 to May 1969 (after McCave, 1971b).

migrating small sand waves on the lower part of the lee slopes are probably caused by the periodic south-flowing tidal current. Although this is regionally less competent than the north-flowing current, it may locally be the more competent on the 'lee' side of large sand waves, just as sand waves have asymmetry counter to the regionally most competent current on the steep slope of a sand bank. Furthermore, the short-lived, occasional net southerly non-tidal flows of water will also exert some influence (Section 4.5.3). Reactivation surfaces due to this flow will be expected. They may be analogous to the master bedding of Allen (1980a, Class V shown in Fig. 5.5(d)). However, the tidal currents for this ground will be less strong than in the theoretical case considered by Allen so that erosional episodes will generally not be so marked.

Twenty out of a total of twenty seven cores from the southern part of this region show cross-stratification to be present (Houbolt, 1968, Enclosure 2), while parts or the whole of some of the cores showed only the churning action of echinoderms. Four out of the five cores, in which the cross-bedding dip direction is known, face in the expected direction of north-easterly net sand transport. The cross-bedding dip values of 15–20° indicate that the majority of the cross-bedding is produced by the small sand waves atop the larger sand waves which seem to have gentler slopes thereabouts.

The northernmost of the large sand waves have planar, north-facing lee slopes showing an abrupt contact with their troughs (Fig. 3.5(d)). The shape of the lee slope, with a gradient of down to only 3°, may imply the presence of planar cross-bedding parallel with its surface. Such bedding would presumably be analogous to that seen in a Dutch estuary (Fig. 5.7), and possibly analogous to that shown in a boomer profile from the Chesapeake and Delaware Canal (USA) for sand waves 1 to 3 metres high, with cross-bedding sloping at about 3 to 12° (Moody and van Reenen, pers. comm., 1966). However, it is possible that the interfaces visible on the profile in the latter example result from muddy layers along master-bedding surfaces, and that smaller scale, higher angle cross-strata are present between these layers. Periods of sand transport by tidal currents on their own will be separated by periods when there is rounding of the crests of the small sand waves and possibly even by the temporary disappearance of these small sand waves. This will be indicated by flat bedding and destruction of bedding by the fauna.

The sinuosity of many sand waves, together with the oblique orientation of small sand waves on top of the large ones (Section 3.3.3(g)) and the presence of sand ripples as well, suggest that the cross-stratification will frequently be three-dimensional and complex.

(b) *Zone of small sand waves*

Available side-scan sonar coverage shows that the small sand waves occurring on their own occupy a narrow band (Fig. 5.8) extending along the northern side of the zone of large sand waves. Their wavelength is known to range from 1 to 10 m and their heights are up to about 1 m. Lee slope angles are expected to be commonly up to 30°, in keeping with those of other small sand waves (Section 3.3.3(d)).

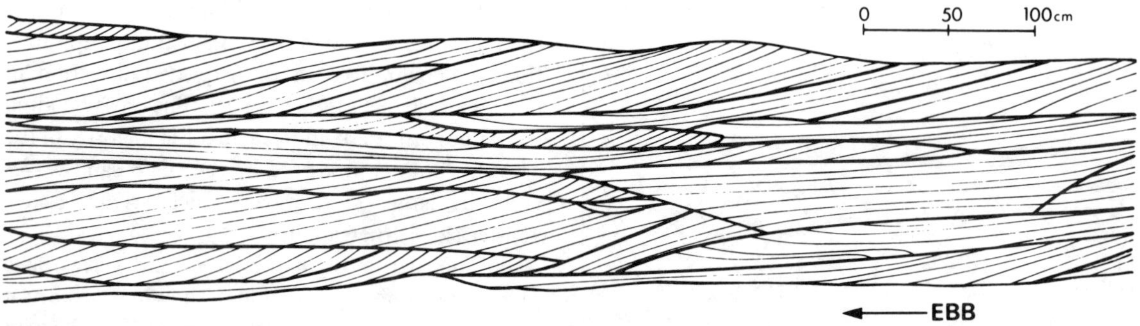

106 Offshore Tidal Sands

![Cross-bedded sands diagram]

← EBB

Fig. 5.7 Cross-bedded sands showing only the influence of the ebb tidal current, in a longitudinal section of a Dutch estuarine channel (after Terwindt, 1971a). This cross-bedding style may be typical of the northernmost large sand waves in the sand sheet west of Holland, although some bimodality in dip directions must be expected there because of occasional reversals in sand transport direction.

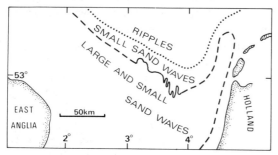

Fig. 5.8 Approximate limits of the zone of small sand waves in the Southern Bight of the North Sea (after Johnson and Stride, 1969).

The median diameter of the sand in this part of the sand sheet is generally coarser than about 0.15 mm and finer than 0.25 mm.

(c) *Outer zones*

The next outer, finer sand zone is normally expected to have asymmetrical ripple marks, as seen in similar situations elsewhere (e.g. Channon and Hamilton, 1976). Structures should possibly include ripple-drift bedding formed during occasions of peak sand transport.

The outermost zone of even finer sand is expected to have nearly symmetrical sand ripples formed largely by storm waves. Short cores taken further seaward beyond this zone show the existence of silty sands merging into muds. The small differences in the abundance of sand in a core are in keeping with the temporal variations in current strength that affect the Southern Bight of the North Sea. The deposition rate of muddy sediments in the Outer Silver Pit is about 2 mm per year (Veenstra, 1965; Zagwijn and Veenstra, 1966) which is less than in the German Bight (Section 5.2.5).

5.2.5 GERMAN BIGHT SAND TO MUD SHEET

A narrow sand wave zone is present along the south side of the German Bight (Reineck, 1963). Box cores and can cores (Reineck, 1963 and 1976) show that cross-bedding is common in this zone and has a spread of dip directions. These directions are rather different from those in the shore zone nearby. The currents to the north of the isle of Norderney reach the same peak speed as those found over the northernmost large sand waves of the Southern Bight of the North Sea. In the region of stronger tidal currents north of Wangeroog Island (Chowdhuri and Reineck, 1978) and in the Jade Estuary (Fig. 5.5(a)) nearby (Reineck, 1963), herringbone cross-bedding is present.

In the south-eastern corner of the German Bight there is only a narrow sand sheet separating the Suderpiep sand bank zone from the mud deposits up to 20 m thick which extend for about 30 km seaward (Fig. 5.9). The ground has been explored by numerous box cores and piston cores (Reineck and Singh, 1973). These show a seawards transition from the medium to fine sands of the Nordegründe sand banks into a sheet of finer sediments in which silty

Offshore tidal deposits   107

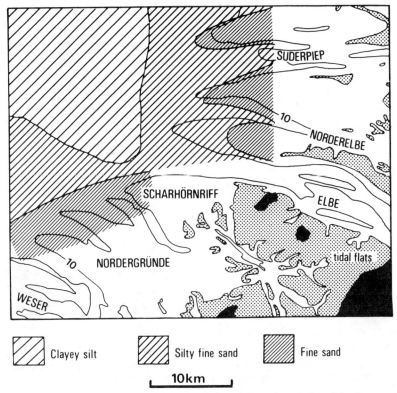

Fig. 5.9 Transition from the estuarine sand bank facies into the offshore sheet facies of the German Bight (after Reineck and Singh, 1973).

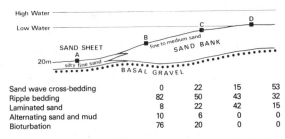

Fig. 5.10 Schematic section extending across Fig. 5.9 from the sand bank facies of Nordergründe into the narrow and restricted development of the sheet facies between the sand banks and extending out into the deeper German Bight. It shows the relative abundance of bioturbation and other structures found in box cores (after Reineck and Singh, 1973).

fine sands give way to clayey silt (Fig. 5.10). The cross-bedding resulting from the presence of sand waves on the sand banks and the abundance of laminated sand decreases to seaward, while the abundance of sand ripples, mud-sand laminae and bioturbation increase to seaward. The short-term temporal variation in grain size in the fine grained sediments is striking (Gadow and Reineck, 1969). It shows how non-tidal storm events can hereabouts temporarily override the effects of normally dominant tidal currents. Cores contain numerous thin layers made of silty fine sand to fine sandy silt that alternate with clayey silt within the top few metres of this 20 m thick sheet. The layers of sandy material can be more than 2 cm thick just west of the sand bank zone but thin out westwards towards the island of Helgoland, just beyond the northwest corner of Fig. 5.11. Correlation between cores shows that some sand layers can be followed for at least 15 km (Reineck, Gutmann and Hertweck, 1967, Fig. 5).

The outward spread of fine sand has been attributed to storm activity, when a temporary wind-induced current aided the tidal current (Gadow and

108   Offshore Tidal Sands

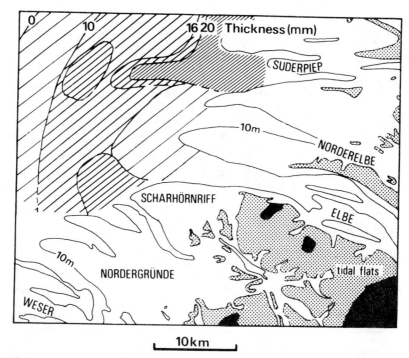

Fig. 5.11 Sand rich layers, in the sheet facies of Fig. 5.10, decrease in thickness (mms) to seaward. The sand is derived from sand banks (defined by the 10 m contour) and tidal flats along the edge of the German Bight and is removed and transported by storm-induced water movements coinciding with ebb-tidal currents (after Gadow and Reineck, 1969).

Reineck, 1969). It is more likely, however, that storm waves would be needed as well to keep the sandy material in almost continuous suspension, so that it would move with the current speed along its transport path of up to 30 km from its supposed source on the seawards tips of the German Bight sand banks (Reineck and Singh, 1972). Radionuclide dating of a 50 cm long section of core shows that the deposition rate varied between about 1.8 mm per year and about 18 mm per year, with an average for the section of 7.7 mm per year. Deposition is discontinuous, reflecting the number and severity of storm surge events (Dominik, Forstner, Mangini and Reineck, 1978). The limitation of this well developed depositional pattern along the edge of the German Bight (and its apparent absence beyond the northern edge of the Southern Bight of the North Sea, where sand gives way to mud, also) is in keeping with the shallower water (respectively 20–30 m instead of 20–40 m) and stronger tidal currents (75 to 100 cm/s instead of 50 to 75 cm/s) of the former. In addition, the peak storm-induced currents of up to 100 cm/s (Gienapp, 1973) and wave energy would be much greater along the inner edge of the German Bight, where there are numerous sand banks reaching up to, or almost up to, low tide level. The similar stratified sediments found in the north-eastern part of the Irish Sea (Pantin, 1978) are close to shore, also. Such material has also been found in a local deep known as Cork Hole, that is located near a sand bank in the Thames Approaches.

5.2.6 IRISH SEA SAND TO MUD SHEET

The transition from the sand wave zone to zones of fine sand to mud up to 30 m thick in the north-western (Belderson, 1964; G. F. Caston, 1976) and north-eastern parts of the Irish Sea (Belderson and Stride, 1969; Pantin, 1978; Cronan, 1972) is far more abrupt in comparison with that found west of Holland. Cores up to 4 m long in the sand and mud deposits of the north-eastern part of the Irish Sea have revealed many temporal changes in grain size (H. Pantin, 1978 and pers. comm. 1979). However, full use of the data is made difficult because of uncertainties about the correlation of layers from core to core. Taking the simplest view of possible correlations it seems that on the western side of the region of deposition there were three episodes when sand was spread out there more widely, with intervening periods when muddy sands were more common. The sand probably reached the deposition sites along the existing transport path from the southwest. Further eastwards mud gives way to sand near to the English coast. A few cores thereabouts show thin alternate laminae of relatively sandy and muddy sediments (Pantin, 1978, Plates 4.3 and 4.4). This deposit resembles material in the German Bight (mentioned above), for which storm-induced currents are suggested as aiding the tidal currents. A study of the magnetic fabric of the more muddy sediments in the Irish Sea (D. Frederick, 1971, pers. comm.) raises hopes of revealing their direction of arrival during deposition.

5.2.7 REGIONAL CROSS-BEDDING DIP DIRECTIONS WITHIN THE SAND SHEET FACIES

It is expected that there will be a northerly prevalent direction of dip for much of the cross-strata in the sand sheet of the north-eastern part of the Southern Bight of the North Sea, in keeping with the direction of net sand transport. These dip directions may, however, range up to 45° or so to either side of this prevalent direction because of the sinuous plan view of the crests of many of the large sand waves (e.g. Belderson, Kenyon, Stride and Stubbs, 1972, Fig. 48; Terwindt, 1971b). In addition, some southerly cross-bedding dips are also expected (Section 5.2.4). These suggestions are supported by the measured dip directions in five core samples (Houbolt, 1968).

The inferred dip directions for cross-bedding of five significant modern sheet sands located in the Irish Sea are shown in Fig. 5.12. The regional variation in direction points to the need for caution in using such data out of its local geographical context. Thus, if all of these cross-bedding dip directions were put together on to a single rose diagram, as they might if the dip directions had been obtained from available outcrops of sedimentary rocks of the same age, it would be difficult to draw meaningful

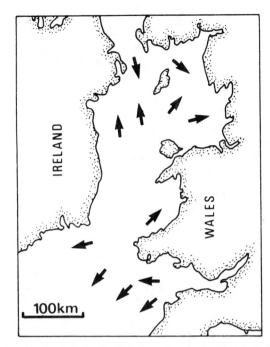

Fig. 5.12 Likely dip directions of cross-strata of the late Holocene sand sheets in the Irish Sea, based on the known asymmetry of the sand waves and tidal currents.

conclusions about regional sand transport directions. It would need to be appreciated that the geographical variability in dip directions was due to tidal currents flowing in widely different directions in adjacent locations

5.2.8 SAND PATCHES

Patches of sand on a lag gravel floor are common features in regions (Fig. 4.4) where near-surface tidal currents reach mean spring peak values of less than about 50 cm/s. They are aligned with their long axes either more or less parallel or transverse to the peak tidal flow (Sections 3.3.4 and 3.4.4). Although there may be some sand movement in these sand patches during storms it is probably justifiable to view the patches as deposits. Those sand patches that receive little additional sand probably have a high preservation potential. This is because they occur in regions of comparatively weak tidal currents where a slight decrease in peak current strength could initiate mud deposition. Their characteristic tabular profile has been observed on sub-bottom profiles, by the authors, beneath a mud blanket north-west of Ireland. Tabular sand patches (erroneously called sand ribbons) are also present at the edge of the Grande Vasière, south-west of Brittany. These are partially buried by mud at times. It is proposed that there is some intermingling of the sand and mud as a result of sand transport and bioturbation (Delanoë and Pinot, 1980).

The sand patches still being supplied with sand are expected to grow into a complete sand sheet, as they seem to get progressively wider towards the supply source (Belderson and Stride, 1966).

The transverse sand patches, which have been sampled in the Celtic Sea, are known to be rich (20–70%) in molluscan valves and other carbonate fragments and must be growing as a result of the fauna living in the sand patch or in the neighbouring lag-gravel floor. Seen as a deposit the transverse sand patches of the Celtic Sea are commonly 2 m thick (max. 4 m), 500 m long and have an irregular to crescentic plan view. The steep edges of these sand patches (possibly even slip faces) may indicate that the sand will be cross-stratified. The sand in the available 7 samples is well sorted, fine to medium grade (average median diameter 0.28 mm), with a slight negative skew (Kenyon, 1970). In general the sand of the sand patches in the English Channel is similar, ranging in composition from almost pure quartz to almost pure shelly material, as in the case of the sand sheets.

'Silt balls' up to about 1 cm diameter have been recognized in the sand patches in the English Channel south of Plymouth. The balls examined were made of sand containing 15 to 20% of silt, giving an overall median diameter of 0.26 mm, compared with 0.34 mm for the surrounding clean sands (Flemming and Stride, 1967). Such balls were also present in the sand patches found near to Land's End (Channon, 1971). Balls of sand up to a few centimetres wide, containing up to 10% silt, have also been found in the lag gravels separating some of the sand patches of the Celtic Sea (Kenyon, 1970). Similar sized muddy bodies are also present in the sheet of muddy sand located nearer to the Bristol Channel (Belderson and Stride, 1966). The origin of the silt balls and mud bodies is uncertain but the following possibilities should be considered: mud clasts made from the break-up of layers of mud that had temporarily lain on migrating sand waves; local amounts of mud accumulating in pits above burrows; aggregates of faecal pellets; or mucilage balls made by crustaceans to cover up the entrances to their burrow systems.

5.2.9 SAND WAVES FORMED BY TIDAL LEE WAVES

The final type of sand deposit discussed here consists of a group of exceptionally large, almost symmetrical sand waves that extend parallel with the general direction of the edge of the continental shelf of La Chapelle Bank, to the west of Brittany, in a water depth of up to about 200 m. They are up to 5 km long, 1 km apart and 7 m high (Stride, 1963b), with slopes of about 4° (Cartwright, 1959, Fig. 2). They seem to be underlain by sand (Belderson, Kenyon and Stride, 1970). This sheet is likely to be preserved but its internal structure is not known. The sand waves have been attributed to the action of tidal lee-waves (Section 3.3.3(i)).

5.2.10 FACIES MODEL OF AN OFFSHORE TIDAL CURRENT SAND SHEET

The sand sheets can be the more extensive of the two main sand facies of the offshore tidal current environment of continental shelves. Sand sheet size and shape are variable, but areas of 20 000 km^2 are not uncommon. Individual sheets can be widely separated from one another (Fig. 5.3). All sand sizes can be represented in the deposits but in practice the range of grain size in an individual sheet can be quite restricted, depending on current strength and grain size availability (Section 5.2.3). An essential feature of the sheets is that they show grain size gradients, which can be longitudinal or transverse to the peak tidal flow, such that the grain size is in equilibrium with the local current strength. Sorting of the sand is very good, especially for fine sand.

The structure of the sand sheet can show an abundance of cross-bedding. The character of this will vary progressively along directions longitudinal or transverse to peak tidal flow. Some tentative

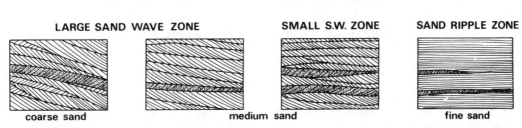

Fig. 5.13 Four much simplified hypothetical sections spaced along the line of the net sand transport-deposition path (arrow) of a sand sheet laid out largely by tidal currents. They show some of the types of structures to expect but are not necessarily all at the same scale. Tidal rythmic bedding and wave and current ripple-bedding as well as probable mud laminae should be present, but are not shown. Some cross-bedding resulting from an additional storm-induced current flowing to the left is emphasized by heavy shading. This effect is probably more evident in the zones of small sand waves and sand ripples. Such effects (derived largely from structures in sedimentary rocks) are described in Johnson (1978) and Walker (1979).

conclusions are given below. A better appraisal will have to await the taking of more samples. Unipolar cross-bedding dips pointing in the regional net sand transport direction are thought to be dominant, but there will inevitably be a spread of directions about the mean value because of the sinuosity of the sand wave crests. Bipolarity of dips in cross-stratification is not thought to be common except in some sand waves. It is expected to be much more abundant where the symmetrical sand waves of a bed-load parting are present at one end of a sand sheet. In general, the nature of the cross-bedding will differ progressively along a sand transport-deposition path (Fig. 5.13) in relation to the progressive weakening of peak tidal current speed along it, and the resulting differences in bed morphology. In the coarser sands there is expected to be master-bedding with dips of less than 20° representing lee slopes of large sand waves, together with high-angle cross-bedding of the small sand waves with dips of up to 30°. Some of these dips are expected to be opposite to the net sand transport direction. Beyond this zone the master-bedding produced by the larger sand waves should have gentler dips of about 5°, with fewer signs of reversed dip directions, such as in Fig. 5.7. Beyond again there should only be high-angle cross-bedding due to small sand waves. The next outer zone should be flat or ripple-bedded, while the outermost zone of fine sand should tend to be flat-bedded or have some symmetrical ripple bedding. This simple hypothetical model should, of course, make more allowance for the changes resulting from cyclic variations in tidal current strength, as well as the superimposed effects of the non-tidal water movements (Sections 4.5, 4.6 and 5.2.4). Muds lie beyond. In general, mud is not expected to be common in the sand wave zones but can be present at localities where mud is abundant in suspension on occasions. Some flaser bedding may then also occur.

The vertical sequence of deposits should reflect the most significant temporal changes in peak tidal current strength. During a ten year period, say, there can be substantial but temporary shifts in the positions of the boundary between each bedform zone. As stated in Section 4.6 the outer limit of tidal current strength needed to form small sand waves (e.g. in the Southern Bight of the North Sea) must fluctuate and will move furthest outwards during periods of exceptionally strong tidal currents in summer (and at other times when aided by unidirectional currents). The result is that the outermost of the small sand waves, up to a metre high perhaps, must be formed there at the expense of the deposit with sand ripples which occurred there previously. When the strong currents abate the crests of these small sand waves will be slowly subdued by wave and faunal disturbance while ripple-bedding will bury the lower, remaining, portions of the high angle cross-bedding of the small sand waves.

Within the zone of small sand waves there are also indications that some large sand waves may grow temporarily (e.g. Caston and Stride, 1973), and then be rounded off when currents weaken. This could lead to preservation of layers of low angle cross-bedding in a zone dominated by high angle cross-bedding. However, the considerably longer lag time required to make large sand waves, in what was previously a region of small waves, suggests that this effect is neither frequent nor widespread.

Fig. 4.24 gives an impression of the temporal changes in position of the outer limit of generation of small sand waves that can be expected on a sand sheet during a ten year period, say. Occasional strong non-tidal water movements both enhance and detract from the tidal effects so that the sequence of changes will be less obviously cyclical.

Viewed over a longer time scale the basal portions of the offshore tidal sand sheets should be deposited preferentially in any local deeps that were present in the sea floor. The supply of quartz sand made available as a result of a marine transgression should be finite, unless the supply can be maintained from land. Where there is a shortage of quartz sand successive deposits should become progressively more calcareous. If the basin widens or deepens in time (or if there is renewed marine transgression) there should be a progressive change into more muddy facies.

Deposition rate will vary in sympathy with the tidal cycles, with the incidence of other water movements, and for other reasons. Thus, the rate may well be highest during the filling of any local deeps in the floor, and subsequently while there is still plenty of sand available for deposition. The completeness of the record of events will then gradually deteriorate because of the progressive lessening in deposition rate. However, as deposition occurs during the migration of bedforms across most of the sand sheet the change from a high to low (vertical) mean deposition rate is rather subtle. Even for a low deposition rate the bedforms will be migrating so that major water movement events may be recorded locally by dint of wiping out a part of the previous record. The same partial loss of pre-existing record will occur if a particularly deep sand wave trough migrates across a sand sheet. The forward migration of a large sand wave that is located near to its outer geographical limit of stability will give a record of events in its growth layers of cross strata, the thickness of layers giving a measure of the relative scale of the events. The synchronous episodes of migration of adjacent sand waves will give repeated records of the same events. Recognition of this repetition in the rocks would enable more effective use to be made of the exposed sections of cross-bedded sands.

The facies model given above differs from those produced by other workers by being tied to the known water movements of the present sea, whereas the others were attempts to interpret marine sedimentary rocks in terms of modern sandy bedform zones but without making enough reference to the water movements in a modern sea. Indeed, the inferred changes in the location of the fossil bedform zone at any place (Anderton, 1976) are more extreme than anything currently known from modern seas. For example, those authors envisaged the entire sand wave zone (which requires many years to be built) being washed away and replaced by the sand ribbon zone at times. In contrast, another interpretation arising from a study of the rocks envisaged individual sand waves as maintaining their identity as they grow in height and wavelength throughout a marine transgression (Nio, 1976). However, the tidal origin of these structures is not established. Maybe they are the product of unidirectional currents such as seen off South Africa (Flemming, 1980) or even formed on a tidal delta.

Preservation of some of the internal structures of existing bedforms is highly likely, just as some destruction of them by the infauna (Section 6.3) is also quite inevitable (Reineck, 1976; Chowdhuri and Reineck, 1978). Where the initial deposition is occurring on an uneven floor (such as drowned channels or glacial relief) the migration of sand waves could lead to tipping of material into local deeps until they are filled in. In such circumstances the accumulation of sand could be relatively rapid, and some burrowing animals would be largely missing, with the result that primary sediment structures might not tend to be destroyed.

In the zone of large sand waves the paucity of burrowing echinoids (Houbolt, 1968, Enclosure 2; and Section 6.11.1) and other bioturbating elements of the infauna show that any cross-bedding stands a good chance of being preserved (Section 6.3.1). The cross-bedding in the zone of small sand waves is somewhat less likely to be preserved because of the

Offshore tidal deposits 113

banks are already so large that their preservation is inevitable. Besides providing data on grain size, composition and internal structure, these modern sand banks provide invaluable indications of the bedforms and local sand transport directions on them (as well as the resulting structures) that are caused by modern water movements (Section 4.7). Comparison of profiles of early and late Holocene sand banks shows the changes that take place in the outline of

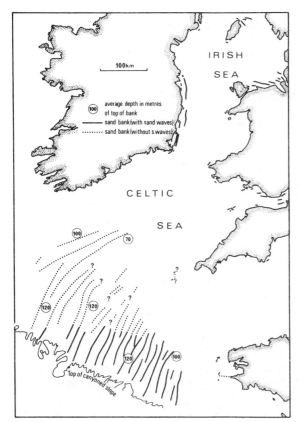

Fig. 5.14 Location of the sand bank facies of the Celtic Sea and neighbouring sea floors.

larger populations of burrowing echinoids and other infauna (Section 6.11.2).

5.3 Sand bank facies

The second of the two most important offshore sand facies of seas dominated by tidal currents consists of the large, discrete banks of sand generally occurring parallel to one another in groups (Figs 5.14 and 5.15). The numerous examples of early Holocene age serve to show their profiles and their range of sizes, but unfortunately few adequate samples of these moribund older sand banks have been taken as yet. Much of the remainder of the information needed about sand banks, in general, has been obtained from the many examples of modern sand banks that are being actively built or maintained by present tidal currents. Some of these younger sand

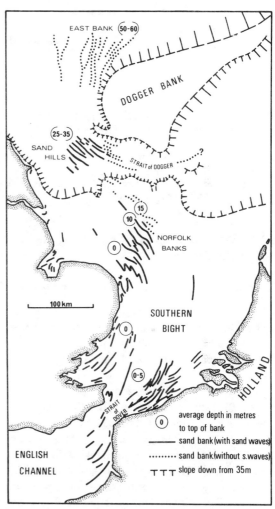

Fig. 5.15 Location of the sand bank facies in the southern part of the North Sea and the eastern end of the English Channel. Those banks with sand waves on them show the mode of active construction by modern tidal currents, whereas the older sand banks (dotted) show various stages of subdued relief indicative of progressive evolution as a deposit.

114   Offshore Tidal Sands

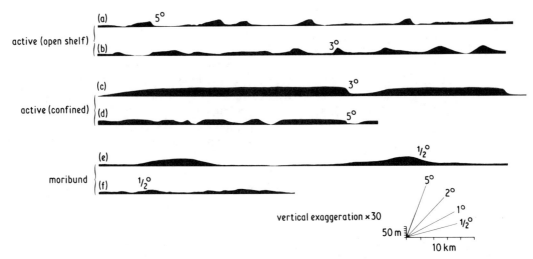

Fig. 5.16 Cross-sections of modern sand banks in the open sea and in estuaries as compared with moribund ones of the early Holocene, (a) Norfolk; (b) Korea Bay; (c) Approaches to Seoul; (d) Thames Estuary; (e) Outer Celtic Sea; (f) Sand Hills, North Sea. When the sand banks in estuaries grow up to sea level their further growth can only be accomplished by spreading laterally. The moribund sand banks have lower slope angles than the active sand banks.

the bank as the growth phase gives way to a potential preservation phase (Fig. 5.16).

5.3.1 EARLY HOLOCENE LOW SEA LEVEL SAND BANK FACIES

(a) *Relative ages of the early Holocene moribund sand banks*

The early Holocene age of some of the sand banks located near to the British Isles is based on several lines of evidence. For example, the Celtic Sea sand banks (Stride, 1963b) are oblique to channels in the underlying floor (Bouysse, Horn, Lapierre and Le Lann, 1976) and so are younger than the Pleistocene low sea levels during which the channels are thought to have been cut. The ostracod fauna found within the top few inches of Labadie Bank (with its least water depth of 70 m on Fig. 5.14) grew in a sea which had about the same seasonal temperatures as are found in the neighbouring part of the Celtic Sea today (Neale, 1974), so a Holocene age for the fauna and sands is probably acceptable. Many of these Celtic Sea sand banks are classed as partly or wholly of early Holocene age as they occur in regions where the present day tidal currents are too weak to build or maintain them. This is confirmed by the absence of sand waves, that are so typical a bedform on the modern sand banks (Section 5.3.2), and by the subdued outline of the banks, as discussed below. Therefore, they must have been formed at lower sea levels than now, when the tidal flow would have been substantially stronger. These sand banks are now mostly in a moribund state prior to burial.

The Celtic Sea sand banks which have a least depth of 120 m should be somewhat older than the ones with a least depth of only 70 m (dotted lines Fig. 5.14). These latter should, in turn, be older than those in the North Sea, whose formation cannot have started until a later stage in the Holocene transgression. The oldest known of these are in the East Bank area, with a least depth of 60–50 m (dotted lines in Fig. 5.15), are believed to have been formed between 12 000 and 9000 years BP (Jansen, 1976). The ones in the 'Straits of Dogger' (35–25 m least depth) can only have commenced forming later on; whereas the outermost of the Norfolk Banks (15 m) can, at most, be only a little older than present day marine conditions. Each group may indicate a period of stable sea level.

(b) *Size and shape of sand banks*

The largest of the moribund sand banks around the British Isles occur in the Celtic Sea within an area of about 20 000 km². Here the longest reaches 110 km, the widest is 15 km and the highest is 55 m. In general, these sand banks broaden towards the edge of the continental shelf. The younger early Holocene sand banks in the North Sea are smaller. The crests of the banks within each series are rather irregular but roughly parallel, and the banks interfinger with one another. Their slopes are everywhere less than one degree. An asymmetrical cross-section is only just perceptible in some of these Celtic Sea sand banks and there are also some signs of very low angle stratification within them.

(c) *Separation of sand banks*

It has been argued that the distance between modern sand banks is roughly proportional to water depth (Off, 1963; Allen, 1968). Early Holocene sand banks occurring in relatively deep water could, therefore, be more widely separated than sand banks occurring in shallow water. In practice, although the Celtic Sea sand banks are, indeed, the most widely separated (up to about 10 km) there is no regular decrease in separation from these to those of progressively shallower parts of the North Sea. It is concluded that, even if there could be a correlation between water depth and bank spacing for some modern sand banks, it does not necessarily apply to sand banks formed during a rapid rise of sea level. This is presumably because the sand banks are so massive that their spacing could not increase fast enough to keep up with increasing water depth, before the tidal currents weakened in strength so much that they were no longer strong enough to move the sand. No such correlation of depth and spacing would be expected for estuarine sand banks because estuaries tend to fill up, so causing a progressive decrease in the width of the channels between them with time.

(d) *Composition of sand banks*

Admiralty Surveys show the composition of many small (lead-line) samples from the Celtic Sea. They reveal that the moribund sand banks are made predominantly of sand, but that some shelly material is also present. In contrast, gravel occurs between the sand banks, while mud is accumulating in places on top of the gravel. There is also mud on the lower flanks of some of these banks.

(e) *Preservation of sand banks*

The four groups of early Holocene sand banks of the Celtic Sea and the North Sea referred to above, show almost conclusively that during a marine transgression this large bedform of sand can be preserved as a discrete feature, isolated from other members of the same group. It is highly likely that offshore sand banks can also be preserved without a rise of sea level. In the Southern North Sea, for instance, tide gauge records show that part of that floor is sinking at up to about 3 mm/year (Jelgersma, 1979), so that preservation of the sand banks is likely in any case, because the tidal currents will gradually decrease in strength.

The preservation process entails not just a decrease in current strength, and a consequent disappearance of sand waves, but is also associated with a decrease in the steepness of the slopes of the sand bank, compared to modern ones (Section 5.3.2). This degradation must be accomplished by removal of material from the upper part of a sand bank and its deposition on the flanks, by storm waves in combination with tidal and non-tidal currents.

5.3.2 LATE HOLOCENE SAND BANK FACIES

(a) *Age of sand banks*

Sand banks with sand waves on them are probably being actively maintained by present day tidal currents. Such sand banks located near to the British Isles are shown in Figs 5.14 and 5.15 as solid lines. The evolution of a few of those in the Southern Bight of the North Sea, for example, may have begun as early as about 8000 years BP (Jelgersma, 1979), when the sea began to spread into that region. However, tidal conditions similar to those of the present day were probably not established there until about 5000 years BP.

(b) *Location and size of sand banks*

The location of modern sand banks depends on their intimate relationship to the existing tidal current regime (Sections 3.4.5 and 4.4.2b) and with obstructions to that flow, in the case of the so-called banner banks. Only the banner banks are solitary. Around the British Isles, at the present time, the offshore sand banks are larger than the ones in estuaries, while the banner banks are the smallest (Table 3.2, Section 3.4.5). None of these are yet as large as the early Holocene ones of the Celtic Sea (Section 5.3.1).

(c) *Shape of sand banks*

Many of the sand banks of European tidal seas and those occurring on Georges Bank, USA (Jordan, 1962), in the Gulf of California (Meckel, 1975) and in the Southern Barents Sea, USSR, are asymmetrical in cross-section, with one side sloping as steeply as 6°. Off (1963) drew attention to the worldwide occurrence of such sand banks on navigational charts.

One group of ten offshore sand banks (the Norfolk Banks) follows the line of the English coast northeast of East Anglia (Fig. 5.15). Three of them (Leman, Well and Swarte) have a relatively wide, rounded southern end while their northern ends tend to be narrow and rather pointed (V. N. D. Caston, 1972). Banks of this series that are located successively further seawards tend to become shorter and lower (Houbolt, 1968). In the same seawards direction the maximum angle of their relatively steep north-eastern slopes tends to decrease from between 5° or 6° in the inshore banks to about 2° for the more offshore Broken and Swarte Banks. This change probably results from the known outward weakening of the tidal currents, rather than indicating increased importance of suspension transport further seaward. Although the median diameter of the surface sand decreases from medium sand with <20% of fine sand on the inshore banks to fine sand on the more offshore banks (Jarke, 1956), this is probably offset as regards suspension transport by the weakening of the tidal current. These banks, are, thus, useful indicators of the early stages of degradation that ultimately gives way to the gentle slopes of the moribund sand banks, discussed above (Section 5.3.1).

A significant difference between the offshore and estuarine sand banks arises when sand continues to be supplied to an estuary so that the congestion increases. In such a case the sand banks become flat topped by growing increasingly in width, as in the eastern Yellow Sea (Fig. 5.16), until they eventually merge into a sheet-like form. Some of the banner banks are more nearly pear shaped rather than elongate (Section 3.4.5(e)).

(d) *Texture and composition*

The known median grain sizes of sand in the late Holocene offshore sand banks range from 0.2 mm in the Southern Bight of the North Sea (Houbolt, 1968) to 1.19 mm on Georges Bank, USA (Stewart and Jordan, 1964). The estuarine sand banks are finer grained, lying between about 0.1 and 0.2 mm in the German Bight (Reineck, 1963), Thames Estuary (Prentice *et al.*, 1968) and Solway Firth (Wilson, 1965), for example. The difference in grain size of the sands of different regions must reflect what has been available, as well as the strength of the water movements that were responsible for transportation. The grain size of the sands can be less than would be expected in terms of current strength, especially in natural sediment traps such as estuaries. For example, in the upper part of the Severn Estuary the sand has a median diameter of 0.15 mm yet the tidal current is strong enough to move particles of 12 mm as bed-load (Hamilton, Somerville and Stanford, 1980). The sorting of such sands is good but has not been shown to reach the same high degree of sorting as in the case of sand sheets of the same grade (cf. Houbolt, 1968, Figs 13 and 34 with Völpel, 1959, Fig. 10) around the British Isles. The sand of the banks on Georges Bank is said to be as well sorted as on neighbouring beaches (Schlee, 1973). In situations where sand has been transported up a longitudinal velocity gradient the sand could be only moderately well sorted. Mud layers are present in the sand banks of estuaries and in some of those of the open sea (Section 5.3.3). Mud clasts, due to the partial break up of the mud layers, are also present (e.g. cores MBO 55 and 56 of Fig. 13 in Houbolt, 1968).

Few samples have been taken from banner banks. They show that there is a wide range in sand size and in shell content (Hails, 1975; Jones *et al.*, 1965).

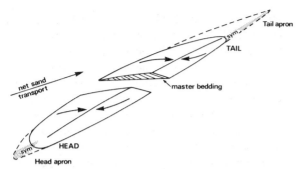

Fig. 5.17 Block diagram of a modern sand bank. The arrows show the regional and local net sand transport directions as shown by sand waves. The rounded head and pointed tail and associated aprons of sand waves (including symmetrical ones) on the outlying flat floor are after Caston, G. F. (1981). Master-bedding lies parallel to the steeper slope of the bank (after Houbolt, 1968) and has a gradient of up to 6°. High angle cross-bedding due to sand waves is also present but is not shown (but see Fig. 5.19). Sand can be added along the whole of one side of the sand bank, from head to tail, but may be lost only from the tail apron (except possibly for some sand lost in suspension from the crest during peak water movements).

5.3.3 INTERNAL STRUCTURE OF OFFSHORE AND ESTUARINE SAND BANKS

Scattered sub-bottom profiles and cores provide valuable indications of the internal structure of some modern offshore and estuarine sand banks. These data can be amplified by inference, both from the large and small-scale surface relief of these sand banks and what is known of the water movements affecting them. Most of the available data refer to sand banks of the North Sea but should have much wider relevance both to other regions and to ancient examples.

The reflection profiles provide some useful indications of the main interfaces within Well Bank which dip at 6° (Fig. 5.17), in Smiths Knoll, located in the North Sea to the north-east of East Anglia (Houbolt, 1968), in West Barrow bank in the Thames Estuary (Prentice *et al.*, 1968) and in Bassure de Baas and Battur Bank at the eastern end of the English Channel (Lapierre, 1975). In each case the available data show that the apparent dip of the interfaces lies approximately parallel with the steeper side of the sand bank (although there are not enough profiles to show the exact orientation of the interfaces). There

Offshore tidal deposits 117

are also hints of such interfaces in an Australian sand bank (Cook and Mayo, 1977).

The cross-stratification that shows up in such profiles is due to the presence of adjacent sediment layers with contrasting acoustic impedance. Such a contrast would be in keeping with the presence in the sand of soft mud layers. Such layers are up to about 1 cm thick and are found in cores (up to 70 cm long) from the southern half of Well Bank (Houbolt, 1968, Enclosure 2). The failure to detect interfaces in reflection profiles of other sand banks with asymmetrical profile is probably because they do not have mud layers. The layers of mud in a sand bank are presumed to originate during occasions of suspension transport with high mud concentration and thus are geographically rather localized. It is notable that for the East Anglian sand banks, including Well Bank and Smiths Knoll, a narrow band of particularly muddy water passes over them (Fig. 5.18) in winter (Joseph, 1955; Lee and Folkard, 1969).

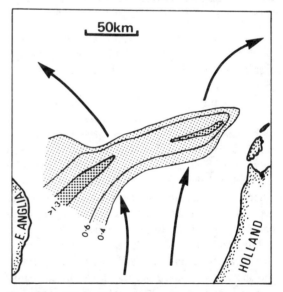

Fig. 5.18 The band of muddy water that, at times, lies over the Norfolk sand banks in the Southern North Sea. The contours equivalent to mud concentration were derived from extinction coefficient measurements made during a February and March (Joseph, 1955). The mud is largely derived from destruction of Pleistocene deposits of the East Anglian coast. It is transported by the slow net drift of water to the east–north–east. Note that this path is approximately at right angles to the peak tidal flow that transports the sand fraction (arrows).

118   Offshore Tidal Sands

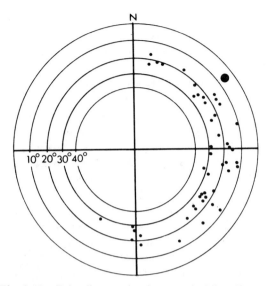

Fig. 5.19 Polar diagram showing cross-bedding dips and dip directions derived from six cores taken on the relatively steeper north-east facing flank of the modern Well Bank, off East Anglia (after Houbolt, 1968). The large dot refers to the steep slope of the sand bank.

The mud layers may result from entrapment of suspended mud by the bottom boundary layer, as suggested in general terms for the continental shelf (Section 5.1(c)).

Those mud layers that become preserved on the steeper slope of an asymmetrical sand bank are likely to be patchy where they have been laid on sand waves. This is because only the mud laid on the steep lee slope of each sand wave is likely to be buried by sand when the sand wave again moves forward, whereas any mud laid on the gentle slope of the sand wave will be eroded, perhaps ending up as mud clasts incorporated into the cross-bedding.

The gentle dips of the large-scale cross-stratification (master-bedding) would hardly be detectable in the short cores taken from the modern sand banks sampled by Houbolt (1968). However, clearly defined cross-bedding with dips of 10–30° was present in the cores and showed a wide scatter in directions (Fig. 5.19), implying the presence of herringbone cross-stratification at times. The favoured explanation is that these dips indicate cross-bedding due to the sand waves present on the steeper slope of the bank. Indeed, many of the dips in the cores from the steeper slope of Well Bank point towards the southern end of the bank, consistent with the known presence of small sand waves moving in that direction along the base of the slope and of others moving up more nearly towards the crest of the bank. Such sand wave crest orientations are shown on sonographs of the nearby Leman and Ower Banks (Caston and Stride, 1970; Caston, 1972). However, the cross-bedding that dips approximately towards the north-east, in the regional net sand transport direction probably represents occasions when the local southerly-going sand waves on the lee slope of the bank have been temporarily reversed. These occasions may well be short-lived as sand waves facing that way have not yet been detected on echo-sounding surveys. They are likely to be associated with enhanced removal of sand from the crest of the bank aided by storm waves at a time when the tidal current was flowing strongly north-east.

The dip directions shown by short cores in six sand banks at the south end of the Southern Bight of the North Sea (Houbolt, 1968, Enclosure 2) show a similar pattern to that found in the sand banks off Norfolk. Of particular interest, in this more southerly region, are the predominant easterly dips shown in the cores from the Outer Ruytingen Bank. These are in keeping with the unusual anticlockwise veering of sand waves on that bank (Figs 4.8 and 4.25). The preservation of this pattern implies the lateral migration of this bank in the direction faced by its steeper south-easterly facing slope, as would be expected.

In summary, cores from the sand banks of the Southern Bight of the North Sea (including South Falls, Sandettie, Outer Ruytingen, West Dyck and the Norfolk Banks) show that the prevalent dip directions for steeply dipping cross-bedding are opposite to the directions of the regionally stronger tidal currents. Thus, because they are formed by the migration of sand waves on the lee side of the bank the dips of those cross-strata are opposite to the dips resulting from regional net sand transport within a sand sheet associated with the same transport path (Section 5.2.4). However, the morphology of sand waves on the heads of some sand banks (Caston, 1981) implies that there should be some preservation thereabouts of the high-angle cross-bedding that dips in the direction of the regional net sand transport.

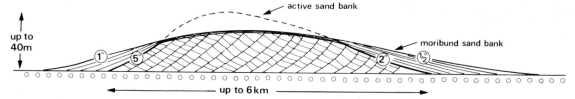

Fig. 5.20 Idealized hypothetical cross section showing changes in the profile and structure of an offshore sand bank that are expected to occur during the transition from active to moribund state. Note that the slopes are exaggerated. The master-bedding within the bank has a true dip of only a few degrees.

Deep channels that cut obliquely through the whole thickness of estuarine sand banks (e.g. Robinson, 1963) or the shallow channels (swatchways) that cut only the tops of some of them, are prone to migrate and leave a record of cut and fill (e.g. Nio, van den Berg and Siegenthaler, 1979). In some deposits the cross-bedding of the associated sand waves dips in the direction of either the ebb or the flood tidal current, whilst elsewhere both directions are represented at times (De Raaf and Boersma, 1971; Terwindt, 1971a). A notable feature of the cross-bedding due to sand waves of the inshore channels of Holland (Visser, 1980), Germany (Wunderlich, 1978, Plates 2 and 3) and off East Anglia (in Corton Road, Lowestoft), is the presence of mud layers amongst the sand layers.

The internal structure of banner banks has not yet been determined. It is expected that they will also be cross-bedded but probably in a more complex manner. This is because the banks may at times 'flutter' from side to side while being held to their point of 'attachment'.

5.3.4 FACIES MODELS OF OFFSHORE AND ESTUARINE TIDAL SAND BANKS

Some tidal sand banks have been shown to grow by deposition of low angle cross-strata parallel to their steeper side. This process is also presumed to occur in many other sand banks within which no relatively strong reflectors are visible on sub-bottom profiles. The bedding represented on sub-bottom profiles is usually shown in publications with a highly exaggerated vertical scale. In reality the true dips are only a few degrees, so that the bedding is unlikely to be recognized as cross-stratification in limited rock exposures. However, what will mainly be seen in outcrops of the sand bank facies are high angle cross-strata that originate by the migration of sand waves present on the steeper side of the bank. Klein (1977b, Fig. 80) produced a hypothetical model of the expected vertical sequence in preserved coalescing, subtidal, tide-dominated sand banks in which these large scale and smaller scale cross-strata are confused. The confusion arises because (a) the large-scale cross stratification is represented on his figure at an angle of 49° rather than its true angle of about 5° (after allowing for vertical exaggeration), and (b) because the sand waves that had been present on the sand bank were thought to represent a late-stage reworking episode only. However, they are known to be present during the whole period of bank construction and should be responsible for the presence of smaller-scale, high angle cross-bedding throughout the whole body of an active sand bank.

(a) *The offshore sand bank facies model*

For an offshore tidal sand bank to be preserved it should progress from an active state to a moribund state and then finally to a buried state. This will occur with progressively decreasing current strength, which generally implies increasing depth of water, but could also be caused by departure from resonance (Section 2.3.2). Fig. 5.20 indicates the likely changes in shape and internal structure. This change will be associated with the disappearance of sand waves. The crest of the bank will be somewhat lowered and rounded, and a veneer of winnowed, ripple-bedded sands will be formed. The material removed from the crest will be deposited as ripple or flat-bedded fine sands on the lower slopes and beyond the base of the original sand bank. The gradient of the steeper slope of the bank will progressively degenerate from about

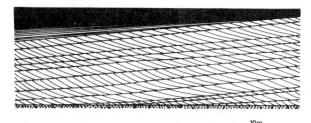

Fig. 5.21 Idealized hypothetical cross-section through the steeper side of an offshore sand bank to give an indication of the nature of the internal structure. The original slope is overlain with flat or ripple-bedded sand and then by mud (black). The slopes are drawn to about true scale.

5° to perhaps 1°, and the more gentle slope will change from about 2° to perhaps 0.5°. In some cases the asymmetry of the original sand bank in cross section may disappear altogether when buried by ripple-bedded sands. Mud clasts representing broken up mud layers may be more or less abundant, depending on how rapid is the onset of a permanent mud blanket.

Fig. 5.21 represents an idealized hypothetical transverse section across a part of the steeper, prograding flank of a preserved tidal sand bank. The dips on both the large-scale cross-strata (master-bedding) resulting from bank growth or migration (maximum of about 5°) and the smaller-scale cross-stratification related to sand wave migration (a maximum of about 30°) are presented at about their true angles.

A number of points should be noted:

(1) The dip of the master bedding will decrease from a maximum of about 5° towards the basal gravel layer, following the observed concave profile of the sand bank in cross-section.

(2) On a cross-section normal to the long axis of the bank the apparent dip of the cross-stratification of sand waves should increase towards the crest of the bank as the sand wave crests become progressively more parallel to the bank crest.

(3) The cross-stratification due to sand waves may include reactivation surfaces and occasional reversals in dip direction.

(4) Mud-rich layers (composed of laminae or separate mud clasts) will vary in their abundance from absent to common.

(5) Episodes of mud deposition may be preserved as mud-rich laminae within foreset-bedding of any sand waves present at the time of mud deposition. Thus the mud laminae, while being confined to a coset, may be separated by a variable distance from those on contemporaneously adjacent sand waves.

(6) Bioturbation should become more prominent in the overlying rippled-sand stage and especially in the mud-blanket state of preservation.

(7) The dip of beds within the rippled or flat-bedded sand zone should be less than the dip of the master-bedding within the cross-stratified zone.

(8) A transverse section across the other (originally more gentle) flank of the bank should indicate master-bedding dipping downwards from that flank towards the bank axis.

The distinction between the sand bank facies and the sand sheet facies (Section 5.2.10) may prove difficult to make in sedimentary rocks lacking adequate exposures. For instance, in a rock exposure equivalent to only a small area of Fig. 5.22 the master-bedding may not be recognized as large-scale stratification because of its gentle dip (on a section parallel to the bank axis its apparent dip will approach zero). At exposures near the base of the sand bank this dip may also be so very gentle (due to the concave slope towards the contact with the underlying basal gravel) as to pass unnoticed. Likewise, the vertical sequence from cross-bedded sands, to ripple or flat-bedded sands and to muds in time, may also apply to the sand sheet facies. Thus, for a correct interpretation of the main sand facies type the emphasis must often be on the overall relationship of the sand bank with its surroundings. It is expected that the dips of the cross-strata due to sand waves will generally be higher in the sand bank facies than in the case of the sand sheet facies, because of the potentially higher sand transport rates on sand banks.

It is not known whether these offshore sand banks can coalesce when sand is super-abundant, as occurs in the case of the estuarine sand bank facies discussed below.

(b) *The estuarine or embayment sand bank facies model*

With a plentiful supply of sand, particularly where it becomes trapped in an estuary or larger embayment,

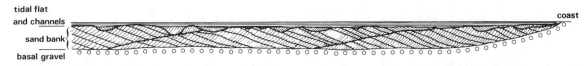

Fig. 5.22 Idealized hypothetical cross-section showing three coalescing sand banks in an estuary that is being infilled with sand at constant sea level. The top surface of these has been cut by tidal flat channels and overlain by tidal flat deposition. Note that the slopes are exaggerated. The master-bedding within the sand banks has a true dip of only a few degrees.

the sand banks build up to the water surface and may then grow laterally until they coalesce. Fig. 5.22 represents an idealized hypothetical transverse section through such a deposit. This is, of course, a simple version of what, in nature, will generally be much more complex. It is assumed that the sand banks build or migrate uniformly in the same direction, and that they are approximately straight-crested. With more complicated 'S' or 'V' shaped sand banks that are related to ebb–flood channel systems the resulting internal structure of both the low and high angle cross-bedded sands will be correspondingly more complex.

A number of points should be noted:

(1) In Fig. 5.22 the dips of the large-scale cross-strata (parallel to the steeper side of each sand bank) are exaggerated from the true angle of not more than about 5°, and the gentle slope of each adjoining bank over which they are accreting are exaggerated from a true angle of about 2°.

(2) The initial spacing of the sand banks may depend on the initial water depth (for a stable sea level).

(3) There will be an upwards transition from inter-bank channels, dominated by lateral migration of the sand banks, to tidal flat channels which may migrate and meander either way towards the top of the sequence. Trough cross-bedding related to these tidal flat channels will be found to a greater or lesser depth into the section depending on both the maximum depth of the channels and the thickness of the sank bank.

(4) With a constant sea level, the thickness of sand beneath the cover of tidal flat deposits will be about equal to the original low tide water depth. With a rising sea level, the sand banks should continue to stack one over the other until a stage is reached where the tidal flat sedimentation becomes predominant over sand bank growth and migration.

(5) The upward transition in sequence from sand bank facies to intertidal facies will tend to take place first at the head and sides of the estuary and then spread seawards until the estuary is infilled.

(6) The distinction between the estuarine sand bank facies overlain by tidal flat facies and the offshore sand sheet facies overlain by tidal flat facies may often not be clear from limited rock exposures.

5.4 Sediment and faunal indicators of shape, depth and exposure of continental shelves

Modern tidal current deposits provide criteria which give some hope of recognizing inner and outer limits of ancient continental shelves, as well as their relative degrees of exposure to the open ocean. The first criterion from modern continental shelf seas is their preferential association with glauconite formation (Odin, 1973, Fig. 1). Glauconite is found, for example, around south-west England (George and Murray, 1977) as well as on many other continental shelves. Secondly, tidal currents strong enough to move sand are largely confined to continental shelves and estuaries. Thirdly, the inner margin of tidal seas can be indicated by a variety of deposits, including those of tidal flats and sand banks, the relative development of which depend on tidal range, amongst other factors (Hayes, 1976). The water depth in a modern estuary can be matched by the height of associated sand banks, as those that reach up to about mean tide level can then only grow out laterally with a flattish top surface (Fig. 5.16(c) and (d)). They can show both longitudinal and transverse bedforms and are subject to channelling. Some almost flat topped, but discrete sand banks are also seen further offshore for the same reason of limited water depth, such as Buiten Ratel near Belgium (Section 15 in Enclosure 2 of Houbolt, 1968). Fourthly, the minimum water depth (as well as the probable proximity of a coast) is also shown by the height of the geographically isolated 'banner' banks (Figs 5.14 and 5.15). These are particularly useful as they can occur on gravel floor swept by strong tidal currents lying between the coast and the sand sheet further offshore. Fifthly, the broad band of equilibrium grade sands around the western and southern edge of the North Sea (Fig. 5.3), for example, gives an overall indication of the whereabouts of a coast. In this case, the sheet deposits lie seaward of all the sand banks. In contrast, for the Celtic Sea, south of Ireland, there would be no such indication however. Sixthly, in a sea dominated by tidal currents, a regional grain size gradient in facies of the same age may (but not always) indicate geographical differences in water depth. However, this is of limited value as can be seen by comparing the distribution of sand and gravel (Fig. 5.3) with the bathymetry (Fig. 1.2). Basically, such gradients show the geographical distribution of average near-bed peak tidal current speed and this is affected by the tidal regime in the whole sea (Chapter 2). An obvious exception is that sand trapped in estuaries and in bed-load convergences may well be much finer than would be expected in terms of current strength alone.

The outer edge of a continental shelf can be indicated by a markedly different fauna, as compared with that of the adjacent continental slope deposits and that of the inner part of the continental shelf (Section 6.15). Proximity to the shelf edge can also be shown by the presence of the very large symmetrical sand waves with rounded crests formed by tidal lee waves, although these require special conditions (Section 3.3.3(i)). The outer edge of the continental shelf in high latitudes can have been especially disturbed by icebergs during glacial times (Plate 1.2), whereas a shallower inner part of the shelf can preferentially show drowned and in-filled valleys (Fig. 1.3).

The offshore tidal current deposits of modern seas have not yet provided criteria for recognizing mid-continental shelf water depths in rocks of past tidal seas. Overall, the most positive indicators of actual water depth may well be the longitudinal sand banks which can grow up to sea levels, although they will lose some height before their preservation.

It has also been argued that the spacing and heights of sand waves indicates the depth of the sea, but it is now more generally acknowledged that this cannot be demonstrated on available data (Section 3.3.3(c)).

The marked differences in the shelly faunas of the Bristol Channel and the Southern Bight of the North Sea, for analogous zones defined by current strength and bedforms, provide a means of recognizing a sea that is open to the west as against one open to northern waters (Section 6.15).

5.5 Longer term evolution of the deposits

The 10 000 year time span of the Holocene is long enough for appreciable thicknesses of sediments to accumulate in some regions, but is a short period in geological terms. For this reason it is worth attempting to extend the existing evolutionary trends of the modern deposits and to make some speculative comments about their slightly longer term evolution. This should lessen the difficulty of comparing past and present offshore tidal current sedimentation. Sea level is taken as constant for the period considered.

Present indications are that the existing estuaries and the inner parts of wider embayments will continue to be filled in by material that has been eroded from coasts or from outlying portions of sea floor and moved along present day transport paths or is arriving by littoral drift. It has already been pointed out that the linear sand banks of the Thames Approaches have grown up to just above low tide level and are now growing laterally and developing flat tops. Their broadening will continue (with muddy deposits accreting in the inner reaches of old estuaries) to the extent already found around the inner part of the German Bight, or in a more extreme form in the approaches to Seoul where they seem to be as much as 30 km wide. As deposition continues such sand banks may become so large that they converge to form a sheet with tidal flats on top. In contrast, the narrow sand banks of Korea Bay in the Yellow Sea are inferred to be extending seaward into deeper water and may not widen much in time.

In the more open parts of seas swept by strong tidal currents the bulk of the two main sand facies will continue to increase, so long as sediment is available and the tidal currents remain sufficiently competent. Erosion in such potential source areas for sand as bed-load partings will tend to deepen and roughen the floor substantially thereabouts, while deposition elsewhere will lessen the water depth and may change the sea floor roughness. In time these changes will affect the tidal current regime so that the sedimentation pattern itself will change (Johnson and Belderson, 1969). Such changes will be particularly marked for localities where the tidal currents are close to resonance at some stage.

Sand will continue to be lost from the land (Section 4.1.1) but may not reach large areas of sea floor that are more distant from it. For these latter areas shelly debris could gradually become dominant and because of this change of substrate the fauna itself could change (Section 6.20). The shelly material is already abundant on parts of the ocean facing edge of the European continental shelf (Fig. 6.2). The depositional model emphasized by some workers of a near-shore sand prism building out into the sea is not seen as a model for general application to offshore parts of seas swept by strong tidal currents, although a band of sand, say, can be laid down parallel with a coast in some parts of a sea (Fig. 5.3).

The continued downwarping of the western edge of Europe will provide additional space for the accumulation of offshore tidal current deposits. The same will apply in the North Sea basin because of its continued sinking (Jelgersma, 1979). Preferential deposition will take place in such sites as the southern edge of the mid-North Sea graben, if it continues to sink. The grain size of the associated sands, and their bedform zones of deposition will, of course, depend on the sand transport paths and the tidal current strengths. These tidal currents, in turn, will be affected by the changes in basin shape and dimensions. The sinking of the Southern Bight of the North Sea is likely to lead to preservation of more of the cross-stratified deposits due to sand waves than would be the case if much of the sand had time to be moved out to the ultimate position of grain movement under present day conditions.

5.6 Sand and gravel deposits of non-tidal marine currents

A few examples of the sand and gravel deposits of non-tidal marine currents must be mentioned, as there is scope for confusing them with the offshore tidal current sand and gravel deposits. The confusion can arise because many of the bedforms of tidal seas mentioned above are a function of grain size and current strength (Sections 3.2.1 and 3.3.3) rather than being an indicator of whether the current that was responsible for them was unidirectional or tidally reversing. For example, the outflowing Mediterranean Undercurrent, when passing through Gibraltar Strait, shows the same sequence of bedforms as would be found for tidal currents: i.e. from bare rock, to sand ribbons and to sand waves with decreasing current strength, together with longitudinal sand bank-like features where there are marked lateral velocity gradients in the current (Kenyon and Belderson, 1973). Similarly, the occasionally strong near-bed currents that flow into the Baltic Sea have also left their mark as an analogous suite of sand bedforms (Werner and Newton, 1975). Sand waves formed by the occasional strong wind-induced currents have also been recognized in the North Sea off the Danish northwest coast (Stride and Chesterman, 1973), while numerical modelling of these currents (Davies, 1976) shows that the sand waves could occur over a larger area. Where strong enough the main ocean currents are also responsible for forming sand ribbons and sand waves along broad stretches of continental shelf, as for example beneath the Agulhas Current, off south-east Africa (Flemming, 1980).

The recognition of deposition by unidirectional, rather than tidal currents might present difficulties when interpreting sedimentary rocks, but these difficulties could probably be resolved by the environmental context, as emphasized in Chapter 7. For example, the deposits of unidirectional currents would not normally be associated with bed-load partings and convergences. However, the potential value of partings is lessened because ocean currents, for instance, can provide somewhat similar deposition patterns. Thus, where an ocean current impinges on a continental shelf and coast at a high angle, the current may split into two diverging parts, one flowing to the left and the other to the right past the adjacent land (as the South Equatorial Current does against South America). The deposits of such a situation could show cross-stratification dipping away to left and right from the region of impingement

Table 5.2 Criteria for discriminating between the modern sand sheet and the sand bank facies, particularly as developed around the British Isles.

	Sand sheet facies	Sand bank facies
Area	Extensive sheets of up to about 400 km × 50 km.	Isolated banks or groups of banks with sizes up to 50 km × 3 km, say, but estuarine banks may merge into a sheet.
Thickness	Up to 10 m, but more when infilling local deeps.	Up to 40 m.
Location	Generally in the open sea.	Frequently (but not always) in large embayments, estuaries, bays and near to coasts and islands.
Grain size	Longitudinal and transverse grain size gradients can range from gravel to silt. Muddy layers can be present in the cross-bedded sand. Grain size of sand and gravel in equilibrium with current speed.	Approximately constant sand grade for each sand bank. Mud layers can be present. Grain size can be finer than current speed would lead one to expect (cf. the case of sand sheets).
Sorting	Very good, especially for fine sand.	Good, but not so perfect as in a sand sheet when considered in bulk terms.
Structures	Geographical sequence (generally longitudinal) in keeping with bedform zones present. Cross-bedding common, largely high angle unipolar and bipolar. Master-bedding dips can decrease from about 15° to about 3° in net sand transport direction. No channelling expected.	Cross-bedding is common throughout the sand bank (except for the overlay of bank degradation deposits). Master-bedding lies parallel to the steeper side of the sand bank with a dip of 6° or less. High angle cross-strata is common, with dip directions spread within a 180° arc. Set thickness can be up to 5 m. Filled channels are common in estuarine sand banks.
Vertical sequence for fixed sea level	Some interfingering of structures of adjacent bedform zones.	More or less uniform, except for mud layers and degradation products.

of the current (Fig. 4.17). A similar effect could be due to an ocean current flowing broadly parallel to a coast but over some stretches being separated from it by a counter current (Fig. 4.18), as occurs off Durban, South Africa (Flemming, 1980), or elsewhere by large oppositely directed eddies on one or both sides of a headland. However, other lines of evidence should enable such situations to be detected.

It is predicted that (for transport paths of equal length, with sediment of the same initial grain size and currents of the same strength) a unidirectional current could produce a more poorly sorted deposit than that of a tidal current. For the latter the periodic backwards and net forwards transport of sand would usually be equivalent to a far longer unidirectional path and should surely give rise to a higher degree of sorting and rounding for the tidal sand deposits, as happens for many beach sands which have experienced periodic transport due to waves. This is presumably the reason for the excellent sorting of the sand in some of the known offshore tidal current sand sheets and for claims that the grains are well rounded.

5.7 Main conclusions

1. The scale of activity of strong currents as well as their longitudinal and transverse velocity gradients are shown by the grain size, sedimentary structures, as well as the numbers and type of fauna (Chapter 6) of the deposits.

2. The dominance of tidal currents over other water movements is shown by (a) the presence of the

two main offshore sand facies of contrasting types, the sand sheet facies and the sand bank facies; (b) unipolar to bipolar cross-stratification with a tendency for predominant directions in the two facies to be opposite even when the two facies are adjacent to one another; (c) with ideal conditions sand bundle thickness in cross-strata may show monthly spring–neap–spring tidal cycles; (d) the upper part of sand bundles of cross-strata due to the stronger tidal flow are partly eroded during the weaker reverse tidal flow (reactivation surfaces); (e) sand grains are particularly well sorted and rounded; (f) the presence of tidal flat deposits nearby.

3. The criteria for distinguishing between the two main offshore tidal current facies are set out below in Table 5.2.

4. The structures of the sand facies are poorly known and should be sampled as a matter of urgency.

Chapter 6

Shelly faunas associated with temperate offshore tidal deposits

6.1 Introduction

Shelly fossils and trace fossils occurring in supposed tidal current deposits are found in marine sedimentary rocks of all ages from the Lower Cambrian onwards (Chapter 7). Although it has been realized for some time that the most successful palaeoenvironmental reconstructions are made using the combined palaeoecological and lithological approach to the interpretation of sedimentary basins and sedimentary successions (Hecker, 1965), little attempt has so far been made to relate faunas on continental shelves to bed transport paths or to particular bedforms. Closer investigation of shelly faunas and their associated sediments in the geological record and their comparison with modern analogues associated with known bedforms should facilitate a more certain recognition of the environments of formation of tidal current deposits of the geological past.

The purpose of this chapter is to provide a few examples of faunas associated with particular bedforms and zones of different sediment mobility, to indicate some of the biological factors of importance to the geologist in studies of these faunas, and to show how faunal differences can provide much useful information to aid the interpretation of ancient tidal current deposits. The sequence of faunal changes along bed transport paths with longitudinal velocity gradients is also demonstrated. It is important to emphasize, however, that the faunas themselves do not provide any actual proof of tidal current activity as such, only activity related to currents. The examples given are mostly taken from the continental shelf seas around the British Isles and on Rockall Bank and include data taken from the literature as well as from the results of Institute of Oceanographic Sciences (IOS) investigations. The species names used are generally those in current use. Where an earlier name is much more widely known, this is given in brackets immediately after the current name. The taxonomic authority for each name in current use is given in Appendix 6.1.

Studies of shelly faunas in Holocene offshore tidal current deposits demonstrate the characteristic types and species of animal associated with each facies or bedform. They also reveal important regional differences in the faunas to be found in otherwise identical bedforms occurring in markedly different geographical locations on the continental shelf. In north-west European waters, for example, there are differences between the faunas living in sand waves on ocean facing continental shelves such as that west of Scotland or in the Celtic Sea, and those in an adjacent, but more enclosed portion of a shelf sea such as the Southern North Sea.

On a smaller scale, subtle variations in the nature

of the sediment constituting a particular bedform such as changes in water content and to a certain extent, in grain-size, may be detected by observing differences in either the faunal density or the spatial distribution of species which are widespread on, or in, the top few centimetres of the sediment.

The nature and diversity of the fauna also gives an indication of the degree of disturbance of the sediment by tidal and other currents. Bedforms that rarely move, generally support a relatively rich fauna that includes both burrowing and epifaunal species which tend, in sands, to be dominated by suspension feeders. For progressively finer grain sizes, the proportion of deposit feeders increases (Craig and Jones, 1966; Johnson, 1964; Saila, 1976). On the other hand, bedforms that move more frequently support a very sparse fauna both in terms of few species being present and in low population densities. Few burrowing species and no epifaunal species are represented (Sections 6.10.4, 6.11.1 and 6.11.4). Any bivalves present generally have thick, robust shells. Shells and fragments derived from the few thin shelled species living in the mobile sediment are very rarely preserved. Indeed, the virtual absence of fossils from many marine sandstones in the geological record may be an indication, in addition to the structure of the deposit (Sections 5.2.4 and 5.3.3), that they may be deposits formed from once-mobile bedforms such as sand waves (Section 7.5).

Faunal differences between bedforms can also be expressed as differences in the bottom shear stress (Warwick and Uncles, 1980; and Section 6.10.1).

The detailed lists of species recorded from a particular bedform or zone may only be directly applicable to similar bedforms or zones in equivalent locations on other parts of the north-west European continental shelf. The more general conclusions as to the nature of the fauna – the faunal diversity, faunal density, presence or absence of infaunal or epifaunal species, presence or absence of robust or delicate species, etc. – should be applicable to tidal current dominated continental shelves in other parts of the world. Similar faunal communities in different parts of the world generally contain species belonging to the same genera or families (Campbell and Valentine, 1977; Thorson, 1957).

More use of published data would have been preferred. However, difficulties do arise when trying to relate biological and geological observations that have been made separately and independently in a particular area. This problem becomes particularly difficult in regions where there is marked geographical variation in the nature of the sediment and where uncertainties about position fixing make it impossible to locate the biological sample within a particular bedform or facies. Difficulties also arise when comparing results obtained with different sampling equipment (Warwick, George and Davies, 1978; Section 6.11). Accordingly, the results which are of most use to the geologist come from integrated, contemporaneous studies of currents, sediments, bedforms, and living and dead faunas.

In spite of this, it is probable that the literature contains many other useful observations and examples of faunas which are relevant to any study of tidal current sedimentation but which have been omitted from this chapter for one reason or another.

An approach of particular importance to the geologist in the study of the inter-relationships between faunas and their associated sediments is the investigation into the processes of preservation of both body fossils and traces (Section 6.3). Much of our knowledge of this comes from the work of Richter (1928) who introduced the term 'actuopalaeontology' for such investigations. Further work by Häntzschel and others and the book by Schäfer (1962) with its English edition (1972) have added considerably to our knowledge in this field. The concepts and ideas expressed in that volume and the processes described, although largely based on work on tidal flats and inshore waters in the North Sea, are equally applicable to the middle and outer parts of the continental shelf in other parts of the world.

Physical and chemical aspects of the environment may be deduced from the mineralogy and other properties of the sediments (Craig, 1966). Knowledge of the structure of the population in terms of recruitment and growth rate is also essential in any study of population dynamics (Craig and Oertel, 1966).

6.2 Faunal associations

Knowledge of faunal associations and the relationships of marine animals to their associated sediments

Table 6.1 Classifications of faunal associations on the continental shelf and upper continental slope using the most significant sources. Three associations largely restricted to intertidal and shallow sub-tidal conditions are also included.

Jones (1950) and Holme (1966)	Petersen (1918) and Spärk (1935)	Thorson (1957)
Boreal Shallow Sand Association	*Tellina tenuis* Community (Spärk, 1935)	*Tellina* Communities
Boreal Shallow Mud Association	*Macoma* Community	*Macoma* Communities
Boreal Shallow Rock Association	*Mytilus* Epifauna *Mytilus edulis* Community (Spärk, 1935)	—
Boreal Offshore Sand Association	*Venus* Community *Tellina* Sub-community (Warwick and Davies, 1977, after Thorson, 1957) *Spisula* Sub-community (Warwick and Davies, 1977, after Thorson, 1957)	*Venus* Communities
Boreal Offshore Muddy Sand Association	*Echinocardium filiformis* Community *Abra* Community	*Syndosyma* Community *Amphiura* Communities
Boreal Offshore Mud Association	*Brissopsis–Chiajei* Community	*Amphipoda–Amphioplus* Community
Boreal Offshore Gravel Association	*Venus fasciata* Community (Spärk, 1935) and *Modiolus* Epifauna	*Venus* Communities (*Venus fasciatum*) *Spisula elliptica-Branchiostoma* Community
Boreal Offshore Muddy Gravel Association (Holme, 1966)	—	—
Boreal Deep Mud Association	*Brissopsis–sarsii* Community and *Amphilepis–Pecten* Community	*Maldane sarsi–Ophiura sarsi* Community
Boreal Deep Coral Association	Deep water Epifauna	

has come from many sources and many different approaches. Many of the ideas were first formulated in the nineteenth century by such pioneers as Edward Forbes and Karl Möbius. They continued to develop, and major advances were made by Petersen (1913, 1918) in his work on macrobenthic associations in Danish waters where he defined a number of communities based on groupings of species which regularly occurred together. They were designated using the names of one or two of the important species in the community. Those of relevance to studies in tidal current sedimentation are listed in Table 6.1.

Other studies (including Einarsson, 1941; Ford, 1923; Molander, 1928, 1962; Spärk, 1935; Stephen, 1933, 1934) have also defined communities based on a variety of criteria including geographical location, sediment type, water depth and salinity. These alternative schemes are summarized in Jones (1950).

In a major review of faunal communities throughout the world (Thorson, 1957), it was demonstrated that the same types of bottom are everywhere inhabited by a series of 'parallel' animal communities in which different species of the same genera replace one another as the 'characterizing species' (Table 6.1).

Other factors such as larval ecology, settlement behaviour, the faunal composition, the spatial relationships between species both horizontally and vertically with the sediment, and the differences in the relative proportions of species possessing hard parts in different sediment types are also important (Johnson, 1964).

The trophic structure of palaeocommunities can be inferred from assemblages which represent only small parts of the original living community (Stanton, 1976) although complications arise when the proportion of preservable species in a particular trophic level is different from its proportion in the total community (Stanton and Dodd, 1976). Complications can also arise when shells from faunas living in the same area but at different times become mixed by processes of winnowing and condensation which remove the sediments that originally separated the assemblages of shells (Fürsich, 1978).

The most complete ecological faunal analysis is embodied in the 'holistic concept' of ecological units which comprise four elements (Kauffman and Scott, 1976). First is the total composition of the fauna. This should be as complete as possible and should include microfaunas and large, but related, rare predator species. Second is the structure of the community and this incorporates a full understanding of the interactions between species and the energy flow in all its forms within the system. Third is the nature of the physical environmental factors, such as temperature, salinity, depth, oxygen content, sediment type, etc. Fourth is the nature of the boundary of the community.

Although these ideals are rarely, if ever, realized in either ecological or palaeoecological studies, it is, nevertheless, useful to bear them in mind when undertaking investigations into continental shelf ecology or palaeoecology. Clearly, species interaction in fossil communities can only be investigated where there is an observable relationship between the preserved hard parts such as one species encrusting another, etc. Problems also arise when considering the relative proportions of a fauna that have hard parts or other preservable structures. These proportions vary in different sediment types and from region to region. The finer sediment in each region is associated with the fewest species having preservable hard parts (Johnson, 1964). Evidence from the Irish Sea (Craig and Jones, 1966) suggests that only about 33% of the common infaunal species and 58% of the common epifaunal species have preservable hard parts. However, the percentage of infaunal species will be increased if structures produced by bioturbation are included. Some confusion exists in palaeoecology in the use of the terms community, association and assemblage (Pickerill and Brenchley, 1975). Several schemes have been proposed to clarify this and to define the terms and the relationships involved in a way that is helpful to the palaeoecologist (Pickerill and Brenchley, 1975; Watkins, Berry and Boucot, 1973).

The term community is used by ecologists where inter-relationships between the species present can be demonstrated. In this chapter the term association will be used except where reference is made to data from important published sources where to change terms would introduce further complications.

A classification scheme for faunal associations on continental shelves and continental slopes in the north-east Atlantic based on temperature range, sediment type and water depth was proposed by Jones (1950) and extended by Holme (1966). Five of these associations (the Boreal Offshore Sand Association, the Boreal Offshore Muddy Sand Association, the Boreal Offshore Mud Association, the Boreal Offshore Gravel Association and the Boreal Offshore Muddy Gravel Association) are present on the continental shelf. Faunas in the English Channel (Holme, 1966), the Southern Irish Sea (Dobson, Evans and James, 1971) and in the Malin Sea (Pendlebury and Dobson, 1976) have been assigned to some of these associations. There are two (the Boreal Deep Mud Association and the Boreal Deep Coral Association) which occur on the deeper parts of the continental shelf, the continental shelf edge and on the upper continental slope (Table 6.1). This grouping of associations is potentially of most use to the sedimentologist or palaeoecologist working on continental shelves and continental slopes.

The relationship of the Jones (1950) classification to those of Petersen and Thorson is outlined in Table 6.1. Other significant associations or modifications recognized by other workers are also included. Taxonomic problems arise when comparing species in associations described by several authors at different times. In an attempt to minimize confusion, the names used by the original authors are used in Table 6.1.

Few attempts have so far been made by these or other workers to relate faunal associations to particular bedforms or to make allowances for varying degrees of sediment mobility. Difficulties are encountered in trying to relate particular faunal communities or associations to particular grain sizes or sediment types (Buchanan, 1963). Some of these problems may be resolved when much more detailed attention is paid to the bedforms in the areas in question. The discussion of the faunas in the Western English Channel (Section 6.9) and in the Bristol Channel (Section 6.10) goes some way towards relating the Jones (1950) Associations to the bedforms.

6.3 Bioturbation

Trace fossils are of great value to the palaeoecologist in interpreting the environment of deposition of marine sediments, as they are widely distributed throughout the different marine environments both at present and in the fossil record from the late Pre-Cambrian onwards. They are always found *in situ* and provide a record of animal behaviour and of different responses to changes in the environment (Rhoads, 1975). They reflect the behavioural responses of the trace constructing animal to particular environmental conditions. Preserved trace fossils may provide a final record of some of the environmental conditions under which they were produced (Hertweck, 1972). Although most of our knowledge of the nature and extent of bioturbation in Holocene sediments has come from work on tidal flats and in shallow inshore waters, it is probable that bioturbation is as extensive on the middle and outer parts of the continental shelf where appropriate conditions exist. For this reason, an account of some of the effects of bioturbation relevant to studies of offshore tidal current deposits is given in this section.

The presence or absence of bioturbation within a sediment is related to a number of factors including the rate of deposition. Where the rate of deposition is variable and periods of rapid deposition alternate with times when little or no deposition takes place, the degree of bioturbation present at different horizons provides an indication of the duration of these respective phases (Reineck and Singh, 1973; van Straaten, 1954).

Escape structures made by infaunal species provide important clues to periods of rapid deposition in tidal current deposited sediments, such as those formed during storms (Goldring, 1964; Reineck, Dörjes, Gadow and Hertweck, 1968; Reineck and Singh, 1973).

The degree of bioturbation can be broadly correlated with grain size, and is greatest in fine-grained sediments (Reineck, 1967).

The burrowing activities of infaunal organisms may completely destroy cross-bedding and other primary sedimentary structures which are so important to the geologist in making the correct interpretation of the origin of the sediments. An understanding of the origin, nature and extent of bioturbation in a sedimentary deposit may, however, provide other evidence as to the origin of the sediments of possibly equal or even greater value to that that could be deduced from the sedimentary structures it has destroyed.

In areas where the sediments are mobile, any traces that are produced by biological activities are likely to be destroyed by the next episode of sediment movement (Frey and Seilacher, 1980; Howard, 1975; Rhoads, 1975).

6.3.1 DEPTH OF DISTURBANCE BY BIOTURBATION

Estimates vary as to the depth of disturbance of the sediment by the activities of modern infaunal animals but most studies suggest that the disturbance at any time is greatest in the top 10–15 cm of the sediment (Saila, 1976). However, larger polychaetes (Rhoads, 1967), bivalves such as *Lutraria* sp. (Holme, 1959) and decapod crustaceans such as *Upogebia* sp. and *Callianassa* sp. (Hertweck, 1972) live and disturb the sediments at depths down to at least 20–30 cm below the sediment surface. Burrows of the stomatopod crustacean *Squilla* sp. are recorded down to at least 60 cm depth within the sediments off Sapelo Island, Georgia, USA (Hertweck, 1972). Tubes of the burrowing sea anemone *Cerianthus* sp. are also found down to depths of 60 cm in the same area (Hertweck, 1972).

In intertidal sediments some species such as the bivalves *Ensis siliqua* and *Mya arenaria* live up to 30 cm below the surface (Holme, 1954; Reineck, 1958; Stanley, 1970; J. B. Wilson, 1967). On the Dutch Wadden Sea tidal flats, the polychaetes *Heteromastus filiformis* and *Arenicola marina* live 10–20 cm below the sediment surface (Cadée, 1976, 1979).

Deep burrowers – such as active crustaceans and deep burrowing bivalves – are difficult to sample (Warwick *et al.*, 1978). For example, burrows up to 90–120 cm deep and up to 4–5 cm in diameter in the sands off Newport Bay, California, are probably formed by a species of stomatopod crustacean (MacGinitie, 1939).

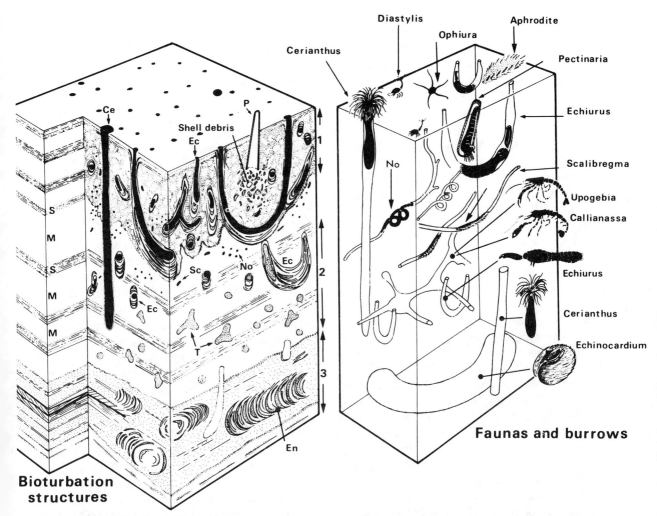

Fig. 6.1 Block diagrams to show the different levels of infaunal colonization and the bioturbation structures produced by polychaetes, echiuroids, crustaceans, anthozoans and irregular echinoids, also the burrows and animals responsible for the bioturbation. In the upper zone (1) the alternate layers of sand (S) and mud (M) are completely destroyed by the echiuroid *Echiurus* sp. Shell debris is accumulated at depth by *Pectinaria* sp. In zone (2) bioturbation is produced by crustaceans such as *Callianassa* sp. and *Upogebia* sp. Some earlier structures produced by *Echiurus* sp. are present. In the lowest zone (3) the characteristic 'orange leaf' bioturbation structures produced by the irregular echinoid *Echinocardium* sp. are present. Molluscs have been excluded (after Reineck, Gutmann and Hertweck, 1967).

Sediments are reworked by deposit feeding polychaetes, bivalves, crustaceans, echinoids and holothurians (Cadée, 1976, Table 2). Complete bioturbation of layers of various thickness can be achieved by many invertebrate species of polychaete, gastropod, bivalve, ophiuroid, echinoid, crustacean and echiuroid (Reineck and Singh, 1973, Table 13). A good example of this can be seen in the Pleistocene Rhodes Formation on the Aegean island of Rhodes (Bromley and Asgaard, 1975), where cross-bedded sands in units approximately 30 cm thick which are undisturbed by bioturbation alternate with generally thicker units in which the bedding has been completely destroyed by the burrowing activities of the irregular echinoid *Echinocardium cordatum* (Bromley and Asgaard, 1975, Fig. 1) (Section 6.13).

The activities of burrowing bivalves also turn over the sediment to a greater or lesser extent. Deep burrowers are rarely disturbed and they therefore seldom need to reburrow (Stanley, 1970). On the other hand shallow burrowing and nestling species do disturb and turn over the sediment (Rhoads, 1967).

Densities of the polychaete *Arenicola marina* on the Dutch tidal flats locally exceed 100 per m^2. The average density over the whole of the Wadden Sea is 17 per m^2 and this corresponds to a layer of sediment 6–7 cm in thickness being totally reworked annually in the whole Dutch Wadden Sea (Cadée, 1976). Densities of the polychaete *Heteromastus filiformis* range up to 9000 per m^2 with an average for the Wadden Sea of 500 per m^2. This corresponds to a layer of sediment 4 cm thick being reworked annually (Cadée, 1979).

On the Pacific coast of North America the decapod crustacean *Callianassa californiensis* can burrow down to a depth of 75–80 cm below the surface. Most of its burrowing activity is at depths of 45–50 cm and the sediments in this zone can be entirely turned over within a few months (MacGinitie, 1935).

In the inshore waters off Sapelo Island, Georgia, USA, extensive areas of the sediment are almost totally reworked and the structures produced by particular burrowers cannot be recognized. Where the degree of bioturbation is less, the activity can sometimes be attributed to one or two specific organisms (Howard and Reineck, 1972).

The populations of contemporaneous animals, all related to the *same* sediment-water interface but present at different depths within the sediment (Fig. 6.1), are also quite distinct, with different groups of species living at each depth (Reineck, Gutmann and Hertweck, 1967; Seibold, 1974; Thamdrup, 1935; J. B. Wilson, 1967).

Deep vertical burrowing is characteristic of nearshore and particularly intertidal environments. Shallower horizontal burrowing to depths of 10 cm is better developed in sediments on level bottoms further offshore (Rhoads, 1967).

6.3.2 TYPES OF BIOTURBATION

Different animals can affect the sediments on continental shelves in different ways. These vary considerably in the degree to which the primary structure of the sediment is affected and a distinction is made between internal disturbance and surface disturbance of the sediment. Some have only a minor effect on the structure of the sediment but are of importance in geochemical processes at the sediment–water interface.

The structures and trace fossils resulting from these animal–sediment interactions have been the subject of numerous papers. For an entry into the extensive literature, the reader is referred to the books and symposium volumes by Crimes and Harper (1970, 1977); Frey (1975); Hantzschel (1975) and Reineck and Singh (1973).

Those structures with preservation potential which may be present on the middle and outer continental shelf can be referred to six broad groups (Frey and Seilacher, 1980), while the Benthic Boundary Layer Working Group on Organism–Sediment Relationships cited nine principal ways in which the interaction can take place (Webb *et al.*, 1976). These two approaches, which are applicable to tidal current deposits, are correlated in Table 6.2.

Trace fossil assemblages – ichnofacies – provide a valuable record of animal activities within the sediment and are of particular importance to the sedimentologist as a further line of evidence about the conditions of deposition of the sediment. Seven ichnofacies have been recognized (Frey and Seilacher, 1980) of which six are marine (Table 6.3). Some of these assemblages occur in areas where tidal current sedimentation takes place and should, therefore, be present in rocks of presumed tidal current origin.

Lenses or layers composed of shell fragments can be formed by the burrowing activities of polychaetes. On the Dutch tidal flats the polychaete *Arenicola marina* accumulates gastropod (*Hydrobia* sp.) shells and shell fragments at the bottom of its burrow. These shell accumulations are 20–30 cm below the sediment surface (Cadée, 1976; van Straaten, 1952). Similar biogenic graded bedding has been reported from Barnstable Harbor, Massachusetts, USA where coarser debris remains at the feeding depth 20–30 cm below the sediment surface as a result of the activities of the polychaete *Clymenella torquata* and at 5–6 cm below the surface by the activities of

Table 6.2 Correlation between trace fossil types and the different effects of animals on sediments.

Trace fossil (*Lebensspuren*) types (Frey & Seilacher, 1980)	Effects of animals on sediments cited by Benthic Boundary Layer Working Group (Webb et al., 1976)
Resting traces (*Cubichnia*). Shallow depressions made by animals that temporarily settle onto, or dig into, the substrate surface. Emphasis is upon reclusion.	—
Crawling traces (*Repichnia*). Trackways and surface or interstratal trails made by organisms travelling from one place to another. Emphasis is upon locomotion.	The disturbance of the sediment surface by active epibenthic surface dwellers such as fish or crustaceans during feeding or movement.
Grazing traces (*Pascichnia*). Grooves, pits and furrows, many of them discontinuous, made by mobile deposit feeders or algal grazers at or near the sediment surface. Emphasis is upon feeding behaviour analogous to 'strip mining'.	The disturbance of the sediment surface by deposit feeders such as gastropods browsing for food which can also change particle size by faecal deposition.
Feeding structures (*Fodinichnia*). More or less temporary burrows constructed by deposit feeders; the structures may also provide shelter for the organisms. Emphasis is upon feeding behaviour analogous to 'underground mining'.	Burrowing polychaetes and echinoderms operating within the sediment and disturbing and sorting it.
Dwelling structures (*Domichnia*). Burrows, borings or dwelling tubes providing more or less permanent domiciles, mostly for hemisessile suspension feeders or, in some cases, carnivores. Emphasis is upon habitation.	Burrowers such as crustaceans transporting sediment upward and horizontally during the construction and maintenance of their burrows. The role of tube building polychaetes such as the Sabellarids – *Lanice*, *Pectinaria* – in concentrating particular sediment grains or grain sizes during the construction of their tubes. The transport of sediment grains upwards and the circulation of water downwards by polychaetes such as *Pectinaria*. The circulation of interstitial water by infaunal suspension feeders such as polychaetes which may move sediment grains.
Escape structures (*Fugichnia*). Traces – Lebensspuren – of various kinds modified or made anew by animals in direct response to substrate degradation or aggradation. Emphasis is upon readjustment, or equilibrium between relative substrate position and the configuration of contained traces.	—
—	The conversion of suspended solids in the water column into deposit faeces by epifaunal suspension feeders such as tunicates.
—	Shell secreting animals such as molluscs converting dissolved ions into shells.

Table 6.3 Trace fossil assemblages and their characteristic environments (adapted and simplified from Frey and Seilacher, 1980).

Trace fossil assemblages	Environment
Trypanites ichnofacies. Variously shaped burrows, anastomosing systems of borings (bryozoan, sponge). Raspings and grazings (chitons, limpets, echinoids). Low diversity but may be abundant.	Hard substrates; rock surfaces or shells.
Glossofungites ichnofacies. Vertical, cylindrical or U shaped or sparsely ramified dwelling burrows, some with protrusive 'spreiten' developed through growth of the animal. Fan shaped forms of *Rhizocorallum*. Low diversity but may be abundant.	Firm substrates; stable coherent substrates in low energy settings or higher energy semi-consolidated substrates offering resistance to erosion.
Skolithos ichnofacies. Vertical, cylindrical or U shaped dwelling burrows, some with protrusive or retrusive 'spreiten' developed as escape structures in response to sedimentation or erosion. More or less vertical *Ophiomorpha* burrows. Low diversity but may be abundant.	Shifting substrates; relatively high energy conditions, clean, well sorted sediments subject to abrupt erosion or deposition.
Cruziana ichnofacies. Surface and intrastratal crawling traces. Inclined U burrows mostly with protrusive 'spreiten'. Variable *Ophiomorpha* and *Thalassinoides* burrow systems. High diversity and abundance.	Well sorted silts and sands or interbedded muddy and clean sands in moderate to low energy environments affected by tidal currents and disturbed by major storm waves.
Zoophycos ichnofacies. Simple or relatively complex grazing traces and shallow feeding structures with planar or inclined 'spreiten', arranged in sheets, ribbons or spirals. Low diversity but may be abundant.	Quiet water conditions on continental shelves and in deeper water in weak current areas. Sediments rich in organic matter but somewhat deficient in oxygen.
Nereites ichnofacies. Complex grazing traces and patterned feeding and dwelling structures with nearly planar 'spreiten'. Crawling grazing traces and sinuous faecal castings.	Deep quiet water with little tidal influence interrupted by intermittent turbidity flows or areas of very slow deposition.

the polychaete *Cistenides gouldi* (*Pectinaria gouldi*) (Rhoads, 1967; Rhoads and Stanley, 1965).

The opposite process has been reported from Mugu Lagoon, California where the coarser sediments and shell fragments are accumulated as mounds on the sediment surface at the mouths of burrows formed by the crustacean *Callianassa californiensis* during its burrowing activities. Other similar shell layers may be present either at depths within the sediment or on the surface in suitable areas on the continental shelf (Warme, 1967).

6.4 Topics and areas excluded

Descriptions and discussion of shelly carbonates of tropical and sub-tropical continental shelves are not included in this book as they are extensively described in other publications. Eight examples from the Caribbean, Gulf of Mexico, and the continental shelf off northern and western Australia have recently been summarized by Ginsburg and James (1974). For an entry into the extensive literature on these, the reader is referred to the books by Bathurst (1971), Matthews (1974), Milliman (1974), J. L. Wilson (1975) and to the bibliography in Ginsburg and James (1974). Much work has been done on the shallow water ooid shoals of the Bahamas (for an introduction to the literature including more recent papers, see Harris, 1979). Carbonate sediments are also present on the continental slope off the Bahamas (Mullins and Neumann, 1979). Consideration of these shallow inshore and continental slope deposits in tropical waters is, however, beyond the scope of this book.

Similarly, sediments and faunal associations on tidal flats and in shallow sub-tidal near-shore waters in temperate latitudes are also comparatively well known and are therefore mostly excluded from this chapter. The sediments and faunas of Dutch coastal waters, the Wadden Sea, north German and Danish tidal flats have been extensively studied (de Jong, 1977; Eisma, 1966, 1968; Reineck and Singh, 1973; Thamdrup, 1935; van Straaten, 1954, 1961). For a bibliography of papers on tidal flats published up to 1973 see Reineck (1973). Other areas investigated include the Ria de Arosa in north-east Spain (Cadée, 1968).

Estuaries and bays in the United Kingdom in which the sediments and faunas have been investigated include the Tay, studied extensively by the Tay Estuary Research Group (Buller and McManus, 1975; Green, 1975; Khayrallah and Jones, 1975; McManus, Buller and Green, 1980; Mishra, 1968), The Wash (Evans, 1965), the Thames (Greensmith and Tucker, 1969; Prentice et al., 1968), the Exe (Thomas, 1980) and the Solway (Perkins, 1974; J. B. Wilson, 1967).

Discussion of microfaunas and details of the faunas of muddy areas both inshore and on the continental shelf are also largely excluded.

6.5 Temperate water regions studied and their geological importance

Although comparatively little attention has been paid by geologists to carbonate sediments in temperate latitudes (Chave, 1967), a number of continental shelf areas have been studied. Those in the northern hemisphere outside European waters include the New York Bight (Freeland and Swift, 1978), Georges Bank off the north-eastern United States of America, off Nova Scotia (Emery, Merrill and Trumbull, 1965; Müller and Milliman, 1973; Wigley, 1961), the Grand Banks, off Newfoundland (Slatt, 1973), off Alaska (Hoskin and Nelson, 1969), off south-west Iceland (Thors, 1978) and the Barents Sea (Bjørlykke, Bue and Elverhøi, 1978). Regions investigated in the southern hemisphere, include the contintental shelves off southern Australia (Chave, 1967; Conolly and von der Borch, 1967; Wass, Conolly and MacIntyre, 1970), off eastern Australia (Marshall and Davies, 1978), off the south and south-east coasts of South Island and off the northern part of North Island, New Zealand (Cullen, 1967; Summerhayes, 1969a, b; Andrews, 1973; Carter, 1975) and off the west coast of South Africa (Birch, 1977).

Studies of the faunal composition of shell gravels and coarse bioclastic sands on continental shelves in temperate latitudes are of significance to the geologist in the light that they can shed on the origins of some of the more important and extensive bioclastic limestones of the geological record. Accumulations of coarse bioclastic carbonate debris of cold or warm water origin can be of considerable economic importance (Mazzullo, 1980). Their value as reservoir rocks can be increased by diagenetic changes such as the selective dissolution of certain of the faunal elements, so creating pockets of higher porosity. However, porosity in carbonates is generally lower than that found in sandstones (J. L. Wilson, 1975).

In some areas in the Western English Channel, particularly off Brittany, the coarse carbonate gravels are dredged for use as agricultural lime (Section 6.6.1, Plate 6.1).

6.5.1 CARBONATE CONTENT OF SEDIMENTS ON THE CONTINENTAL SHELF AROUND THE BRITISH ISLES

The calcium carbonate content of the sediment around the British Isles shows considerable variation. The shell gravels in the Western English Channel commonly have values of 60–80% (Fig. 6.2). The sands forming the sand patches and the rippled sands on the continental shelf west of Scotland can contain 25–30% carbonate while the coarser shell gravels in the Fair Isle Channel between Orkney and Shetland commonly contain 90% carbonate. In contrast the carbonate content of the active sand banks of the southern North Sea is rarely more than 20% and is generally about 5% or less (Section 6.13).

6.6 Faunas in shallow nearshore waters

As this book is concerned with offshore tidal current deposits, only brief mention is made of a few studies

136 Offshore Tidal Sands

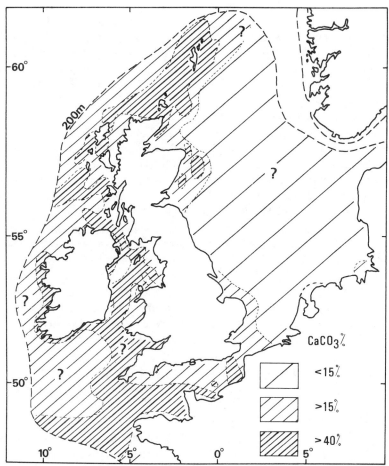

Fig. 6.2 An impression of the abundance of bioclastic debris in the uppermost superficial sediments of the north-west European continental shelf. There is much local variation in abundance. Data are scarce in some regions. Sources: Boillot (1965), Borley (1923), Bouysse, Le Lann and Scolari (1979), Channon and Hamilton (1976), Cronan (1969), Deegan, Kirby, Rae and Floyd (1973), Dobson, Evans and James (1971), Dyer (1970), Farrow, Cucci and Scoffin (1978), Gunatilaka (1977), Hails (1975), Keary (1969), Kenyon (1970), Larsonneur (1972), Lees, Buller and Scott (1969), Lefort (1970), Lovell (1979), Moore (1968), Owens (1979), Pendlebury and Dobson (1976), Pratje (1931) and unpublished data from Institute of Geological Sciences and Institute of Oceanographic Sciences (analysis by N. H. Kenyon).

of areas where either the adjacent offshore faunas are referred to later in this chapter or where deposits of geological importance are located such as calcareous algal gravels (Section 6.6.1).

In inshore waters on the north side of the English Channel and in sandy bays and estuaries, species present include the bivalves *Tellina tenuis* and *Donax vittatus* and the polychaete *Arenicola marina*. The species present are in the Boreal Shallow Sand Association (Table 6.1; Holme, 1961, 1966). In estuaries and shallow subtidal waters where the sediments are finer grained, the bivalves *Macoma balthica*, *Scrobicularia plana*, *Cerastoderma edule* (*Cardium edule*) and *Tellina tenuis* occur (Holme, 1949, 1961, 1966). These species comprise the Boreal Shallow Mud Association (Table 6.1).

In inshore waters of the Southern Irish Sea the Boreal Shallow Sand Association is present in intertidal and shallow subtidal sandy areas and species belonging to the Boreal Shallow Rock Association including the bivalve *Mytilus edulis*, the gastropods *Patella vulgata*, *Nucella lapillus* and *Littorina* sp. and the barnacle *Balanus balanus* are present over much of the shore zone. These species contribute carbonate to the adjacent sediments (Dobson *et al.*, 1971).

In nearshore waters west of Scotland up to eleven facies have been distinguished based on the nature of the floor and the associated faunal types (Farrow, Scoffin, Brown and Cucci, 1979). These range from ophiuroid covered rock surfaces to fine sands with *Turritella communis* shells occupied by hermit crabs and sipunculans, or to muds with pennatulids and burrows of the decapod crustacean *Nephrops norvegicus*. Algae play an important role in the stabilization of the sediment in these shallow nearshore waters. The deepest records of the large brown algae in British waters are at a depth of 30 metres. The extensive *Laminaria* 'kelp forests' off the east coast of Scotland and England are much shallower than this. They can also act as agents of sediment transport as the holdfast may be attached to dead shells, pebbles or small cobbles. If the alga is uprooted during a storm then the thallus and holdfast, with the attached material, may be transported into a different sedimentary environment (Pasternak, 1971). The thallus of large brown algae such as *Laminaria hyperborea* and *Laminaria saccharina* can act as a substrate for such carbonate producers as barnacles, serpulids and bryozoans (Farrow *et al.*, 1979).

6.6.1 TEMPERATE WATER CALCAREOUS ALGAL GRAVELS

Calcareous algae such as *Lithothamnium corallioides*, *Lithothamnium glaciale* and *Phymatolithon calcareum* are important producers of carbonate in shallow nearshore waters off Brittany, in the Western English Channel, to the south and west of Ireland, off the west coast of Scotland, the Orkney and Shetland Isles and off the north-east coast of Spain (Adey and Adey, 1973; Blunden, Farnham, Jephson, Fenn and Plunkett, 1977; Bosence, 1976, 1978, 1979, 1980; J. Cabioch, 1970; L. Cabioch, 1968; Cadée, 1968; Chardy, Guennegan and Branellec, 1980; Farrow, Cucci and Scoffin, 1978; Farrow, Scoffin, Brown and Cucci, 1979; Gunatilaka, 1977; Holme, 1966; Keegan, 1974; Larsonneur, 1971; Lees, Buller and Scott, 1969; Lemoine, 1923; Piessens and Lees, 1977; Pruvot, 1897; J. B. Wilson, 1979c).

Five distinct faunal associations have been recognized among the algal gravels off Co. Galway, Ireland (Bosence, 1979). These algal gravels are known as 'maërl' and support a rich fauna of herbivorous gastropods including *Bittium reticulatum*, *Gibbula cineraria*, *Gibbula magus*, and burrowing or nestling infaunal bivalves including *Venus verrucosa*, *Parvicardium ovale* (*Cardium ovale*), *Paphia rhomboides* (*Venerupis rhomboides*), *Paphia aurea* (*Venerupis aurea*) and *Thyasira flexuosa*. The bivalves *Hiatella arctica* and *Gastrochaena dubia* bore or nestle closely into the larger calcareous algal rhodoliths. Serpulid and spirorbid polychaetes, bryozoans and the bivalve *Anomia ephippium* are present as sessile epifauna attached to bivalve valves and algal rhodoliths. For full details of the species associated with the Co. Galway algal gravels see Bosence (1979) and Keegan (1974). The algal gravels off Roscoff in the Western English Channel also support a rich fauna which is described in detail by L. Cabioch (1968).

Deposits containing fragmented rhodoliths can be fairly extensive in sheltered inshore waters. The gravels supporting the algae are generally fairly coarse (Plate 6.1). Agitation and sorting of the algal debris by waves, and to a lesser extent by tidal currents, produces deposits of varying grain sizes. Algal debris may be transported ashore where extensive beaches are formed which are composed almost entirely of well sorted and graded rhodolith fragments.

6.7 Faunas of the middle and outer continental shelf

The sediments and faunas in the German Bight have been the subject of extensive investigations by the Senckenberg Institute, Wilhelmshaven. For details of these investigations and references to the extensive literature the reader is referred to the book by Reineck and Singh (1973). The faunas described below are from those parts of the continental shelf around the British Isles for which limited suitable data are available and where the faunas can be related

138 Offshore Tidal Sands

Table 6.4 Areas on the continental shelf around the British Isles where faunal data relating to bedform zones and to tidal current strength are used in this chapter. References are to sections in Chapters 3 and 6.

Current strength	Facies	Bedform zone		English Channel	Bristol Channel	Southern North Sea	West of Scotland
Strong ↑	Rock	Bare rock floor or thin veneer of sediment	—	6.9.1	6.10.2	—	—
	Gravel sheet	Gravel waves and furrows	3.4.2	6.9.1	—	—	6.12.1
		Sand ribbons	3.4.4	6.9.2	6.10.3	—	—
	Sand sheet	Large sand waves with gravel between or	3.3.3	6.9.3	6.10.4	—	—
		Large sand waves continuous sand	3.3.3	6.9.3	6.10.4	6.11.1	—
		Small sand waves	3.3.3	—	—	6.11.2	—
		Rippled sand or	3.3.2	6.9.4	6.10.5	6.11.3	6.12.2
↓ Weak		Sand patches (reduced sediment supply)	3.3.4 & 3.4.4	—	—	—	6.12.2 & 6.12.3
Strong to Moderate	Sand bank	Active offshore sand banks	3.4.5	—	6.13	6.13	—
Now weak		Moribund offshore sand banks	3.4.5	—	—	—	—

to particular bedforms associated with a known sediment transport–deposition path.

In each region described, the faunas will generally be discussed in sequence starting in the zone of strongest tidal currents and proceeding into zones of successively weaker tidal currents (Table 6.4). Regions described include the Western English Channel, the Bristol Channel, the Southern North Sea, the continental shelf to the west and north of Scotland and the Norfolk Banks.

Data are only available for some of the bedform zones present in each of these regions. Some bedform zones are present in one region and not in another (Table 6.4).

6.8 Faunas of a bed-load parting

Bed-load partings are by their nature likely to be zones of the sea floor where the sediments (if present) will generally be coarser than those in the adjacent zones of weaker currents on either side. They will frequently be areas of extensive rock outcrop and boulder fields with some coarse shell sand and shell gravel trapped in eroded joint planes and depressions in the rock and in the interstices between the boulders. The rocks and boulders may support an epifauna (Plate 6.2), probably consisting of serpulids, barnacles, byssally attached bivalves, gastropods, regular echinoids and ophiuroids. Some time after death the shells will be further broken down, largely by biological action (Section 6.16), into gravel, sand or smaller sized fragments and will contribute carbonate debris of the appropriate size and shape to be transported away from the bed-load parting in either direction by the two diverging bed transport paths. Bed-load partings and indeed all areas of rocks and boulders in strong current areas can be regarded as potential carbonate 'factories' (Plate 6.2) which provide additional sediment to the transport paths. In some cases such as that of the English Channel Bed-Load Parting (Section 6.9.1) the rock floor has a thin veneer of gravel on it in most of the zone. In contrast, the Southern Bight Bed-Load Parting in the North Sea has a sand floor.

6.9 Faunas associated with bedform zones in the Western English Channel

The faunas of the English Channel are comparatively well known. Extensive investigations have been carried out both in the inshore waters of England and France and in the deeper waters offshore (Holme, 1961, 1966; L. Cabioch, Gentil, Glaçon and Retière, 1977). Particular areas such as the Eddystone area 22 km south of Plymouth (Allen, 1899; Ford, 1923; Smith, 1932), the area off Looe (Vevers, 1951, 1952), the continental shelf off Brittany and the Baie de la Seine (Auffret, Berthois, L. Cabioch and Douville, 1972; L. Cabioch, 1968; Glemarec, 1969a, b; Larsonneur, 1971; Lefort, 1970) have been studied in detail.

6.9.1 FAUNAS FROM THE GRAVEL SHEET

Within the area of the gravel sheet lies the English Channel Bed-Load Parting and the area of the gravel waves and furrows (Section 4.3.2 and Fig. 4.4). The faunas within this area are generally sparse (L. Cabioch *et al.*, 1977; Holme, 1966; N. A. Holme, pers. comm., 1979). Within the area of the bed-load parting, the fauna contains only epifaunal species as there is only a thin veneer of sediment covering the rock in places. Burrows are absent. Species present include the asteroids *Henricia sanguinolenta* and *Solaster papposus*, the holothurian *Cucumaria* sp., the bivalve *Chlamys varia* and hermit crabs inhabiting vacant gastropod shells. The bryozoan *Flustra foliacea*, the hydroid *Nemertesia* sp. and the ophiuroid *Ophiothrix fragilis* are common. The ophiuroid *Ophiocomina nigra* is rare. Burrowing anemones are also present (L. Cabioch, Gentil, Glaçon and Retière, 1975; N. A. Holme, pers. comm., 1979). The floor in places consists of a shell pavement composed of valves of *Glycymeris glycymeris* and *Venus* sp.

Within the zone where furrows (Section 4.3.2) are present in the gravel (Fig. 6.3) the fauna includes the bivalves *Venus verrucosa*, *Paphia rhomboides* (*Venerupis rhomboides*), *Gari tellinella* which nestle in shallow burrows in the gravel, the attached form *Chlamys distorta* and the more active species *Aequipecten opercularis* (*Chlamys opercularis*) and *Limaria hians* (*Lima hians*). *Chlamys varia*, which may be either byssally attached or free living, is also present. Other species include the gastropod *Buccinum undatum*, the bryozoan *Flustra foliacea*, the ophiuroid *Ophiothrix fragilis* and the asteroid *Asterias rubens* (Holme, 1966). Other bryozoans including *Schizomavella auriculata*, *Schizomavella linearis*, *Porella concinna*, *Cellepora pumicosa* and *Turbicellepora avicularis* and the serpulid polychaetes *Protula tubularia* and *Serpula vermicularis* are also present (L. Cabioch *et al.*, 1977). These species occur throughout the Channel where appropriate substrates are present. The fauna within the gravel sheet can be assigned to the Boreal Offshore Gravel Association (Holme, 1966) although the number of species of that Association present is much less than is present in the sand ribbon zone (Section 6.9.2) and on the gravel floor associated with the zone of large sand waves (Section 6.9.3).

6.9.2 FAUNAS FROM THE SAND RIBBON ZONE

The faunal diversity gradient on the gravel floor of the Western English Channel within the sand ribbon zone increases towards the west. In the eastern part of the zone species present on the gravel floor between the sand ribbons include the bivalves *Venus verrucosa*, *Paphia rhomboides* (*Venerupis rhomboides*), *Acropagia crassa* (*Tellina crassa*), *Chlamys varia* and *Chlamys distorta*. The ophiuroid *Ophiothrix fragilis* is also present. Further west within the sand ribbon zone, the gravel floor fauna is much more diverse and includes the bivalves *Striarca lactea* (*Arca lactea*), *Circomphalus casina* (*Venus casina*), *Paphia rhomboides* (*Venerupis rhomboides*), *Gari tellinella*, *Glycymeris glycymeris* and the polychaete *Chaetopterus variopedatus*. Other species sometimes present include the bivalves *Chlamys varia*, *Limaria hians* (*Lima hians*), *Clausinella fasciata* (*Venus fasciata*), *Dosinia exoleta*, *Galeomma turtoni*, *Spisula elliptica*, and the ophiuroids *Amphiura filiformis*, *Ophiura albida*, *Ophiothrix fragilis* and *Ophiocomina nigra*. These species are characteristic of the Boreal Offshore Gravel Association (Holme, 1966). Dead shells of species such as *Palliolum tigerinum* (*Chlamys tigrina*), *Gouldia minima* (*Gafrarium minimum*) and *Limaria loscombi* (*Lima loscombi*) are common. In

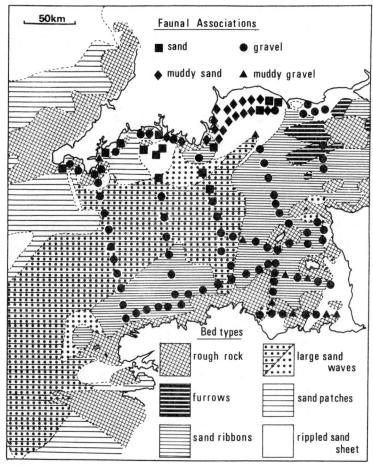

Fig. 6.3 Bed types off south-western England and western France (from IOS data; see also Fig. 4.4) and living faunal associations in the Western English Channel (after Holme, 1966).

some areas within the sand ribbon zone, particularly in the southern half of the English Channel, the gravel fauna differs slightly and can be assigned to the Boreal Offshore Muddy Gravel Association (Fig. 6.3). Additional species present include the burrowing callianasid crustaceans *Upogebia deltaura* and *Upogebia stellata* and the bivalve *Nucula nucleus* (Holme, 1966).

In the Baie de la Seine and in the Western English Channel off Roscoff, samples and photographs of the gravel floor show numerous valves of *Glycymeris glycymeris*, *Paphia rhomboides* (*Venerupis rhomboides*), *Chlamys varia*, *Aequipecten opercularis* (*Chlamys opercularis*), *Modiolus barbatus*, the echinoids *Psammechinus miliaris* and *Echinocyamus pusillus* and the gastropod *Buccinum undatum* (L. Cabioch, 1968; Larsonneur, 1971) associated with the pebbles.

The empty shells of gastropods such as *Buccinum undatum* are frequently utilized by the hermit crab *Pagurus bernhardus*. The crabs often scratch the columella of the shell leaving a characteristic trace (Boekschoten, 1966). The availability of gastropod shells of a suitable size is one of the major controls on the size of hermit crab populations (Kellogg, 1976). Hermit crabs are responsible for transporting and thus dispersing gastropod shells over much wider areas than those occupied by the original living gastropod population (L. Cabioch, pers. comm.). Dead shells of *Buccinum undatum* are found in the

Western English Channel. As it is an eastern species (Holme, 1966), it is possible that some of these shells may have been partly transported westwards in the Channel by successive populations of hermit crabs.

The pebbles in the gravels within the sand ribbon zone north of Roscoff (Fig. 6.3) support an epifauna of hydroids, sponges, serpulid polychaetes and bryozoans. Important carbonate producing species include the bryozoans *Parasmittina trispinosa* and *Reptadeonella violacea* and the serpulid polychaete *Hydroides norvegica* (L. Cabioch, 1968).

Little is known about the fauna living in the sand ribbons themselves. It is probable that very few species are present due to the frequency of disturbance of the sand by the strong tidal currents characteristic of the sand ribbon zone. Evidence from Fouveaux Strait, south of South Island, New Zealand, shows that in the sand ribbon zone the settlement of the bivalve *Ostrea sinuata* is largely controlled by the nature of the floor. The oyster thrives on the medium to fine sand, pebble and gravel floors. Spat which fall on the coarse sands are unable to become cemented and will be buried by the mobile sand, while those that settle on the very coarse gravels are probably too easily dislodged from the uneven surface (Cullen, 1962).

6.9.3 FAUNAS FROM THE ZONE OF LARGE SAND WAVES

Over much of the zone, the sand waves are isolated and widely separated with extensive areas of gravel floor between them. This gravel supports a Boreal Offshore Gravel Association fauna that is similar to that present in the zone of sand ribbons (Section 6.9.2) but the numbers of particular species present such as the polychaete *Chaetopterus variopedatus*, and the bivalves *Circomphalus casina* (*Venus casina*) and *Clausinella fasciata* (*Venus fasciata*) tend to be greater than that found in the gravels in the sand ribbon zone.

Where the sand cover between the sand waves is more continuous, additional species including the bivalves *Dosinia lupinus*, *Chamelea gallina* (*Venus gallina*) and *Lutraria angustior* are present and the fauna is more related to the Boreal Offshore Sand Association (Holme, 1966).

Some of the characteristic western species such as the bivalve *Tellina pygmaea* (Holme, 1966) from the gravel and sand in the sand wave zone are also found in the eastern part of the English Channel in the sand wave zone near Dover Strait. It is probable that this represents eastward movement of larvae as a result of the net eastward flow of water along the English Channel.

6.9.4 FAUNAS FROM THE ZONE OF RIPPLED SAND

Rippled sands are present off Plymouth and in Lyme Bay (Great West Bay) (Figs 4.4 and 6.3). The fauna can be assigned to the Boreal Offshore Sand Association (Holme, 1966) but it is much more diverse and is somewhat different in faunal composition to that of the sand floor of the zone of sand waves. Species present include the bivalves *Spisula elliptica*, *Aequipecten opercularis* (*Chlamys opercularis*), *Nucula nucleus*, *Palliolum simile* (*Chlamys similis*), *Gouldia minima* (*Gafrarium minimum*), *Chamelea gallina* (*Venus gallina*), *Arctica islandica* (*Cyprina islandica*), *Corbula gibba*, *Abra prismatica*, *Lucinoma borealis*, *Acanthocardia echinata* (*Cardium echinatum*) and *Lyonsia norwegica*, the gastropods *Turritella communis*, *Colus gracilis* and *Aporrhais pespelecani* and the ophiuroids *Ophiura albida* and *Ophiura texturata* (Holme, 1966).

In Lyme Bay (Great West Bay) the fauna is somewhat different and includes species assigned to the Boreal Offshore Muddy Sand Association (Holme, 1966). The species present include the bivalves *Mactra stultorum* (*Mactra corallina*), *Phaxas pellucidus* (*Cultellus pellucidus*), *Thyasira flexuosa*, *Myrtea spinifera*, *Spisula elliptica*, *Spisula subtruncata*, *Dosinia lupinus*, *Chamelea gallina* (*Venus gallina*), *Mysia undata*, *Gari fervensis*, the gastropod *Turritella communis*, the ophiuroids *Amphiura filiformis* and *Ophiura texturata*, the echinoid *Echinocardium cordatum* and in some locations, the callianassid crustacean *Upogebia stellata*.

6.10 Faunas associated with bedform zones in the Bristol Channel

In the Bristol Channel there is a geographical succession of zones from east to west, of deep erosion

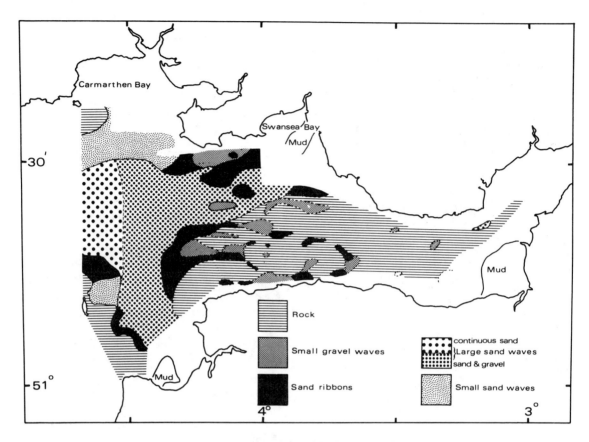

Fig. 6.4 Bed types of the Bristol Channel based on Institute of Oceanographic Sciences studies and Warwick and Davies (1977).

(mainly bare rock floor), of sand ribbons, of sand waves, and of rippled sand or flat floors beyond, corresponding to the progressive decrease in peak tidal current strength (Section 4.3.2).

The faunas associated with these zones have recently been investigated (Warwick and Davies, 1977; Warwick and Uncles, 1980). The faunas recorded have been assigned to communities based on Petersen (1913) and Thorson (1957) (Table 6.1). Many of the species recorded during the study are soft bodied errant forms which have little or no preservation potential and will not be discussed here. The species mentioned below generally include only those which could be potential fossils. The location of the bedforms and their associated faunal communities is shown in Figs 6.4 and 6.5.

6.10.1 BENTHIC FAUNAS IN RELATION TO TIDAL BOTTOM STRESS

A correlation between the tidally averaged magnitude of M_2 bottom stress and the faunal associations has been established along the sand transport path in the Bristol Channel (Warwick and Uncles, 1980). The maximum bottom stress at mean spring tides, which may be the more relevant quantity for the fauna is similarly correlated, being everywhere 3.7 times the M_2 mean stress.

Values for the tidally averaged bottom stress (Fig. 6.6) along the transport path range from at least 19 dynes/cm² on the rock floor colonized by the Reduced Hard Bottom Community (Section 6.10.2) to 6 to 7 dynes/cm² in the zone of sand waves

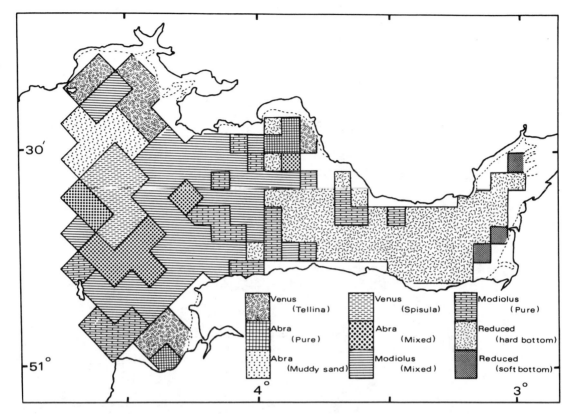

Fig. 6.5 Main macrofaunal communities of the Bristol Channel (after Warwick and Davies, 1977).

colonized by the *Spisula* Sub-community (Section 6.10.4). In the rippled muddy sands in the bays (Section 6.10.5), where the *Tellina* Sub-community is found, the tidally averaged bottom stress is generally less than 2.5 dynes/cm^2 (Fig. 6.6).

On more exposed parts of the continental shelf the values would be slightly less as the smaller tidal currents and tidal stress would be compensated by the stronger non-tidal currents and waves during storms.

Available evidence suggests that suspension feeders are the dominant forms in zones of high bottom stress and deposit feeders dominate in zones of low bottom stress (R. Warwick, pers. comm., 1980). Faunas in areas of high bottom stress also tend to be dominated by one species whereas the faunal diversity is much higher in areas of low bottom stress (Warwick, 1980, and pers. comm., 1980).

6.10.2 FAUNAS FROM THE ROCK FLOOR

A variety of bottom types are present ranging from rock outcrops and boulders to gravel with some admixture of sand. In the Bristol Channel, east of Swansea Bay, on the ground where there is strong tidal scour, the faunal diversity is reduced. The fauna present comprises the Reduced Hard Bottom Community (Warwick and Davies, 1977) which consists of twenty-three species. Those with the greatest preservation potential, and hence importance to the geologist, include the bivalves *Hiatella arctica* and *Nucula tenuis*, the asteroid *Asterias rubens*, the ophiuroid *Ophiothrix fragilis*, the hermit crab *Pagurus bernhardus* and the polychaete *Melinna cristata*. The small bivalve *Sphenia binghami* attaches itself by byssus threads inside *Hiatella* boreholes.

Other species present include the tube building sabellarid worms *Sabellaria spinulosa* and *Sabellaria alveolata* which form extensive 'reefs' in places. Sabellarid 'reefs' and other structures occur in temperate and tropical regions between latitudes 72°N and 53°S (Kirtley and Tanner, 1968). In European

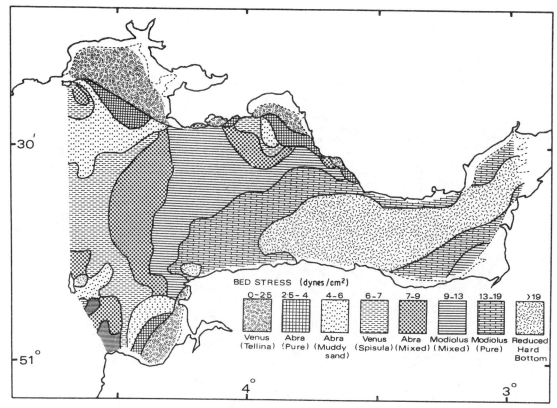

Fig. 6.6 Main macrofaunal communities of the Bristol Channel and the associated range of tidally averaged bottom stress (after Warwick and Uncles, 1980).

waters, in addition to the Bristol Channel, they occur extensively in the Southern North Sea (Richter, 1920, 1927; Schäfer, 1962), where they are known to fishermen as 'riffs' or 'ross'. They also occur in the Irish Sea (Herdman, 1920) and in the Bay of Mont Saint-Michel in the Western English Channel.

Further west in the Bristol Channel, the Reduced Hard Bottom Community is replaced by the more diverse *Modiolus* Community (Fig. 6.5). Where the bottom consists of rock outcrops or boulders the faunal list for the *Modiolus* Community contains forty-six species. Towards the sand ribbon zone (Section 6.10.3) where there is some admixture of sand and gravel in addition to the rock outcrops, the fauna is much more diverse and a further fifty-four Mixed Bottom species (of which twenty are also characteristic of other communities) may be added to the list (Warwick and Davies, 1977). Those species present with good preservation potential include the

bivalves *Nucula tenuis*, *Chlamys varia*, *Modiolus modiolus*, the gastropods *Calliostoma zizyphinum*, *Buccinum undatum*, *Diodora graeca* (*Diodora apertura*), the echinoid *Psammechinus miliaris*, the asteroid *Asterias rubens*, and the ophiuroids *Ophiura albida* and *Ophiothrix fragilis*.

6.10.3 FAUNAS FROM THE SAND RIBBON ZONE

Faunas in the sand ribbon zone (Fig. 6.4) include species in both the *Abra* and *Modiolus* Communities. The faunal composition of the *Abra* Community depends partly on whether the bottom is consolidated or whether some mud or muddy sand may be present (Fig. 6.5). In the sand ribbon zone, the muddy sand and mud species (Warwick and Davies, 1977, Table 2) will be absent. Species present include the infaunal bivalves *Abra alba*, *Spisula solida*, the polychaetes *Scalibregma inflatum*, *Scoloplos armiger*, *Notomastus latericeus*, and *Pectinaria koreni*. The polychaete *Lumbriconereis latreilli*, and the amphipod crustacean *Ampelisca spinipes* are particularly abundant in the sand ribbon zone often averaging over 100 individuals per square metre.

Gravel and mixed bottom faunas within the sand ribbon zone

Within the sand ribbon zone in the strong current areas where the floor consists of pebbles and gravel with shell debris but where little or no sand is present, species common to both *Venus* and *Abra* Communities are present in association with *Modiolus* Community species. Some species are particularly abundant on these mixed bottoms and include the echinoid *Echinocyamus pusillus*, the amphipod crustaceans *Ampelisca spinipes* and *Stenothoë marina* and the sipunculan *Phascolion strombi* which inhabits dead *Turritella communis* and scaphopod shells (Warwick and Davies, 1977).

6.10.4 FAUNAS FROM THE ZONE OF LARGE SAND WAVES

The zone of large sand waves in the Bristol Channel sand sheet has a sparse fauna. Epifaunal and delicate species are virtually absent. The fauna is characteristic of the *Abra* and *Modiolus* Communities where there is gravel between the sand waves (cf. Section 6.9.3), and of the *Spisula* Sub-community (a subdivision of the *Venus* Community, Table 6.1) where the sand waves form a continuous cover (Figs 6.4 and 6.5). The *Spisula* Sub-community includes the bivalve *Spisula elliptica*, the polychaetes *Nephthys cirrosa* and *Paraonis lyra*, the crustacean *Gastrosaccus spinifer* and the cephalopod *Sepiola atlantica*. Other species which may be present occasionally include the bivalves *Ensis siliqua*, *Chamelea gallina* (*Venus gallina*) and *Abra alba*, the ophiuroid *Ophiura albida* and the echinoid *Echinocardium cordatum* (Warwick and Davies, 1977). Although the species recorded differ from those in the sand wave zone of the sand sheet facies in the Southern North Sea (Section 6.11.1), the fauna is similar in that epifaunal and most delicate species are absent, as is to be expected in the unstable sediment typical of an area of sand waves. Indeed *Spisula elliptica* is the only species commonly recorded that has any significant preservation potential. It generally lives for up to two and a half years (Warwick and George, 1980).

The faunas from the outer zones of this sand sheet beyond the zone of large sand waves have not yet been studied, but it is possible that some elements of this fauna will be analogous to those described in Section 6.10.5.

6.10.5 FAUNAS FROM THE RIPPLED MUDDY SANDS IN BAYS

Fine sands, some of which are muddy, are found in the bays on the north and south sides of the Bristol Channel. In Carmarthen Bay and in the outer part of the Bristol Channel the *Tellina* Sub-community (a second subdivision of the *Venus* Community, Table 6.1) is present. Characteristic species include the bivalves *Tellina fabula*, *Pharus legumen*, the asteroids *Astropecten irregularis*, *Asterias rubens* and the ophiuroid *Ophiura texturata*. Other species present include the amphipod crustaceans *Bathyporeia guilliamsoniana* and *Pontocrates arenarius*.

The presence of the bivalve *Donax vittatus* in the bays is thought to be related to the shore community in the tidal zone (Warwick and Davies, 1977). It is also present, however, in the zone of rippled sand in the Southern North Sea (Section 6.11.3).

6.11 Faunas associated with bedform zones in the Southern North Sea

Details of the faunas in the Southern North Sea sand sheet associated with the zone of large sand waves, the zone of small sand waves and the zone of rippled sand (Sections 4.3.2 and 5.2.4) used in this chapter have been largely derived from the extensive surveys by Davis (1923, 1925). Further data derived from Dutch trawling studies some fifty years later (van Noort, van Leeuwen and Creutzberg, 1979a–d) have also been incorporated. Problems arise when trying to compare the results of surveys in the same area taken at different times using different sampling methods (Ursin, 1960). Particular problems arise when comparing the results from grab sampling with those of dredging or trawling. For example, the asteroid *Asterias rubens* is obtained in most dredge or trawl samples in the Central and Southern North Sea (Ursin, 1960), sometimes in enormous numbers (van Noort *et al.*, 1979a–d) whereas it was never recorded during the extensive grab surveys (Davis, 1923, 1925).

6.11.1 FAUNAS FROM THE ZONE OF LARGE SAND WAVES

Available evidence suggests that the large sand waves with small sand waves on them (Sections 4.3.2 and 5.2.4) support a very meagre shelly fauna. In that part of the zone east of 3°E, Davis (1925) recorded only seven species from the area (Table 6.5) with a maximum of four recorded from any one station. If only those species with the highest preservation potential are considered, the shallow burrowing bivalve *Donax vittatus* is the only species present of any significance. Densities of 5–10 per m² were recorded from most stations. *Donax vittatus* is a strong shelled, active, bivalve commonly found in the sands of the surf zone on the lower shore (Section 6.10.5). It is also found in disturbed clean sands offshore. It is therefore well adapted to living in the unstable sands of these large sand waves.

Table 6.5 Numbers of species of polychaetes, molluscs, crustaceans, echinoderms and others present in the large sand wave, small sand wave and rippled sand zones, Southern North Sea (faunal data from Davis (1925)).

Group	Zone of large and small sand waves	Zone of small sand waves only	Zone of rippled sand
Polychaeta	2	6	16
Mollusca	3	10	21
Crustacea	0	3	13
Echinodermata	1	6	10
Others	1	0	4
Total no. of species	7	25	64

Densities of bivalve species vary from year to year and depend on the success of the spatfall (Warwick and George, 1980; Warwick *et al.*, 1978). *Donax vittatus* grows rapidly and can live for two to three and a half years (Warwick *et al.*, 1978; Ansell and Lagardère, 1980). Predation by the gastropod *Lunatia catena* (*Natica catena*) is the principal cause of death (Negus, 1975; Warwick *et al.*, 1978). As the *Donax vittatus* shells are robust they contribute most of the carbonate debris in the sand wave zone.

Other species occasionally present include the bivalves *Chamelea gallina* (*Venus gallina*), *Tellina fabula*, the asteroid *Astropecten irregularis* (Table 6.6) and the polychaetes *Nephthys caeca* and *Sthenelais limacola*. The sand eel *Ammodytes* sp. was recorded at one station. The relatively high degree of disturbance of the surface sediments in the sand wave zone by both tidal currents and storm waves probably accounts for the very low faunal diversity.

Results from the later survey (van Noort, van Leeuwen and Creutzberg, 1979a–d) also indicate a meagre fauna from this zone. Species recorded included the bivalves *Donax vittatus*, *Spisula elliptica* and *Ensis minor*, the ophiuroids *Ophiura albida* and *Ophiura texturata*, the decapod crustaceans *Philocherus trispinosus*, *Crangon vulgaris* and *Crangon allmani* and the hermit crab *Pagurus bernhardus*.

The echinoid *Echinocardium cordatum* is present in the large sand waves off the Dutch coast. Bioturbation resulting from its feeding activities (Section 6.3.1) frequently destroys cross-bedding in the sand waves (Houbolt, 1968).

6.11.2 FAUNAS FROM THE ZONE OF SMALL SAND WAVES

In contrast to the zone of large sand waves (Section 6.11.1) the faunal diversity in the zone of small sand waves is slightly higher (Tables 6.5 and 6.6). Potentially preservable species that are present include the bivalves *Tellina fabula*, *Gari fervensis*, *Phaxas pellucidus* (*Cultellus pellucidus*), *Thyasira flexuosa*, the irregular echinoids *Echinocyamus pusillus* and (less commonly) *Echinocardium cordatum* and *Echinocardium flavescens* (Davis, 1925). *Donax vittatus* was not recorded from this zone. This omission is likely to be a sampling problem rather than a true absence, however, as only a few of Davis's sample stations lie within this rather narrow zone. The numbers present at any one time are also dependent on the success of the spatfall in the previous years. The presence of the echinoid *Psammechinus miliaris* at one station is likewise thought to be fortuitous and

Table 6.6 Occurrence of certain mollusc and echinoderm species in the large sand wave, small sand wave and rippled sand zones, Southern North Sea (faunal data from Davis (1925)). Species names used are those in Appendix 6.1 and include some changes from those used in Davis (1925).

Species	Large and small sand waves	Small sand waves only	Rippled sand
Gastropoda			
Cylichna cylindracea	absent	rare	rare
Lunatia alderi	present	present	present
Bivalvia			
Abra prismatica	absent	absent	present
Chamelea gallina	rare	absent*	rare
Clausinella fasciata	absent	absent	rare
Donax vittatus	common	absent*	common
Dosinia exoleta	absent	absent	rare
Ensis ensis	absent	rare	common
Gari depressa	absent	absent	rare
Gari fervensis	absent	present	rare
Lutraria lutraria	absent	absent	fairly common
Mactra stultorum	absent	rare	abundant
Nucula nitidosa	absent	present	fairly common
Phaxas pellucidus	absent	present	present
Spisula solida	absent	rare	rare
Spisula subtruncata	absent	absent	common
Tellina fabula	rare	common	common
Thyasira flexuosa	absent	rare	rare
Asteroidea			
Astropecten irregularis	rare	absent*	rare
Ophiuroidea			
Amphiura chiajei	absent	present	rare
Amphiura filiformis	absent	present	present
Ophiura albida	absent	absent	present
Ophiura texturata	absent	rare	absent
Echinoidea			
Brissopsis lyrifera	absent	absent	rare†
Echinocardium cordatum	absent	rare	abundant
Echinocardium flavescens	absent	rare	present
Echinocyamus pusillus	absent	present	present
Psammechinus miliaris	absent	rare*	absent
Holothuroidea			
Cucumaria elongata	absent	absent	rare
Total no. of species present	5	17	27

* Abundance estimate probably not reliable as number of samples is insufficient.
† This is probably a temporary occurrence (Ursin, 1960).

does not indicate that this species is restricted to the zone. *P. miliaris* is not common in the Southern North Sea. Its density is less than 1 per m² (Ursin, 1960). The listing of *Echinocyamus pusillus* at some stations probably indicates the presence of some coarser sediments in places. Other species recorded include the gastropod *Cylichna cylindracea* and the polychaetes *Ophelia limacina* and *Glycera unicornis*.

Additional bivalve species recorded during the later survey (van Noort, van Leeuwen and Creutzberg, 1979a–d) include *Nucula nucleus*, *Spisula subtruncata* and *Chamelea gallina* (*Venus gallina*).

6.11.3 FAUNAS FROM THE ZONE OF RIPPLED SAND

The rippled sands (Section 4.3.2) support a rich fauna of molluscs, echinoderms, polychaetes and crustaceans. Sixty-four species are present over much of the area (Table 6.5). The dominant species are the bivalve *Mactra stultorum* (*Mactra corallina*) and the echinoid *Echinocardium cordatum*. The bivalves *Ensis ensis*, *Lutraria lutraria*, *Spisula subtruncata*, *Donax vittatus*, *Tellina fabula* and the ophiuroids *Amphiura filiformis* and *Ophiura albida* are also common (Table 6.6). Densities of *Mactra stultorum* (*Mactra corallina*) vary considerably throughout the area. In particular locations which (assuming comparable densities in the sediment between adjacent stations) may be up to 370 km^2 in extent, densities of 700 to 9500 per m^2 were reported (Davis, 1925). Birkett (1954) recorded a maximum density of *Mactra stultorum* of 2775 per m^2. Densities of *Echinocardium cordatum* recorded by Davis (1925) range from 5 to 10 per m^2. Other species recorded include the polychaetes *Nephthys caeca*, *Scoloplos armiger*, *Chaetopterus variopedatus*, *Notomastus latericeus*, *Pectinaria koreni* and *Pectinaria auricoma*, the amphipod crustaceans *Bathyporeia pelagica* and *Ampelisca brevicornis*, the decapod crustaceans *Callianassa subterranea*, *Pagurus bernhardus* and *Portunus depurator*. The asteroid *Astropecten irregularis*, the ophiuroids *Ophiura texturata* and *Amphiura chiajei*, the echinoid *Echinocyamus pusillus* and the bivalve *Nucula nitidosa* were recorded in the southern part in the transition area close to the zone of small sand waves.

Echinocardium cordatum can tolerate considerable variation in grain size within the zone of rippled sand. The large aggregations of adults present are derived from populations of larvae which aggregate prior to settlement (Buchanan, 1966).

6.11.4 FAUNAL DIFFERENCES FROM THE SAND WAVE ZONE TO THE ZONE OF RIPPLED SAND

The dramatic increase in faunal diversity along the sand sheet from the zones of large and small sand waves to the zone of rippled sand in the Southern North Sea (Section 5.2.4) is well illustrated by the increase in the number of species present in each zone from seven to sixty-four (Table 6.5) based on the data presented by Davis (1925). Many of the polychaete and small crustacean species present have a very low preservation potential. The increase in the faunal diversity can nevertheless be seen equally in those species with a high preservation potential such as the molluscs and echinoderms (Table 6.6). Differentiation of the two zones based on faunal and other sedimentological evidence should, therefore, be possible in the fossil record.

The small infaunal bivalve *Phaxas pellucidus* (*Cultellus pellucidus*) is recorded from the small sand wave and rippled sand zones (Davis, 1925) and it is widespread in the sands of the North Sea (Petersen, 1977). The predatory gastropod *Lunatia alderi* (*Natica alderi*) is also recorded from all three zones.

In a more recent study of North Sea faunas (Petersen, 1977), twenty-two species of bivalves were listed from the area that includes the zone of rippled sand. Of the fourteen relatively common species, only four – *Nucula tenuis*, *Dosinia lupinus*, *Thyasira flexuosa* and *Thracia phaseolina* – were not also recorded by Davis (1925). Petersen (1977) also recorded seven rare species (all with densities of less than 1 per m^2 and recorded from only a few stations) which were not listed by Davis (1925). These include the bivalves *Mysella bidentata*, *Arctica islandica* (*Cyprina islandica*), *Timoclea ovata* (*Venus ovata*), *Tellina tenuis*, *Thracia convexa* and *Cochlodesma praetenue*. As the area investigated by Petersen (1977) included only the zone of rippled sand and no new data were obtained from either the zone of large sand waves or the zone of small sand waves, these additional species from the zone of rippled sand are not included in the numbers listed in Table 6.5 which is based solely on data in Davis (1925). The results from the later survey (van Noort *et al.*, 1979a–d) also show an increase in the faunal density and diversity between the zone of large sand waves and the zone of rippled sand.

6.12 Faunas associated with bedform zones on the Atlantic continental shelf between Brittany and Scotland

Data on the faunas associated with the shell gravels in the strong current zones of the Fair Isle Channel between the Orkney and Shetland Islands, and on the rippled sands, sand patches and lag gravels in the weaker tidal current areas on the middle and outer parts of the continental shelf to the west and north of Scotland, have been obtained during Institute of Oceanographic Sciences investigations over the past decade.

6.12.1 FAUNAS FROM THE GRAVEL SHEET ZONE, FAIR ISLE CHANNEL

Over much of the region of stronger tidal currents in the Fair Isle Channel between the Orkney and Shetland Islands the sediments consist of shell gravels and coarse shell sands. Bivalve species which are common in these gravels include *Palliolum tigerinum* (*Chlamys tigrina*), *Circomphalus casina* (*Venus casina*), *Clausinella fasciata* (*Venus fasciata*), *Timoclea ovata* (*Venus ovata*), *Spisula elliptica*, *Gari costulata*, *Goodallia triangularis* (*Astarte triangularis*) and *Glycymeris glycymeris*. *Modiolus modiolus* is also common where the gravels are coarser. Its shells support an epifauna of barnacles, serpulid polychaetes and bryozoans (Plate 6.3). Gastropods present include *Rissoa parva*, *Hinia incrassata* (*Nassarius incrassatus*), *Bulbus islandicus* (*Amauropsis islandica*) and *Lunatia montagui* (*Natica montagui*) (Plate 6.3).

6.12.2 FAUNAS FROM THE RIPPLED SAND ZONE

The rippled sand zone of the outer parts of the continental shelf in the weak current areas west of Brittany, in the western side of the Celtic Sea, and west of Scotland support a characteristic fauna containing elements which appear to be largely restricted to that environment and can, therefore, be regarded as 'indicator species' for rippled sand zones on open ocean-facing continental shelves when they occur together.

(a) *Principal corbonate producers* – Ditrupa and Caryophyllia

The most important of these 'indicator species' is the serpulid polychaete *Ditrupa arietina*. Its distinctive tubes have been dredged in large quantities from areas of rippled sand and also from sand patches on the continental shelf around the British Isles. Off France, these areas of rippled sand on the outer part of the continental shelf are referred to as the 'sables à alènes' (the shape of the *Ditrupa* tube resembles a curved needle) because of the abundance of the *Ditrupa* tubes (Glémarec, 1969a, b, 1973). *Ditrupa arietina* tubes are frequently confused with shells of the scaphopod *Antalis* sp. (*Dentalium* sp.). However, its shape and structure are quite different from those of *Antalis* sp. (*Dentalium* sp.) (Plate 6.4).

Reliable data on the density and distribution of living *Ditrupa arietina* in the sand sheets and sand patches are hard to obtain. Dead tubes appear to be much more widely distributed. Evidence from grab and dredge samples from individual sand patches on the continental shelf west of Scotland does, however, indicate that *D. arietina* lives in discrete patches where the density can be as high as 1600 per m^2. The ratio of living to dead tubes in one sample was almost 1 to 1. Living *D. arietina* appear to be most common in the sand sheets and sand patches although they also occur in the fine sands associated with the infilling of iceberg plough marks on the edge of the continental shelf and upper part of the continental slope (Plate 1.2). Although very fresh looking tubes are present in grab samples obtained from the gravels between the sand patches, no live *D. arietina* have so far been obtained. It is not yet possible, therefore, to say whether or not *D. arietina* can also live successfully on a gravel floor.

Estimates of the numbers of empty *D. arietina* tubes, based on grab samples from the rippled sands on the continental shelf west of Scotland, suggest that densities of at least 3000 complete dead tubes per m^2 are not uncommon. *D. arietina* fragments and complete tubes commonly account for 50–70% and sometimes up to 94% by weight of the total carbonate present in the sand (Section 6.18.2, Plate 6.5).

The *D. arietina* tubes act as a substrate for the

settlement and growth of other animals including the coral *Caryophyllia smithii*. Planulae that settle and develop on tubes containing living worms are kept at the sediment surface by the worm and derive support during the critical phases of early growth (J. B. Wilson, 1976). Evidence from the distribution of *Caryophyllia smithii* attached to *D. arietina* on the continental shelf west of Scotland (J. B. Wilson, 1975) and from the Celtic Sea suggests that it is largely restricted to the zone of rippled sand on an ocean facing continental shelf. The areas of rippled sand on the continental shelf west of Scotland are associated with depressions on the floor and may be fed by the advance and gradual coalescence of sand patches from the adjacent shelf. *Caryophyllia* on *Ditrupa* is also recorded from the continental shelf off Brittany (Glémarec, 1969a, b). It is also the narrow based form similar to that on the continental shelf west of Scotland and in the Celtic Sea (but listed as *Caryophyllia clavus* by Glémarec).

After the settlement and early growth of the *Caryophyllia smithii* the tubes tend to become buried in the sediment and after the death of the worm the tubes are attacked by boring algae and fungi (J. B. Wilson, 1976). *D. arietina* tubes which are not colonized by *Caryophyllia smithii* also become buried in the sediment presumably by the lateral migration of sand on the surface of the sand sheet. The tubes become discoloured by the algae and fungi (Plate 6.4) and eventually they break into several segments which in turn may break longitudinally into halves. The >2 mm fraction from the sand sheet is in places composed almost entirely of such fragments (Plate 6.5). Evidence from radiocarbon dating of both fresh tubes and bored, discoloured tubes indicates that the discolouration of the tubes by algal and fungal boring takes place in just a few years.

Other species supported by *D. arietina* tubes which are sometimes present in the zones of rippled sand include the serpulid *Hydroides* sp., the barnacle *Verruca stroemia*, and the foraminifera *Cibicides refulgens* and *Acervulina* sp. and several species of bryozoans. The majority of empty *D. arietina* tubes have patches of discolouration which indicate the points of attachment of foraminifera or encrusting bryozoans which are no longer attached to the tube. As the attachment of the bryozoan depends on the adhesive properties of a multi-polysaccharide periostracum (Taverner-Smith and Williams, 1972) rather than direct attachment of the skeletal carbonate to the tube, it is possible that they are more readily detached than are corals. In the comparatively few cases where bryozoans are present on the unbroken empty tubes, they are generally attached to one side only implying that empty *D. arietina* tubes on the surface of the sea floor remain undisturbed by bottom currents or burial by migrating sand ripples for at least long enough for the colony to become established and to grow. Estimates based on measured bryozoan growth rates imply stability of the order of 100 days from settlement. The broken portions of tube, which can roll more easily, have bryozoan colonies such as *Pyripora catenularia* and *Alderina imbellis* consisting of lines of individuals which completely encircle the tube or run along its length.

The empty tubes also act as habitats for the sipunculan *Golfingia minuta*, the amphipod crustacean *Siphonoecetes* sp. and as temporary refuges for certain errant polychaetes. In one sample from the continental shelf west of Scotland, 14% of the empty tubes examined contained sipunculans and 3% contained amphipods.

The eunicid polychaete *Hyalinoecia tubicola* is also an important member of the rippled sand zone fauna. Densities of up to 40–50 per m^2 have been obtained from the continental shelf west of Scotland. As the tube is thin and non-calcareous, however, it is unlikely to survive burial or transport and is therefore unlikely to be preserved. Other portions of tubes from species such as *Hyalinoecia bilineata* could perhaps be preserved, however, as the tube in this species is strengthened with gravel and often has small pieces of *Ditrupa arietina* tube and other shell fragments embedded in it. *Ditrupa arietina* fragments are also sometimes incorporated into the tubes of terebellid polychaetes (McIntosh, 1913). The scaphopod *Antalis entalis* (*Dentalium entalis*) is also present in the sand patches and it occurs in 25% of the samples taken north of Brittany. Its occurrence on the continental shelf west of Scotland appears to be broadly comparable to this although empty shells and fragments are present in most sand sheet and sand patch samples. The empty scaphopod shells are often colonized by the sipunculan *Phascolion strombi*. They are carried on the sediment surface by the sipunculan and act as a substrate for the attachment of *Caryophyllia smithii* and *Hydroides norvegica* (J. B. Wilson, 1976). Other empty scaphopod shells are utilized by hermit crabs.

(b) *Principal carbonate producers – Molluscs, Echinoderms and Bryozoans*

Molluscs recorded from the areas of rippled sand off Brittany include *Circomphalus casina* (*Venus casina*), *Astarte sulcata*, *Palliolum simile* (*Chlamys similis*) and *Lunatia montagui* (*Natica montagui*) (Glémarec, 1969a, b).

Shells, valves and fragments from thirteen gastropod species and twenty-nine species of bivalves (Plate 6.6) have been recorded from the sand sheets on the continental shelf west of Scotland. The more important of these include the gastropods *Aporrhais serresiana*, *Lunatia alderi* (*Natica alderi*), *Lunatia montagui* (*Natica montagui*), *Turritella communis*, *Acteon tornatilis* and *Cylichna* sp., the bivalves *Portlandia philippiana* (*Yoldia tomlini*), *Cardiomya costellata* (*Cuspidaria costellata*), *Palliolum simile* (*Chlamys similis*), *Pandora pinna*, *Palliolum tigerinum* (*Chlamys tigrina*), *Abra prismatica*, *Tridonta elliptica* (*Astarte elliptica*), *Timoclea ovata* (*Venus ovata*), *Dosinia lupinus* and the scaphopod *Antalis* sp. (*Dentalium* sp.). Predation by *Lunatia* (*Natica*) is widespread in the sand sheets. Up to 20% of the *Ditrupa* tubes and many of the bivalve valves and gastropod shells in samples from the continental shelf west of Scotland were bored by *Lunatia* sp. (*Natica* sp.).

The irregular echinoid *Echinocyamus pusillus* is present but not common on the sand sheets. Up to 30 per m^2 have been recorded from the continental shelf west of Scotland. Dead tests are present in most sand sheet samples, however. Fragments from the tests of larger irregular echinoids are also fairly common in the sediment. The presence of valves and fragments from the commensal bivalve *Tellimya ferruginosa* (*Montacuta ferruginosa*) associated with these fragments implies that some probably came from *Echinocardium cordatum*.

The bryozoan *Cellaria fistulosa* is occasionally found if a suitable substrate for attachment such as hydroid thecae and tubes of the polychaete *Chaetopterus variopedatus* are present. It can also root itself into the sand.

Inarticulate and articulate brachiopod valves and fragments, plates from the barnacle *Balanus balanus*, and valves and fragments of the bivalve *Anomia* sp. are occasionally found in the sediment. It is possible that these have been derived from isolated cobbles or large shell fragments on the surface of the sand as there is, in most cases, little evidence of abrasion. They could, however, also be derived from the gravel or cobbles adjacent to the sand sheet. Fragments from regular echinoids are rare and when present are generally very worn and eroded, suggesting transport from at least the adjacent gravel.

(c) *Faunal evidence for the edge of a rippled sand zone*

Towards the edge of the rippled sand zone of a sand sheet and also the edge of a sand patch where the sand cover on the cobble or gravel floor is thin, large boulders may protrude through the sand. The exposed surfaces of these boulders support an epifauna of encrusting and erect bryozoans, serpulids and inarticulate brachiopods (Fig. 6.7). The cobbles covered by the sand patch carry the worn and eroded remains of the epifauna that they supported before they became buried by the sand. Generally all that can be seen of this former epifauna are some eroded portions of attached serpulid tubes, barnacle basal discs and the cemented valves of the inarticulate brachiopod *Crania anomala*. Most of the delicate encrusting bryozoan growth on these cobbles has been removed by abrasion from the sand grains.

6.12.3 FAUNAS ASSOCIATED WITH GRAVELS IN WEAK CURRENT AREAS WEST OF SCOTLAND

Examination of pebbles and granules from the gravels associated with the sand patches on the continental shelf off Scotland from areas where the near surface mean spring peak tidal currents are less than 50 cm/s shows that most of the smaller clasts do not support a significant epifauna. This suggests that intermittent movement of sand grains even in areas of weak tidal current activity may be sufficient to

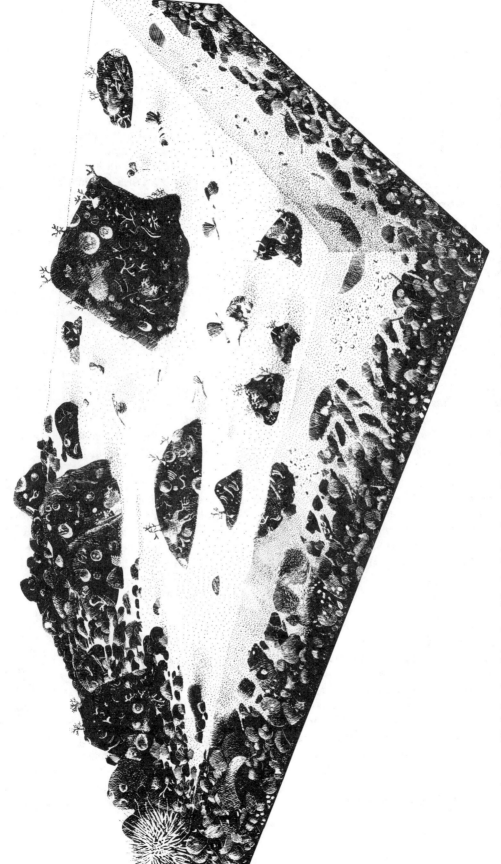

Fig. 6.7 Block drawing of the edge of the rippled sand zone of a sand sheet on the continental shelf west of Scotland. The cobble and gravel floor supports an epifauna of encrusting and erect bryozoans, inarticulate brachiopods, serpulids, attached bivalves and the occasional regular echinoid. Towards the edge of the sand sheet where the sand cover is thin, the larger boulders still protrude above the sand surface and support a similar living epifauna. The remains of the more robust and durable elements of the former epifauna on the now buried cobbles can still be seen and include eroded serpulid tubes and the remains of cemented valves of the inarticulate brachiopod *Crania anomala*. The sand supports a fauna including the serpulid polychaete *Ditrupa arietina* and the coral *Caryophyllia smithii*.

prevent the settlement of some epifaunal larvae. However, the larger cobbles and small boulders associated with these gravels which are resting on the gravel surface do support a fauna of encrusting bryozoans and serpulid tubes. Molluscs associated with the gravel include the bivalves *Tridonta elliptica* (*Astarte elliptica*), *Astarte sulcata*, *Timoclea ovata* (*Venus ovata*), *Bathyarca pectunucloides* and the gastropods *Lunatia montagui* (*Natica montagui*), *Cytharella coarctata* (*Mangelia coarctata*) and *Teretia anceps* (*Philibertia teres*).

6.13 Faunas of active sand banks

Evidence from gravity cores (Houbolt, 1968) and box cores (Reineck, 1963) indicates the presence of the irregular echinoid *Echinocardium cordatum* in many of the sand banks off Norfolk, in the eastern part of the English Channel and in the German Bight (Section 3.4.5). *E. cordatum* lives at the bottom of a vertical mucus-lined burrow (Schäfer, 1962). Bioturbation by *E. cordatum* resulting in the characteristic 'cut and fill' feeding trails (Reineck, 1963; Reineck et al., 1968; Schäfer, 1962) effectively destroys any cross-stratification present in the sediment within the burrow in the zone 15–20 cm below the surface (Fig. 6.1, Section 6.3.1). Some offshore populations of *E. cordatum* in the western part of the North Sea live in shallower burrows about 2 cm below the sediment surface (Buchanan, 1966). Feeding activities into the surrounding sediment can extend up to 50–60 cm from the burrow. When the echinoid moves on, the burrow is abandoned and it fills with bedded sand and other detritus (Schäfer, 1962). With high densities of *E. cordatum*, virtually all the sediment in the feeding zone – which can be up to 15–20 cm below the surface – is disturbed by bioturbation over a wide area (Section 6.3.1).

Some cores from the relatively steep east side of Well Bank (Houbolt, 1968) indicate complete disturbance of the sediment by *E. cordatum* from the surface to depths of 55–60 cm. On the gently sloping west side of Well Bank, the top of Brown Ridge and in the sands on the deeper floor between the sand banks, the top 10 cm or so of the sediment is cross-bedded and overlies a zone of bioturbation by *E. cordatum* (Houbolt, 1968). As this bioturbation is at the depth at which contemporaneous disturbance by *E. cordatum* could be expected, it is probable that this disturbance is related to the present surface and the preservation of the cross-bedding in the top 10 cm of the cores shows that no vertical burrows were sampled. During severe storms, when the sediments are extensively disturbed by wave action, *E. cordatum* can be washed out of its burrow onto the surface of the sand. Those that survive this disturbance have to burrow again into the sediment.

On the tops of the sand banks, the sediments are also disturbed by the burrowing activities of large populations of sand eels (*Ammodytes marinus* and *Hyperoplus lanceolatus* (*Ammodytes lanceolatus*)). There is some evidence to suggest that this activity takes place primarily during the winter months (Macer, 1966).

Available evidence suggests that the fauna of the Norfolk sand banks is similar to that present in the zone of large sand waves in the Southern North Sea (Section 6.11.1). These sediments generally contain about 5% carbonate debris consisting mainly of bivalves and fragments (Houbolt, 1968). These are probably derived from molluscs living in the sediments of the bank. The low carbonate percentage again indicates the paucity of the fauna living in these active sand banks.

Up to 20% carbonate was occasionally recorded in the sediment of the Flemish sand banks while in the Outer Gabbard sand bank, mollusc debris accounted for up to, and in some cases over, 50% of the sediment (Houbolt, 1968).

Coarser lag deposits are found between the sand banks which contain bivalve valves and large fragments of sabellarid colonies (Houbolt, 1968).

The sand floor between the Norfolk banks supports a fauna including the ophiuroid *Ophiura albida*, the polychaetes *Pectinaria koreni*, *Owenia fusiformis*, and *Sabellaria* sp., the bivalves *Spisula solida* and *Abra alba* and the gastropod *Lunatia alderi* (*Natica alderi*) in addition to *Echinocardium* sp. (Davis, 1925).

The inshore linear sand banks in the Bristol Channel support an impoverished fauna (Tyler and Shackley, 1980). None of the species present have a high preservation potential. They include the mysid crustacean *Gastrosaccus spinifer*, the amphipod

154 Offshore Tidal Sands

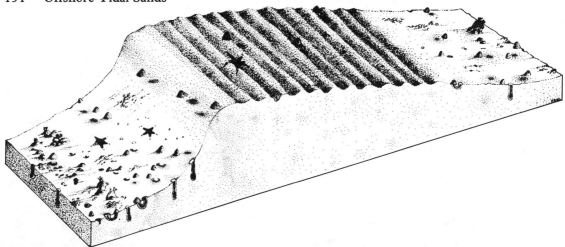

Fig. 6.8 Block drawing of a stable sand wave in the Kiel Bight showing colonization of the trough by burrowing polychaetes and the bivalve *Mya arenaria* (after Werner, Arntz and Tauchgruppe Kiel, 1974).

crustacean *Pontocrates arenarius* and the polychaete *Nephthys cirrosa*. *Gastrosaccus spinifer* and *Nephthys cirrosa* are also present in the zone of large sand waves in the Bristol Channel (Section 6.10.4). *Pontocrates arenarius* is present in the zone of rippled sands in the bays (Section 6.10.5).

The gravel floor adjacent to the Bristol Channel sand banks supports a fauna including the polychaete *Sabellaria* sp. indicating much more stable conditions.

The limited faunal data available from the Norfolk sand banks and the Bristol Channel sand banks suggests that in each region the faunas in active sand banks are generally somewhat similar to those in the nearby zones of active sand waves.

There are insufficient data as yet to draw any conclusions as to relative differences in faunal diversity between active sand banks and sand sheets with large sand waves on them. The ability to distinguish between them may therefore depend almost entirely on internal structure (Sections 5.2.4 and 5.3.3).

6.14 Faunal evidence for stability of sand waves

The faunal diversity is low where sand waves are particularly active (Sections 6.10.4, 6.11.1) and the faunal densities of the few species that are present are also low.

The active sand waves on Warts Bank, south-west of the Isle of Man in the Irish Sea, support an impoverished fauna including the sand eels *Hyperoplus lanceolatus* (*Ammodytes lanceolatus*) and *Gymnammodytes semisquamatus* and the hermit crab *Pagurus bernhardus*. The crests of these sand waves move between 5 and 10 cm per day on average during the summer months with movement of up to 74 cm recorded for three sand wave crests over one flood tide (Jones, Kain and Stride, 1965). Migration rates such as these clearly discourage the development of an infauna within the sediments forming the bank.

If, however, the sand waves are only moving occasionally, or if the sand waves are separated by wide areas of gravel floor, an infauna may become established in the troughs. For example, the tube building terebellid polychaete *Lanice conchilega* has been reported from sand wave troughs in the German Bight (J. Ulrich, pers. comm.). Another example from the virtually non-tidal Baltic Sea has been described from the very shallow waters in the Kiel Bight, where there are populations of the bivalve *Mya arenaria* with densities greater than 100 per m^2 in the troughs between the sand waves which are moved by only occasional unidirectional currents (Werner, Arntz and Tauchgruppe Kiel, 1974). On some sand waves, which have been stable for a sufficient period of time, the population became established over the whole surface of the sand wave with densities of *Mya arenaria* of 300–400 per m^2 being recorded from the troughs (Fig. 6.8). The age class structure of the population reflects the length of time since the sand waves last moved.

Although this example is taken from a predominantly non-tidal area, the principle should nevertheless be applicable to sand wave zones in tidal current areas to a limited extent.

In areas where the sand waves are separated by lag gravels or boulder and cobble beds the presence of epifauna on the cobbles could also indicate slow migration of the sand waves. If, however, no epifauna is present this would be good evidence of the frequent movement of the sand waves.

6.15 Faunas as environmental indicators

The use of faunas as environmental indicators is of the greatest importance to the geologist who is interpreting the conditions of deposition of particular sedimentary formations of past seas. The literature on the use of faunal evidence in environmental and palaeoenvironmental interpretation is very extensive (for an entry into the literature see the books by Ladd, 1957; Imbrie and Newell, 1964; Hecker, 1965; McKerrow, 1978; Rigby and Hamblin, 1972).

Several examples of faunal differences between open shelf environments and partially enclosed shelf sea environments for modern tidal seas have been given in this chapter. Carbonates in shallow nearshore waters west of Scotland and Ireland and in the western half of the English Channel frequently contain calcareous algae (Section 6.6.1) whereas sediments in the nearshore waters of the North Sea do not. The fauna of the rippled sand zone of the North Sea (Section 6.11.3), differs from that found on the open oceanic shelf (Section 6.12.2). The faunal differences between the sand wave zones in the Bristol Channel (Section 6.10.4) and the Southern North Sea (Section 6.11.1) may reflect the differences in distance to the open ocean represented in these two locations. Further examples of the use of faunal assemblages to determine differences between faunas in adjacent sand transport paths (Section 6.15.1), to determine the proximity of an open ocean (Section 6.15.2) and to indicate the location of the continental shelf edge and upper continental slope (Section 6.15.3) are given below.

6.15.1 FAUNAL DIFFERENCES BETWEEN ADJACENT SAND TRANSPORT PATHS

In areas such as the south-east African continental shelf (Section 4.4.7, Fig. 4.18) where the bed-load parting in a predominantly non-tidal region coincides with the boundary of a major eddy, the faunas associated with the transport paths on either side of the bed-load parting should be somewhat different. Such faunal differences between transport paths diverging from a bed-load parting detected in the fossil record could be used to provide evidence of the palaeo-circulation in ancient shelf seas.

In tidal areas the faunal differences between adjacent sand transport paths should be less distinct with more species common to both transport paths. Faunal differences between the Western English Channel sand transport path and the Eastern English Channel sand transport path are partly dependent on circulation patterns, stratification, etc. in the overlying water column. The faunal diversity in the Western Channel is much greater than that in the Eastern English Channel (L. Cabioch et al., 1977; Holme, 1961, 1966). Species present throughout the English Channel include the bivalves *Paphia rhomboides* (*Venerupis rhomboides*), *Tellina donacina*, *Glycymeris glycymeris*, *Aequipecten opercularis* (*Chlamys opercularis*), *Ensis ensis*, *Ensis arcuatus*, *Nucula hanleyi*, *Nucula nitidosa*, *Thyasira flexuosa*, *Abra alba*, *Gari fervensis*, *Phaxas pellucidus* (*Cultellus pellucidus*), *Mactra stultorum* (*Mactra corallina*), *Lutraria lutraria*, *Lutraria angustior*, *Dosinia exoleta*, *Gari tellinella* and *Tellina pygmaea*, the gastropods *Hinia incrassata* (*Nassarius incrassatus*) and *Lunatia alderi* (*Natica alderi*), the echinoids *Echinocardium cordatum*, *Echinocyamus pusillus* and *Spatangus purpureus*, the ophiuroids *Ophiothrix fragilis*, *Ophiura albida* and *Ophiura texturata*, the bryozoans *Schizomavella auriculata*, *Schizomavella linearis*, *Porella concinna*, *Parasmittina trispinosa*, *Cellepora pumicosa* and *Turbicellepora avicularis* and the serpulid polychaetes *Protula tubularia* and *Serpula vermicularis* (L. Cabioch et al., 1977; Holme, 1961, 1966). Some of these listed are more common on one side of the Channel than on the other (Holme, 1966).

Species restricted to the Western English Channel include species grouped as Western, West Channel,

Sarnian and Cornubian (Holme, 1961, 1966) and include the bivalves *Arca tetragona*, *Palliolum tigerinum* (*Chlamys tigrina*), *Limaria hians* (*Lima hians*), *Limaria loscombi* (*Lima loscombi*), *Arctica islandica* (*Cyprina islandica*), *Dosinia lupinus*, *Circomphalus casina* (*Venus casina*), *Chamelea gallina* (*Venus gallina*), *Clausinella fasciata* (*Venus fasciata*), *Solecurtus chamasolen*, *Astarte sulcata*, *Lucinoma borealis*, *Myrtea spinifera*, *Gouldia minima* (*Gafrarium minimum*), *Gari costulata*, *Pododesmus patelliformis* (*Monia patelliformis*), the gastropods *Turritella communis*, *Hinia pygmaea* (*Nassarius pygmaeus*), the articulate brachiopod *Terebratulina retusa* (*Terebratulina caputserpentis*) which is largely restricted to the southern side of the Channel, the ophiuroid *Ophiocomina nigra*, the bryozoans *Cellaria fistulosa*, *Cellaria sinuosa*, *Cellaria salicornioides*, *Palmicellaria skenei*, *Porella compressa*, *Hornera lichenoides* and the coral *Caryophyllia smithii* (L. Cabioch et al., 1977; Holme, 1961, 1966).

Only a few species are restricted to the Eastern English Channel. They include the bivalves *Mya truncata* and *Saxicavella jeffreysi*. The bryozoan *Flustra foliacea* is commonest in the eastern part of the Channel although it is also present towards the west (L. Cabioch et al., 1977; Holme, 1966). Dead shells of the bivalve *Spisula elliptica* are common in the Western English Channel, although it is mostly found live at present in the Eastern English Channel (Holme, 1966).

6.15.2 THE PROXIMITY OF THE OPEN OCEAN

The faunal assemblages from the areas of rippled sand on the middle and outer parts of the continental shelf to the west of Scotland, described in Section 6.12.2 may be regarded as indicative of the presence of an ocean facing continental shelf. Further evidence of the proximity to, or the isolation from, an open ocean may be obtained from consideration of the ratio of planktonic to benthonic foraminifera and the relative size and the diversity of planktonic foraminifera in bottom sediments (Murray, 1976). On the continental slope, the proportion of planktonic foraminifera present in the sediments is much greater than it is closer inshore. In the Celtic Sea, for example, values of the planktonic/benthic ratio were greater than 60% on the outer continental shelf but only 10% on the middle part of the continental shelf. There is also a marked reduction in the size of the planktonic foraminifera present from the outer part of the continental shelf to the middle continental shelf. In general, the upper part of the continental slope is characterized by wide ranges of sizes of both adults and juveniles and a high planktonic/benthic ratio (greater than 70%). The outer part of the continental shelf has a slightly lower diversity with juveniles and adults of some species and juveniles only of others and a planktonic/benthic ratio in the range 40–70%. The middle part of the continental shelf has low diversity with a high proportion of juveniles and a planktonic/benthic ratio in the range 10–60%. The inner part of the continental shelf has very low diversity with almost all juveniles and a planktonic/benthic ratio of less than 20% (Murray, 1976).

6.15.3 THE EDGE OF THE CONTINENTAL SHELF

The outer edge of a continental shelf can be indicated by a faunal assemblage which differs from the assemblage on the adjacent continental slope and from the assemblage on the inner part of the continental shelf. For example, the continental shelf edge and the upper part of the continental slope to the north-west of Scotland, to the west of Ireland and the flanks of large shoal areas such as Rockall Bank, Faeroe Bank, etc. support a fauna which includes the deep-water coral *Lophelia pertusa* (*Lophelia prolifera*) (Teichert, 1958; J. B. Wilson, 1979a). On the upper continental slope west of Scotland and on Rockall Bank it is most common at depths of 220 m–250 m and occurs as discrete 'patches' which may be up to 40 to 50 m across. Individual colonies within the 'patch' are generally 1 to 1½ m in height (Plates 6.7 and 6.8). As the coral patch develops, the boring activities of clionid sponges detach portions of living colonies and ultimately break down whole colonies. The resulting coral debris acts as a substrate for a rich fauna of bryozoans, serpulids, etc. as well as a substrate for further growth of the coral itself (J. B. Wilson, 1979a, b). These coral 'patches' therefore, provide substrates for epifaunal communities on the edge of the continental shelf and upper part of the continental slope where other substrates such as rock outcrops or boulders are relatively uncommon. In the deeply dissected canyon heads on the edge of the continental shelf west of the Celtic Sea, rock outcrops and coarser sediments do occur.

Other species generally found on the continental shelf edge and upper part of the continental slope to the west and north of Scotland include the echinoids *Spatangus raschii* (J. B. Wilson, 1977, Fig. 12), *Echinus elegans* and *Cidaris cidaris*. The spines of *Cidaris* themselves provide a substrate for colonization by bryozoans, the barnacle *Verruca stroemia* and the goose barnacle *Lepas* sp. Other echinoids including *Echinus acutus* and *Echinocyamus pusillus* are also found at the edge of the continental shelf but they also occur elsewhere on the continental shelf and continental slope.

Of the asteroid species present, only *Pseudarchaster parelii* can be regarded as an indicator species for the edge of the continental shelf. None of the ophiuroids present are restricted to the edge of the continental shelf.

The bivalve *Astarte sulcata* is very common on the continental shelf edge but it also occurs in small numbers in the gravels associated with sand patches in the weak current areas and in the zones of rippled sands on the outer continental shelf (Section 6.12.3). Other continental shelf edge bivalves found to the west and north of Scotland include the epibyssate forms *Bentharca nodulosa*, which is generally attached to pebbles, *Limopsis minuta* which may be attached to pebbles or to coral debris, and *Limopsis aurita* which lives either partly buried in the sediment or just at the surface, loosely attached by its single byssal strand (Oliver and Allen, 1980a, b). Other bivalves include *Acropagia balaustina* (*Tellina balaustina*), *Cardiomya costellata* (*Cuspidaria costellata*), *Chlamys sulcata*, *Manupecten alicei* (*Chlamys alicei*) and *Propeamussium hoskynsi* (*Chlamys hoskynsi*).

The gastropod fauna at the edge of the continental shelf west and north of Scotland is very diverse and includes *Troschelia bernicienesis*, *Colus howsei*, *Solariella amabilis*, *Jujubinus clelandi* (*Cantharidus clelandi*), *Cylichna alba* and *Typhlomangelia nivalis*. A selection of bivalves and gastropods characteristic of the edge of the continental shelf, including these and other species, is illustrated in Plate 6.9.

It should be noted that the edge of the continental shelf to the west and north of Scotland differs from that to the south-west of Ireland in that it is not gullied and dissected by submarine canyons. Any downslope transport of sediment must therefore be minimal.

The association of these coral, echinoderm and mollusc species on the edge of the continental shelf to the west and north of Scotland appears to be diagnostic of such conditions in that area. In other parts of the world, however, certain of these species occur in areas other than the edge of the continental shelf or upper part of the continental slope. For example, *Lophelia pertusa* (*Lophelia prolifera*) occurs in much deeper water in the Bay of Biscay and off south west Ireland, in the Straits of Florida and on the Blake Plateau (Joubin, 1922; Neumann, Kofoed and Keller, 1977; Stetson, Squires and Pratt, 1962). The bivalves *Bentharca nodulosa*, *Limopsis aurita* and *Limopsis minuta* occur at 20 m depth in the Norwegian fjords and in progressively deeper waters further south towards the Bay of Biscay where they are found at depths down to 600 m–1050 m (Oliver and Allen, 1980a, b).

6.16 Factors determining the faunal composition of death assemblages in shell gravels

The fossil record provides the geologist with the preserved remains of the hard parts of some members of the faunas that inhabit the bedforms and deposits on tidal current dominated continental shelves. Many factors affect the faunal composition of these assemblages of hard parts. A brief account of some of those of relevance to the present study is given below.

6.16.1 PREDATION ON SHELL BEARING INVERTEBRATE FAUNAS

Predation is a major factor in the release of shells and shell fragments into the sedimentary environment. Echinoderms, especially asteroids, the larger crustaceans such as crabs and lobsters, gastropods and benthic and benthopelagic fish are the principal predators that are active on continental shelves.

Mollusc shells are often damaged by the large

scavenging crustaceans during their feeding activities. In some cases the crustacean 'peels' (Vermeij, 1978, Fig. 2.13) or crushes (Vermeij, 1978, Figs 2.11, 2.12, 2.14) the gastropod or bivalve shell. Sometimes the mollusc survives the attack. Many gastropod shells show evidence of repeated unsuccessful attacks (Vermeij, 1978, Fig. 2.17). Claw, leg and carapace fragments from these predatory crustaceans are occasionally found in the sediment.

Many bivalve valves display the small circular holes indicative of attack by predatory gastropods such as *Lunatia* sp. (*Natica* sp.), *Nucella lapillus* and *Urosalpinx cinerea* (Carriker and Yochelson, 1968; Carriker, Scott and Martin, 1963; Carriker, 1978; Carter, 1968; Schäfer, 1972). Other gastropods such as *Buccinum undatum* use the lips of their shells to force the valves of a bivalve apart. This generally damages the margin of the bivalve shell (Carter, 1968).

The cephalopod *Octopus vulgaris* feeds on gastropods. It gains entry into the shell by boring (Nixon, Maconnachie and Howell, 1980; Wodinsky, 1969).

Asteroids are known to prey on species in most of the major groups of bivalves. Shells are often released undamaged into the sediment by the processes of starfish predation although some valves may be cracked (Carter, 1968).

The quantitative effects of predation by benthopelagic and benthic fish on both infaunas and epifaunas are hard to assess. Several species feed on infaunal bivalves such as *Scrobicularia plana*, *Mya arenaria*, *Lutraria lutraria*, *Ensis* sp. and *Pharus legumen* by cropping the siphons (Ford, 1925). Siphons from the bivalve *Pharus legumen* were recorded in the stomachs of the dab (*Limanda limanda*), the solenette (*Buglossidium luteum*), the sand goby (*Pomatoschistus minutus*) and the dragonette (*Callionymus lyra*) (Warwick et al., 1978). As the siphons can regenerate in most cases, the fish population can be sustained without killing the infaunal bivalve population.

Numerous other species of mollusc, echinoderm and crustacean form part of the diet of many species of benthopelagic fish on the European continental shelf (Carter, 1968; Eisma, 1966; Ford, 1925; Steven, 1930; Wheeler, 1969).

Lemon sole (*Microstomus kitt*) prey on a wide range of bottom invertebrates including the polychaetes *Serpula vermicularis*, *Hydroides norvegica*, *Pomatoceras triqueter*, *Ditrupa arietina*, *Placostegus tridentatus*, *Hyalinoecia tubicola*, the gastropods *Lunatia* sp. (*Natica* sp.), *Iothia fulva* (*Lepeta fulva*), *Emarginula fissura* (*Emarginula reticulata*), young individuals of the bivalves *Ensis* sp. and *Pecten* sp., and the ophiuroids *Ophiura affinis* and *Ophiura albida* (Rae, 1956). The sipunculan *Phascolion strombi* and the hermit crab *Pagurus* sp. are also important elements in the diet of the sole (Rae, 1956). These must have been obtained from the gastropod or scaphopod 'host' shells that were being used by the hermit crabs and the sipunculans.

Stomach contents of fish such as plaice (*Pleuronectes platessa*) and the Dover sole (*Solea solea*) generally contain mollusc shell fragments. Calculations show that an individual plaice may eat approximately 70 molluscs each week and that the number of molluscs eaten by the plaice population in an area of 7650×10^6 m^2 off the Hook of Holland amounted to 98×10^9 molluscs each week (Eisma, 1968).

The echinoid *Echinocyamus pusillus*, ophiuroids including *Ophiura albida*, the scaphopod *Antalis* sp. (*Dentalium* sp.), and the bivalves *Palliolum simile* (*Chlamys similis*) and *Abra prismatica* are all recorded as part of the diet of the haddock (*Melanogrammus aeglefinus*) from the west of Scotland continental shelf (Jones, 1954; Ritchie, 1937). The bivalves *Abra prismatica* and *Abra alba* are part of the diet of the whiting (*Merlangius merlangus*) (Jones, 1954). Many species of mollusc including the gastropods *Lunatia* sp. (*Natica* sp.), the scaphopod *Antalis* sp. (*Dentalium* sp.) and the bivalves *Nucula* sp., *Glycymeris glycymeris*, *Modiolus* sp., *Aequipecten opercularis* (*Chlamys opercularis*), *Limaria* sp. (*Lima* sp.), *Arctica* sp., *Dosinia* sp. and *Donax* sp. and the echinoid *Spatangus purpureus* form part of the diet of the cod (*Gadus morhua*) (Rae, 1967).

Antalis sp. (*Dentalium* sp.), *Ditrupa arietina*, *Echinocyamus pusillus* and *Ophiura* sp. have been recorded from the megrim (*Lepiodorhombus whiffiagonis*), although in this case it is possible that they may have been incidentally acquired from the stomachs of fish eaten by the megrim (Rae, 1963).

Ophiuroids form part of the diet of the lemon sole (*Microstomus kitt*), the plaice (*Pleuronectes platessa*), the Dover sole (*Solea solea*), the cod (*Gadus morhua*), the haddock (*Melanogrammus aeglefinus*) and the long rough dab (*Hippogloides platessoides*) (Lande, 1976; Warner, 1971). Fish stomachs often contain numerous fragmented ophiuroids which, in due course, contribute ophiuroid debris to the sediment.

Fish are also responsible for predation and breakage of epifaunal species attached to hard substrates. For example, on the east coast of the United States, near Fort Pierce, Florida, the sheepshead (*Archosargus probatocephalus*) is responsible for the production of up to 30 g/m² of barnacle plates and fragments per day. The production of barnacle plates by causes other than predation by the sheepshead, as measured from a population covered by a cage that prevented the fish from gaining access to the barnacles, was only 0.8 g/m² per day (Hoskin, 1980).

Direct evidence from the sediments of the presence of fish populations on the continental shelf is limited to the occasional presence of otoliths.

6.16.2 THE ROLE OF BORERS IN THE BREAKDOWN OF SHELLS

Biological agents are mostly responsible for breaking down shell material into smaller grains. In many cases this breakdown is achieved by boring by either chemical or mechanical means (sometimes a combination of both) by a diverse group of organisms including bacteria, fungi, algae, bryozoans, polychaetes, sponges and molluscs (Carriker, Smith and Wilce, 1969; Boekschoten, 1966; Menzies, 1957; Sognnaes, 1963; Warme, 1975). The clionid sponges are the major agents in the breakdown of coral debris both in tropical reef corals (Bromley, 1978; Goreau and Hartman, 1963; Neumann, 1966) and in the deep water corals present on the edge of the continental shelf to the west of the British Isles and on Rockall Bank (J. B. Wilson, 1979a). Clionid sponges are also responsible for the break up of thick shelled bivalves such as *Arctica islandica* (*Cyprina islandica*), *Ostrea edulis* (Boekschoten, 1966; Bromley and Tendal, 1973; Cobb, 1969) and *Glycymeris glycymeris*. The sponge *Cliona celata* bores into the shells of the bivalve *Modiolus modiolus* and destroys parts of the prismatic layer of the shell and exposes the nacreous layer. In extreme cases the sponge perforates the shell of the live *Modiolus modiolus*, causing some distortion of the shell with nacreous papillae formed round the sponge penetration by the *Modiolus modiolus* (Comely, 1978).

Clionid sponges and boring polychaetes inhabiting shells are themselves grazed by echinoids. The rasping action of the jaw apparatus of the echinoid during feeding causes further bio-erosion of the shell and leaves characteristic paired grooves round the clionid or polychaete boring (Bromley, 1975).

Boring algae and fungi are important agents in the breakdown of shell debris on the continental shelf (Boekschoten, 1966; Cavalière and Alberte, 1970; Golubic, 1969; Kohlmeyer, 1969; Wilkinson and Burrows, 1972; J. B. Wilson, 1976, 1979b).

6.16.3 MECHANICAL BREAKAGE AND DISSOLUTION OF SHELLS

Abrasion and mechanical breakdown of shells is only of significance in very delicate and fragile shells. An experimental study in a circular flume showed that 15–30% of the shells of the thin shelled bivalves *Tellina tenuis*, *Tellina fabula* and *Abra alba* remained intact, 50–60% were separated into valves and most of the rest were fragmented. With the more robust shells *Macoma balthica*, *Chamelea gallina* (*Venus gallina*) and *Spisula* sp., 75–85% of the shells were still intact, 5% separated into single valves and only about 5% were broken into coarse fragments (Eisma, 1968).

In another experimental study where different biogenic fragments were agitated in sterile seawater at a constant temperature, the weight loss for delicate ophiuroid sclerites and spicules was 50% while that for much more robust barnacle plates was only 5% (Lefort, 1970).

Available evidence suggests that solution of shell debris on the continental shelf does not take place to any significant extent. This is in keeping with the pH values of about 8 that have been recorded in coastal waters off the Netherlands (Eisma, 1968). However, in certain local areas, such as the Skagerrak between Denmark and Norway, solution does play an important part in the breakdown of shells into fine particulate carbonate (Alexandersson, 1976, 1979).

The presence of otoliths in sediments in the Northern North Sea and off southern Norway which have been weakened by a process of differential solution or leaching has also been recorded (Gaemers, 1978).

6.16.4 DIFFERENCES IN FAUNAL COMPOSITION BETWEEN LIVING AND DEAD FAUNAS

The faunal composition of a temperate water shell gravel reflects to some degree the composition of the living fauna from which it was derived. The interrelationships are complex, however, and vary greatly from one envirionment to another. Generally samples contain more dead species and more dead individuals than live species and live individuals. Thus death assemblages can give more information about the structure and diversity of living populations as they represent the material from a much greater length of time than that represented by a single sampling programme of the living fauna (Warme, Ekdale, Ekdale and Petersen, 1976). An assemblage of dead shells can, thus, be viewed as a cumulative ecological history which reflects the average community ecology much better than sampling the living community on one occasion.

In some instances, such as on tidal flats (J. B. Wilson, 1967) and in calcareous algal gravels (Bosence, 1979), important differences in faunal composition can be recognized between the living and dead faunal assemblages.

Evidence from a comprehensive investigation in the tidal Ria de Arosa estuary on the north-west coast of Spain showed that, although mollusc species present in the living fauna are always present in the dead fauna, there was no quantitative correlation between the living and dead molluscan assemblages (Cadée, 1968).

On the other hand, the faunal composition of the living and dead molluscan assemblages in the channels on the floor of the Kiel Bight, Baltic Sea, North Germany, generally reflected one another (Arntz, Brunswig and Sarnthein, 1976).

The distribution of dead shells and fragments >1 mm along most of the Dutch coast is also very similar to that of the living molluscan fauna (Eisma, 1968).

Species diversities and relative species abundances in living communities appear to be preserved in the associated dead assemblages of empty mollusc shells in Mugu Lagoon, California and on the north-east coast of the Yucatan Peninsula, Mexico, and that even in the high energy sub-tidal sand channel habitat in both Mugu Lagoon and the Yucatan, post mortem transport of molluscan shell debris is insignificant (Warme et al., 1976).

Surveys in the English Channel covering the same ground as earlier investigations demonstrate significant changes in the species composition of the living populations present (Holme, 1966). The distribution of dead shells of particular species in the English Channel is generally more widespread than the occurrence of living individuals. A study of the distribution of dead shells of particular species has confirmed the absence of these species from certain regions in the English Channel (Holme, 1966).

In large areas such as the western part of the North Sea where many extensive and intensive investigations have been carried out, very few were conducted over successive years. Changes in the North Sea faunas with time are only known for a few areas or species over comparatively short periods of time (McIntyre, 1978).

The nature and structure of the carbonate debris itself plays an important role in both temperate and tropical regions in determining the relative proportions of the faunal elements that are preserved. For example, on the Great Bahama Bank in areas of extensive coral reef development, coral fragments generally only account for less than 10% of the carbonate in the sediments adjacent to the reefs (Bathurst, 1971). This may be because the coral generally disintegrates directly into fine aragonite needles which are too small to be identified in the sand fraction (Goreau, in Bathurst, 1971). The calcareous green alga *Halimeda* sp. on the other hand is greatly over-represented in the sand fraction of Bahamian sediments because of its rapid rate of growth and consequent high rate of production of carbonate debris.

Available evidence from the rippled sands on the continental shelf west of Scotland suggests that *Caryophyllia smithii* debris is also under-represented in the carbonate fraction when compared to its importance as a member of the rippled sand community. This is probably also due to the nature of the coral microstructure. Ophiuroids such as *Ophiothrix fragilis* often occur in large numbers with 'forests' of upturned arms (L. Cabioch, 1968; Warner, 1971) but their remains are rare or even absent from adjacent sediments (Farrow *et al.*, 1978).

Robust, thick-shelled molluscs are much more likely to be preserved in shell gravels than are delicate thin-shelled species. For example, valves of the robust bivalve *Cerastoderma edule* (*Cardium edule*) are three times more abundant than those of the bivalve *Macoma balthica* in surface shell beds on the tidal flats in the Solway Firth although *Macoma balthica* is the dominant molluscan species living in the tidal flats (J. B. Wilson, 1967). The thin-shelled bivalve *Tellina tenuis* which lives in the more mobile sands and in the small sand waves (megaripples of J. B. Wilson, 1967) is not represented in the surface shell beds. Fragile shelled species including the bivalves *Tellina fabula*, *Tellimya ferruginosa* (*Montacuta ferruginosa*), *Abra alba* and *Tellina tenuis* are also under-represented in dead shell accumulations on the Dutch coast (Eisma, 1968).

6.17 Age of temperate water carbonates

Very few radiometric age determinations of temperate water biogenic carbonates have so far been published. Up to 50 g of calcium carbonate are usually required to provide sufficient carbon for a reliable ^{14}C age determination. There are only a few species of mollusc which can provide this in a single shell or valve.

Care must taken in interpreting the results of radiocarbon dating. The age of a single shell may not necessarily indicate the age of the deposit in which it was found as it could have been reworked from an earlier deposit. On the other hand, mean ages determined from bulk samples, containing tens or even hundreds of shell fragments, do not give the range of ages of the material present in the deposit. Nevertheless, those determinations that have been made on bulk samples do give a good general indication of the probable age of the deposit and in some cases they prove the persistence, and hence the geological importance, of carbonate accumulations in temperate waters on continental shelves in several different parts of the world.

The carbonates accumulating on the east Australian continental shelf include material deposited since the stabilization of sea level at about 6000 years BP, following the Holocene transgression, in addition to other earlier material formed during the period of lowered sea level (Marshall and Davies, 1978; Maxwell, 1969). Age determination on *Mya arenaria* shells from shell gravels in the Barents Sea, Arctic Ocean, gave ages of 2400 to 8700 BP (Bjørlykke *et al.*, 1978), again indicating a Holocene age for these carbonates.

Early Holocene dates were also obtained from shells in carbonate gravels on the New Zealand continental shelf, off Stewart Island (Cullen, 1970).

6.17.1 AGE OF SHELL GRAVELS ON THE CONTINENTAL SHELF AROUND THE BRITISH ISLES

Radiocarbon dating of bulk samples (consisting of 50 g of shells, valves and fragments in the >2 mm fraction) suggests that the shell gravels on the continental shelf west and north of Scotland and on Rockall Bank have been gradually accumulating during the Holocene (Section 6.18.2). The four mean ages that have been published range from 2118 ± 45 years to 5406 ± 50 years BP (J. B. Wilson, 1979c) and indicate that the residence time for particular faunal elements in these shell gravels from the continental shelf west and north of Scotland and Rockall Bank should be measured in thousands of years. They therefore represent a geologically significant permanent deposit (see also Section 6.19).

6.17.2 RATES OF DEPOSITION

Estimates of carbonate production based on the carbonate content and absolute dates of sediment cores suggest that the thickness of carbonate sediments deposited in tropical waters during the 10 000 years of the Holocene range from 61 to 305 cm (Chave, 1962). Other data for tropical continental shelf carbonates suggest average rates of deposition of 10–100 cm per 1000 years (Matthews, 1974).

Little is known about possible rates of deposition

in cool water carbonate areas. Evidence from the Oligocene of New Zealand suggests accumulation rates of 1–2 cm per 1000 years in bioclastic limestones which display many features that characterize modern temperate water carbonate deposits (Nelson, 1978, Table 2). The radiocarbon ages given above for surface and near-surface bulk samples from the Scottish continental shelf and Rockall Bank tend to support such average rates of deposition for temperate water carbonates. It is important to note, however, that disturbance of the shell gravels by severe storms will tend to mix and hence average the age of the gravels in the top 10 cm or so of the sediment at any time.

6.18 Relative proportions of the major carbonate producers in death assemblages of continental shelf carbonates

The world-wide variations in the faunal composition of continental shelf carbonates in both temperate and tropical waters and their relationships to temperature and salinity have been the subject of a number of recent papers (Lees, 1973, 1975; Lees and Buller, 1972).

Two principal 'associations' of skeletal grains have been recognized (Lees and Buller, 1972). The 'foramol association' is found in temperate latitudes and consists of molluscs, benthonic foraminifera, echinoderms, bryozoans, barnacles, ostracods, calcareous sponge spicules, calcareous polychaete tubes and ahermatypic corals. Calcareous red algae could also be present in water shallower than 46 m around the British Isles (Adey and Adey, 1973). Maërl (Section 6.6.1) is generally found in waters less than 20 m in depth (Adey and Adey, 1973). Calcareous red algae may occasionally be found in waters much deeper than this. For example, *Phymatolithon rugulosum* which is commonly found at depths of 3–21 m (Adey and Adey, 1973) is also present at depths of 90 m on Rockall Bank (Clokie, Scoffin and Boney, 1981).

The actual faunal composition of the 'foramol association' varies widely and not all constituents are present in each location. The 'chlorozoan association' is found in warmer waters and also includes these components to a greater or lesser degree but with the addition of significant contributions of debris from hermatypic corals and calcareous green algae such as *Halimeda* sp. and *Penicillus* sp. The contribution made by barnacle and bryozoan debris to these warm water biogenic carbonates is much less than that found in temperate latitudes (Lees and Buller, 1972).

In warm water areas where the salinity is higher the hermatypic coral components are not present and the association is termed the 'chloralgal association'. In areas, such as the Persian Gulf, where corals can survive even when the salinity exceeds 48‰, a fourth association — the 'extended chlorozoan association' — has been recognized (Lees, 1975).

6.18.1 FAUNAL COMPOSITION OF DEATH ASSEMBLAGES IN SHELL GRAVELS IN THE STRONG CURRENT AREAS, WESTERN ENGLISH CHANNEL AND CELTIC SEA

Shell gravels are widespread over large areas in the Western English Channel and on the outer parts of the continental shelf west of Brittany (Boillot, 1964, 1965; Boillot, Bouysse and Lamboy, 1971; Bouysse, LeLann and Scolari, 1979; Holme, 1966; Lefort, 1970). The sediment is composed of bivalve valves and fragments including *Glycymeris glycymeris*, *Aequipecten opercularis* (*Chlamys opercularis*), *Palliolum tigerinum* (*Chlamys tigrina*), *Gouldia minima* (*Gafrarium minimum*), *Circomphalus casina* (*Venus casina*), *Clausinella fasciata* (*Venus fasciata*), bryozoans including fragments of *Crisia* sp. and *Cellaria* sp., echinoderms, foraminifera, barnacles, gastropods, serpulids (including *Ditrupa arietina*) and scaphopods. There is a very sparse fauna living in these gravels. Species recorded include the echinoid *Echinocyamus pusillus* and the cephalochordate *Amphioxus lanceolatus* (Holme, 1966).

In the Western English Channel off Cornwall and in the outer Bristol Channel off North Cornwall (Channon and Hamilton, 1976) the shell gravels are formed into symmetrical, wave-formed (Section 5.2.2) ripples with wavelengths of 100–125 cm (Flemming and Stride, 1967). The coarse gravel between the ripples is largely composed of bivalve valves and other coarse bioclastic debris. In places

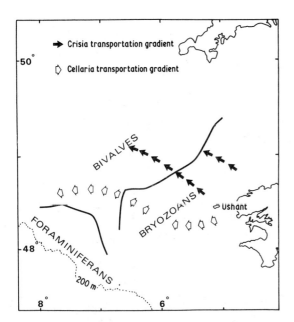

Fig. 6.9 Distribution of dominant faunal elements in the shell gravels in the Western English Channel off Brittany, and the direction of transport of fragments from the bryozoans *Crisia* sp. and *Cellaria* sp. from the inshore waters off Ushant and the Brittany coast (after Bouysse, Le Lann and Scolari, 1979).

the gravel ripples support populations of the brittle star *Ophiocomina nigra* which are dispersed over the surface (J. B. Wilson, Holme and Barrett, 1977). In other parts of the Western English Channel, dense aggregations of the brittle star *Ophiothrix fragilis* are to be found often associated with rock outcrops (Warner, 1971; L. Cabioch, 1968; Larsonneur, 1971; Lefort, 1970; J. B. Wilson, Holme and Barrett, 1977). Sediments adjacent to these aggregations do contain some ophiuroid ossicles. The ophiuroid content of the sediment decreases abruptly, however, away from the rock outcrops. This is largely due to the fragile nature of ophiuroid debris (Larsonneur, 1971; Lefort, 1970, and Sections 6.16.1, 6.16.4). Ophiuroid debris is also important in the outer part of the Western English Channel (Boillot, 1965).

Bivalve fragments are most common in the central part of the Celtic Sea while foraminifera are most abundant towards the shelf edge. Bryozoans are dominant off north-west Brittany. *Crisia* sp. fragments and other lamellar bryozoan debris comprise over 50% of the bioclastic fraction west of the Isle of Ushant off Brittany and decrease west–northwestwards towards the edge of the continental shelf to only 5–25% of the total. *Cellaria* sp. fragments comprise over 50% of the bioclastic fraction north-west of Ushant and north of Brittany and the proportion decreases north-westwards to 5–25% towards mid-English Channel (Fig. 6.9).

6.18.2 FAUNAL COMPOSITION OF DEATH ASSEMBLAGES IN SHELL GRAVELS ON THE CONTINENTAL SHELF WEST OF SCOTLAND

The gravels on the continental shelf to the west and north of Scotland are particularly rich in shelly carbonate debris. In some areas the carbonate content is greater than 95% (Fig. 6.2). Values in the Malin Sea north of Ireland can be up to 78% (Pendlebury and Dobson, 1976).

The absence of non-carbonate clastic components in many of these gravels may be partly due to the winnowing action of the stronger tidal currents which carry the sand sized grains away from the area and partly to the limited supply of non-carbonate clastics to the region. On continental shelves in other colder water areas the relative carbonate content of the sediment is much higher where there is a reduced supply of terrigenous clastic sediment (Milliman, 1974).

(a) *Shell gravels in the strong current zone*

The most extensive deposits of shelly carbonate

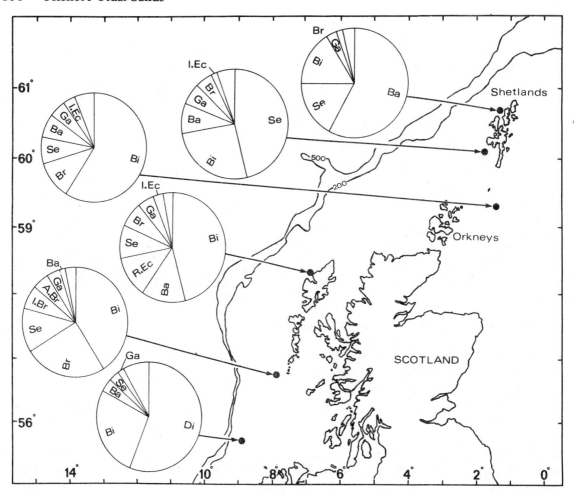

Fig. 6.10 Variation in the faunal composition of the carbonate fraction of coarse sands and gravels on the continental shelf west and north of Scotland. A.Br, articulate brachiopods; Ba, barnacles; Bi, bivalves; Br, bryozoans; Di, *Ditrupa arietina*; Ga, gastropods; I.Br, inarticulate brachiopods; I.Ec, irregular echinoids; R.Ec, regular echinoids; Sc, scaphopods; Se, attached serpulids; blank sector, other minor faunal components (after J. B. Wilson, 1979c).

gravel around Scotland occur in the stronger current areas to the west and north of the Orkney Isles and Shetland Isles and in the Fair Isle Channel between the Orkney and Shetland Isles. Evidence from vibrocoring (Allen, Fannin and Farrow, 1979) shows that these carbonate gravels and sands can be at least 3 metres in thickness. The shell gravels on the continental shelf near the Shetland Isles are mostly composed of calcareous worm tubes, barnacles and bivalve fragments. The barnacles may contribute up to 50–60% of the total carbonate in some areas and polychaete tubes may also provide as much as 40–50%. Bivalves may contribute up to 55% of the total but they are only rarely the main constituent of these inner shelf gravels. Bryozoans can contribute up to 25% while gastropods generally comprise about 5% (Fig. 6.10). Other constituents include inarticulate brachiopod valves, plates, spines and fragments from both regular and irregular echinoids, chiton plates and coral fragments. Certain of these contribute up to 5% of the total carbonate content in some areas. In most cases, however, they rarely contribute more than 1–2% of the total (J. B. Wilson, 1979c).

The source of much of the carbonate in these shell gravels from near the Shetland Isles is thought to be relatively local as there are extensive rock outcrops and boulders on the inner continental shelf which support a rich epifauna including serpulids, barnacles, bryozoans, inarticulate and articulate brachiopods, chitons, regular echinoids, etc.

Some of the bivalve species present such as *Anomia* sp. and *Hiatella* sp. may be locally derived from the outcrops. Others such as *Glycymeris glycymeris* may be nestling in the gravel itself while valves from thinner shelled species such as *Thracia* sp. may have been transported from less disturbed sandy areas.

The shell gravels on the inner continental shelf west of the Orkney Islands and west of the Outer Hebrides off the west coast of Scotland generally have a higher percentage of bivalve shells and fragments – up to 30–55% – than is found in the shell gravels off the Shetland Isles and bivalves are often the dominant constituent. Barnacles and serpulids are also common and each often contributes 15–35% (Fig. 6.10) although locally they may contribute substantially more than this. Again it is thought probable that much of the carbonate is derived from faunas living on the adjacent extensive areas of rock outcrops and boulders.

On the middle part of the continental shelf, bivalves tend to be the dominant constituent often comprising up to 50% of the gravel. Other important contributors are barnacles, calcareous polychaetes and bryozoans. The composition of these gravels from the middle of the continental shelf is variable and again probably reflects local variations in the nature of the fauna in the adjacent source areas. At particular locations very high percentages of particular faunal elements may be present. For example, one mid-continental shelf gravel contains 69% serpulid worms, while in another, inarticulate brachiopods make up 7% and articulate brachiopods 5% (Plate 6.10) of the total carbonate (J. B. Wilson, 1979c), suggesting derivation from an adjacent rocky area or boulder field.

(b) *Shell gravels in the zone of weak tidal currents, outer continental shelf*

On the outer part of the continental shelf to the west and south-west of Outer Hebrides, the faunal composition of the carbonate fraction of the gravels is somewhat different. Although bivalves are generally the dominant constituent, making up to 40 to 50% of the total, the serpulid polychaete *Ditrupa arietina* is also important (Fig. 6.10) and complete tubes and fragments may locally be the dominant constituent of the gravel, making up 50 to 70% of the total. They generally make up 10 to 25% of the total carbonate. These tubes and fragments are largely derived from the adjacent sand patches and rippled sands (Section 6.12.2). Barnacles are also locally important in the gravels towards the edge of the continental shelf where they may comprise up to 25% of the total carbonate. Generally, however, the barnacle content on the outer part of the continental shelf is less than 10%.

In the Malin Sea north of Ireland, bivalves account for 41% of the total carbonate. Other components include barnacles – 19%, polychaetes – 11%, echinoids – 7% and gastropods – 6% (Pendlebury and Dobson, 1976).

6.19 Temporal changes in the faunal composition of shell gravels

Examination of the fragments making up one of these gravels which gave a mean age of 2752 ± 55 years BP (J. B. Wilson, 1979c) showed that the fragments could be divided into two broad general categories – unstained, relatively fresh looking fragments and yellow or golden brown stained fragments. Radiocarbon age determinations made on samples of these showed that the fresh, unstained fragments had a mean age of 2480 ± 55 years BP while the mean age of the stained fragments was 4210 ± 55 years BP. The unstained fraction contained bryozoan fragments, fragments from the bivalves *Spisula* sp., *Anomia* sp., *Timoclea ovata* (*Venus ovata*) and *Chlamys* sp., gastropods including *Trochus* sp. and *Nassarius* sp., some serpulid polychaete tubes and some barnacle plates. In contrast, the stained fraction contained a much higher proportion of barnacle plates – mostly *Verruca stroemia* and *Balanus* sp. – and serpulid tubes than did the unstained fraction, but it contained fewer bryozoan fragments.

These results suggest that changes in the composition of the faunas contributing to these biogenic carbonate accumulations may have taken place during the Holocene. However, care must be taken when making direct comparisons between the faunal composition of the carbonate fraction and the composition of the living fauna which contributed the biogenic carbonate to the gravel (Section 6.16.4). Many complex factors including the growth rates and hence rates of carbonate production, the hydrodynamics of transport and accumulation of particular fragments, the structure and therefore the durability of the fragment (Section 6.16.4), its resistance to abrasion (Section 6.16.3), and the activities of borers (Section 6.16.2) have to be taken into account.

6.19.1 FAUNAL EVIDENCE OF LOWERED SEA LEVEL

Shelly lag gravels related to periods of lowered sea level are exposed in some places on the present sea floor. Radiocarbon dating showed that the mean age of the shell fragments in one of these lag gravels on the continental shelf west of Scotland, which is exposed between two adjacent sand patches at a depth of 130 m is 8335 ± 60 years BP. Although some later Holocene carbonate is present, most of the shell fragments are grey and pitted (Plate 6.11). Many of these are plates from the barnacle *Balanus balanus*. An age determination on a sample of these gave a mean age of 11 560 ± 80 years BP suggesting that they were mostly derived from a barnacle population flourishing during the period of low sea level in the late Pleistocene and probably also during the early stages of the Holocene transgression. The gravel also contains bivalve species such as *Mya truncata* which lived in much shallower waters.

The presence of worn valves of the intertidal bivalve *Macoma balthica* in the sediments of the eastern part of the English Channel and in more inshore waters throughout the Channel (Holme, 1966) is also probably evidence of lowered sea level conditions.

Age determination on supposed littoral shells collected from the continental shelf west of France gave ages of 9000 years BP, 5100 years BP and 3000 years BP at depths of 188 m, 50 m and 32 m respectively (Lapierre, 1969).

6.20 Long term evolution of temperate shelf carbonates

It is difficult to speculate with success on the evolution of continental shelf carbonate-rich deposits. However, one or two concepts are applicable to any process models that are constructed. The cycle begins after a period of lowered sea level with the continental shelf consisting of cobble and boulder deposits with extensive areas of rock outcrops and poorly sorted sands and fine and coarse gravels. As the sea level rises and the transgression takes place, the resulting tidal current activity gradually sorts the sand and gravel into sand sheets which include sand waves where the tidal currents are strong and areas of rippled sand where the currents are weaker. Some of the sand in the areas of weak tidal currents becomes sorted into sand patches where there is inadequate sand available to form a more continuous zone of rippled sand. In areas where there is little or no additional clastic sediment being transported onto the continental shelf from either other parts of the shelf or from the land by rivers, the sand sheets and sand patches remain more or less constant in size. Any additional sediment will be biogenic in origin and the growth of the bedform will depend on the rate of productivity of biogenic sediment.

The exposed cobbles, boulders and rock outcrops are colonized by carbonate producing epifaunal species of erect and encrusting bryozoans, barnacles, serpulid polychaetes, inarticulate brachiopods such as *Crania anomala*, bivalves such as *Anomia* sp. and *Arca tetragona*, gastropods, ophiuroids and regular echinoids such as *Echinus esculentis* (Fig. 6.7, Plate 6.10).

An infauna including bivalves, polychaetes and crustaceans living at different depths within the sediment, scaphopods, gastropods and burrowing irregular echinoids would gradually colonize the sheets of rippled sand and the sand patches. Their burrowing and feeding activities produce bioturbation structures within the sediment (Section 6.3, Fig. 6.1). The fauna in the rippled sand zones becomes highly diverse (Sections 6.11.3, 6.12.2). In the zone of large sand waves where the tidal currents are much stronger only a few species colonize the zone and no infauna develops (Sections 6.10.4, 6.11.1).

As the fauna in the zone of rippled sand develops shell debris gradually accumulates within the sand with the effect of gradually increasing the mean grain size of the sediment. The shell debris accumulates both at the surface by the activities of burrowing crustaceans (Section 6.3.2) and at depths within the sediment related to the depth at which the infaunal bivalves and polychaetes are living (Section 6.3.2). Buried shell and shell fragment layers are thus formed by successive populations of these infaunal bivalves and polychaetes. Through time these layers gradually increase in thickness and extent with the result that it becomes increasingly more difficult for the deeper burrowing bivalves and polychaete species present in the original infauna to penetrate the layers and to remain part of the rippled sand zone fauna. These species gradually die out and are replaced by others which can tolerate the coarser sediments (Keary and Keegan, 1975).

Periods of slight increase in the current strength would tend to winnow away the smaller clastic grains and thus accelerate the overall increase in grain size. Even without the winnowing effect of periodic increases in current strength, the rippled sands or the sand patches will tend to show an upwards increase in grain size as the sediment became more carbonate-rich. The increased grain size will also make the sediment more resistant to erosion and transport (Pasternak, 1971). The coarse carbonate fragments and shells at the surface would themselves be colonized by encrusting bryozoans and serpulids which would in turn add to the accumulating carbonate debris.

At the same time, continued epifaunal production of carbonate on the nearby rock outcrops and boulders would produce accumulations of carbonate gravel in the interstices between the boulders or hollows in the outcrop (J. B. Wilson, 1977, Fig. 5). As this process continues, the hollows and interstices become filled up and the shell debris gradually encroaches onto the surfaces of the rocks and the boulders thus reducing the extent of the surfaces available for further epifaunal colonization. The increased competition for the remaining space will further alter the faunal composition of the rock epifauna. The associated shell gravels will themselves also become colonized by a specialized fauna including shell bearing species which in turn will contribute additional carbonate to the gravel.

Biogenic carbonate debris thus accumulates to form shell gravels in areas of boulders and rock outcrop and also in the sediments of the zones of rippled sand and sand patches. Shelly carbonate thus becomes the dominant constituent of the deposits forming on a continental shelf when other material is not available.

Although this scheme is greatly simplified and few details of the particular carbonate producers present are given, it can nevertheless be used as the basis on which local modifications covering particular situations and conditions can be made.

Periodic influxes of clastic sediment would clearly modify the above scheme by providing further sand for colonization by the infauna and the deposits of the first cycle could be preserved by further rises in sea level and the influx of fine sand or silt. Evidence from cores on the New Zealand continental shelf (Norris, 1972) and the continental shelf west of Scotland (Institute of Geological Sciences data) suggests that shelly material tends to occur in layers separated by thicknesses of clastic sediment containing only scattered shells and shell debris. The shell layers probably represent periods of reduced clastic sediment input in addition to the winnowing mentioned above. Evidence from extensive sampling on the continental shelf west and north of Scotland suggests that the present surface is one on which shell accumulation has been taking place in particular areas for periods of perhaps thousands of years (Section 6.17.1; J. B. Wilson, 1979c).

6.21 Applications to the fossil record

Knowledge of modern warm water carbonate deposition has already been applied to the geological record by other workers. A general model of shallow water tropical carbonate sedimentation has been widely applied (Matthews, 1974; Selley, 1970). It defines five zones from the open sea to the supratidal and incorporates data on sedimentary structures, sediment type and the faunal composition of the bioclastic components (Heckel, 1972; Irwin, 1965; Nelson, 1978, Fig. 1). Nine standard facies belts ranging from marginal basins and open shelf environments to reefs and other organic build-up structures and platform deposits have been established (J. L. Wilson, 1970, 1974, 1975).

Few attempts have so far been made to apply knowledge of modern temperate carbonates to the geological record, nor has much attention been paid to those aspects of carbonate sedimentation particularly related to tidal current activity. These should, however, be fruitful fields for future research. Faunal aspects of certain tidal current dominated deposits in the geological record are given in Chapter 7.

6.22 Main conclusions

1. Shelly faunas can provide a measure of the amount of sediment movement by currents. They do not by themselves provide evidence that this activity is unequivocally related to tidal currents.

2. A distinction can be made between various bedform zones on the basis of the fauna. Stable floors support a rich fauna including both infaunal and epifaunal species. Active bedforms support few species and few individuals of these species.

3. In a zone where there is erosion and strong tidal scour epifaunal species predominate.

4. In a sand sheet there is a marked increase in both faunal diversity and faunal density along a traverse from the zone of large sand waves to the zone of rippled sand.

5. Differentiation between active linear sand banks and a nearby sand sheet covered in large sand waves is not generally possible on faunal grounds as the active sand banks themselves are covered by large sand waves. Both may contain bioturbation structures produced by burrowing echinoids.

6. The occasional presence of an infauna in the troughs between sand waves suggests stability of the bedform in such examples for long enough for the fauna to develop. Should an infauna be present on the faces of the sand wave as well as in the trough, stability for rather longer periods of time is implied.

7. Faunas from bed-load partings may be fairly sparse and may include echinoids, ophiuroids, bryozoans, byssally attached bivalves and barnacles.

8. The dead faunal composition of biogenic carbonate gravels on the continental shelf reflects in broad general terms the living faunal composition of the source of area, although the relative proportions of particular animal groups may be markedly different.

9. Bioturbation structures may provide evidence as to the origin of particular marine sedimentary deposits of equal value to the sedimentary structures that have been destroyed.

10. The few radiometric age determinations on temperate water carbonates that are available show that most deposits contain carbonate that is several thousand years old. They have, therefore, accumulated over that period of time.

11. Since the faunal compositions of biogenic carbonates on the inner, middle and outer parts of the continental shelf are different, faunas may be used to determine the palaeogeographic setting of a particular tidal current sedimentary sequence. They can provide evidence indicating, for example, the probable location of the open ocean or a partially enclosed inner shelf sea or the position of the continental shelf edge and upper continental slope.

12. On a continental shelf with minimal clastic input the carbonate content of the sediment will gradually increase. There will also be a tendency for its grain size to increase. This in turn will alter the composition of the fauna living on the sediment and hence the composition of the later carbonate debris accumulating in the sediment.

Appendix 6.1

A list of species mentioned in Chapter 6 giving name in current use and taxonomic authority is given below. Well-known names for certain species that have now been superseded are given in brackets.

The sources were as follows:

Brunton, C. H. C. and Curry, G. B. (1979), *British Brachiopods*, Synopses of the British Fauna, New Series No. 17. Linnean Society of London and Academic Press London.

Hayward, P. J. and Ryland, J. S. (1979), *British Ascophoran Bryozoans*, Synopses of the British Fauna, New Series No. 14, Linnean Society of London and Academic Press London.

Marine Biological Association (1957), *Plymouth Marine Fauna*, Marine Biological Association of the United Kingdom, Plymouth.

McKay, D. W. and Smith, S. M. (1979), *Marine Mollusca of East Scotland*, Royal Scottish Museum, Edinburgh.

Mortensen, Th. (1927), *Handbook of the Echinoderms of the British Isles*, Oxford University Press, Oxford.

Parke, M. and Dixon, P. S. (1976), Checklist of British Marine Algae – Third revision, *Journal Marine Biological Association UK*, **56**, 527–594.

Ryland, J. S. and Hayward, P. J. (1977), *British Anascan Bryozoans*, Synopses of the British Fauna, New Series No. 10. Linnean Society of London and Academic Press London.

Wheeler, A. (1969), *The Fishes of the British Isles and North-West Europe*, MacMillan, London.

Winckworth, R. (1932), The British Marine Mollusca, *Journal of Conchology*, **19**, 211–252.

Personal communications – P. E. Gibbs, P. J. Hayward, G. Oliver and S. M. Smith.

Rhodophyta
  Lithothamnium corallioides Crouan
  Lithothamnium glaciale Kjellman
  Phymatolithon calcareum (Pallas) Adey et McKibbin
  Phymatolithon rugulosum Adey
Phaeophyta
  Laminaria hyperborea (Gunnerus) Foslie
  Laminaria saccharina (Linnaeus) Lamouroux
Phylum Protozoa
  Order Foraminifera
    Cibicides refulgens (Montfort)
Phylum Porifera
  Class Demospongiaria
    Cliona celata Grant
Phylum Coelenterata
  Class Anthozoa
    Caryophyllia smithii Stokes & Broderip
    Lophelia pertusa (Linnaeus) (*Lophelia prolifera*)
Phylum Annelida
  Class Polychaeta
    Sthenelais limicola Ehlers
    Nephthys caeca (Müller)
    Nephthys cirrosa Ehlers
    Glycera unicornis Savigny
    Hyalinoecia tubicola (Müller)
    Hyalinoecia bilineata Baird
    Lumbriconereis latreilli (Audouin & Edwards)
    Scoloplos armiger (Müller)
    Paraonis lyra Southern
    Chaetopterus variopedatus (Renier)
    Clymenella torquata (Leidy)
    Ophelia limacina Rathke
    Scalibregma inflatum Rathke
    Notomastus latericeus Sars
    Heteromastus filiformis (Claparède)
    Arenicola marina (Linnaeus)
    Owenia fusiformis Delle Chiaje
    Sabellaria alveolata (Linnaeus)
    Sabellaria spinulosa Leuckart
    Pectinaria koreni (Malmgren)
    Pectinaria auricoma (Müller)
    Cistenides gouldi (Verrill) (*Pectinaria gouldi*)
    Melinna cristata (Lars)
    Lanice conchilega (Pallas)
    Serpula vermicularis Linnaeus
    Hydroides norvegica (Gunnerus)
    Pomotoceros triqueter (Linnaeus)
    Protula tubularia (Montagu)
    Placostegus tridentatus (Fabricius)
    Ditrupa arietina (Müller)
Phylum Sipunculoidea
  Golfingia minuta (Keferstein)
  Phascolion strombi (Montagu)
Phylum Brachiopoda
  Class Inarticulata
    Crania anomala (Müller)
  Class Articulata
    Terebratulina retusa (Linnaeus) (*Terebratulina caputserpentis*)
Phylum Arthropoda
  Sub-phylum Crustacea
    Class Cirripedia
      Balanus balanus da Costa
      Verruca stroemia (Müller)
    Class Malacostraca
      Ampelisca brevicornis (A. Costa)
      Ampelisca spinipes Boeck
      Bathyporeia guilliamsoniana (Bate)
      Bathyporeia pelagica (Bate)
      Stenothoë marina (Bate)
      Pontocrates arenarius (Bate)
      Gastrosaccus spinifer (Göes)
      Crangon vulgaris Fabricius
      Crangon allmani Kinahan
      Philocheras trispinosus (Hailstone)
      Callianassa subterranea (Montagu)
      Callianassa californiensis Dana
      Nephrops norvegicus (Linnaeus)
      Upogebia deltaura (Leach)
      Upogebia stellata (Montagu)
      Pagurus bernhardus (Linnaeus)
      Portunus depurator (Linnaeus)
Phylum Mollusca
  Class Gastropoda
    Emarginula fissura (Linnaeus) (*Emarginula reticulata*)
    Diodora graeca (Linnaeus) (*Diodora apertura*)
    Patella vulgata Linnaeus
    Iothia fulva (Müller) (*Lepeta fulva*)
    Solariella amabilis (Jeffreys)
    Calliostoma zizyphinum (Linnaeus)
    Gibbula magus (Linnaeus)
    Gibbula cineraria (Linnaeus)
    Jujubinus clelandi (W. Wood) (*Cantharidus clelandi*)
    Rissoa parva (da Costa)
    Turritella communis (Risso)

Bittium reticulatum (da Costa)
Aporrhais pespelecani (Linnaeus)
Aporrhais serresiana (Michaud)
Bulbus islandicus (Gmelin) (*Amauropsis islandica*)
Lunatia alderi (Forbes) (*Natica alderi*)
Lunatia catena (da Costa) (*Natica catena*)
Lunatia montagui (Forbes) (*Natica montagui*)
Nucella lapillus (Linnaeus)
Urosalpinx cinerea (Say)
Pyrene haliaeeti Jeffreys
Colus gracilis (da Costa)
Colus howsei (Marshall)
Buccinum undatum Linnaeus
Hinia incrassata (Ström) (*Nassarius incrassatus*)
Hinia pygmaea (Lamarck) (*Nassarius pygmaeus*)
Troschelia berniciensis (King)
Volutomitra groenlandica (Beck)
Typhlomangelia nivalis (Loven)
Cytharella coarctata (Forbes) (*Mangelia coarctata*)
Bela nebula (Montagu) (*Mangelia nebula*)
Teretia anceps (Eichwald) (*Philibertia teres*)
Acteon tornatilis (Linnaeus)
Cylichna cylindracea (Pennant)
Cylichna alba (Brown)
Class Scaphopoda
Antalis entalis (Linnaeus) (*Dentalium entalis*)
Class Bivalvia
Nucula nucleus (Linnaeus)
Nucula tenuis (Montagu)
Nucula nitidosa Winkworth
Nucula hanleyi Winkworth
Portlandia philippiana (Nyst) (*Yoldia tomlini*)
Limopsis aurita (Brocchi)
Limopsis minuta (Philippi)
Glycymeris glycymeris (Linnaeus)
Striarca lactea (Linnaeus) (*Arca lactea*)
Bathyarca pectunculoides (Scacchi)
Bentharca nodulosa (Müller)
Arca tetragona (Poli)
Anomia ephippium (Linnaeus)
Pododesmus patelliformis (Linnaeus) (*Monia patelliformis*)
Mytilus edulis Linnaeus
Modiolus modiolus (Linnaeus)
Modiolus barbatus (Linnaeus)
Ostrea edulis (Linnaeus)
Ostrea sinuata Lamarck
Chlamys varia (Linnaeus)
Chlamys distorta (da Costa)
Chlamys sulcata (Müller)
Aequipecten opercularis (Linnaeus) (*Chlamys opercularis*)
Palliolum tigerinum (Müller) (*Chlamys tigrina*)
Palliolum simile (Laskey) (*Chlamys similis*)
Manupecten alicei (Dautzenberg & Fischer) (*Chlamys alicei*)
Propeamussium hoskynsi (Forbes) (*Chlamys hoskynsi*)
Limaria hians (Gmelin) (*Lima hians*)
Limaria loscombi (Sowerby) (*Lima loscombi*)
Astarte sulcata (da Costa)
Tridonta elliptica (Brown) (*Astarte elliptica*)
Goodallia triangularis (Montagu) (*Astarte triangularis*)
Thyasira flexuosa (Montagu)
Myrtea spinifera (Montagu)
Lucinoma borealis (Linnaeus)
Galeomma turtoni (Sowerby)
Tellimya ferruginosa (Montagu) (*Montacuta ferruginosa*)
Mysella bidentata (Montagu)
Arctica islandica (Linnaeus) (*Cyprina islandica*)
Acanthocardia echinata (Linnaeus) (*Cardium echinatum*)
Parvicardium minimum (Philippi) (*Cardium minimum*)
Parvicardium ovale (Sowerby) (*Cardium ovale*)
Cerastoderma edule (Linnaeus) (*Cardium edule*)
Dosinia exoleta (Linnaeus)
Dosinia lupinus (Linnaeus)
Gouldia minima (Montagu) (*Gafrarium minimum*)
Venus verrucosa Linnaeus
Circomphalus casina (Linnaeus) (*Venus casina*)
Timoclea ovata (Pennant) (*Venus ovata*)
Clausinella fasciata (da Costa) (*Venus fasciata*)
Chamelea gallina (Linnaeus) (*Venus gallina*) (*Venus striatula*)
Paphia aurea (Gmelin) (*Venerupis aurea*)
Paphia rhomboides (Pennant) (*Venerupis rhomboides*)
Mysia undata (Pennant)
Donax vittatus (da Costa)
Tellina tenuis da Costa
Tellina fabula Gmelin
Tellina pygmaea Loven
Tellina donacina Linnaeus
Acropagia balaustina (Linnaeus) (*Tellina balaustina*)
Acropagia crassa (Pennant) (*Tellina crassa*)
Macoma balthica (Linnaeus)
Scrobicularia plana (da Costa)
Abra alba (W. Wood)
Abra prismatica (Montagu)
Gari fervensis (Gmelin)
Gari depressa (Pennant)
Gari tellinella (Lamarck)
Gari costulata (Turton)
Solecurtus chamasolen (da Costa)
Pharus legumen (Linnaeus)
Phaxas pellucidus (Pennant) (*Cultellus pellucidus*)
Ensis arcuatus (Jeffreys)
Ensis ensis (Linnaeus)
Ensis minor (Chenu)
Ensis siliqua (Linnaeus)
Mactra stultorum (Linnaeus) (*Mactra corallina*)

Spisula elliptica (Brown)
Spisula solida (Linnaeus)
Spisula subtruncata (da Costa)
Lutraria angustior Philippi
Lutraria lutraria (Linnaeus)
Mya arenaria Linnaeus
Mya truncata Linnaeus
Sphenia binghami (Turton)
Corbula gibba (Olivi)
Hiatella arctica (Linnaeus)
Saxicavella jeffreysi Winckworth
Gastrochaena dubia (Pennant)
Cochlodesma praetenue (Pulteney)
Thracia phaseolina (Lamarck)
Thracia convexa (W. Wood)
Lyonsia norwegica (Gmelin)
Pandora pinna (Montagu)
Cardiomya costellata (Deshayes) (*Cuspidaria costellata*)
 Class Cephalopoda
  Sepiola atlantica Orbigny
  Octopus vulgaris Cuvier
Phylum Bryozoa
 Pyripora catenularia (Fleming)
 Flustra foliacea (Linnaeus)
 Alderina imbellis (Hincks)
 Cellaria sinuosa (Hassall)
 Cellaria fistulosa (Linnaeus)
 Cellaria salicornioides Lamouroux
 Reptadeonella violacea (Johnston)
 Parasmittina trispinosa (Johnston)
 Porella compressa (J. Sowerby)
 Porella concinna (Busk)
 Palmicellaria skenei (Ellis & Solander)
 Schizomavella auriculata (Hassall)
 Schizomavella linearis (Hassall)
 Cellepora pumicosa (Pallas)
 Turbicellepora avicularis (Hincks)
 Hornera lichenoides (Linnaeus)
Phylum Echinodermata
 Class Asteroidea
  Astropecten irregularis (Pennant)
  Pseudarchaster parelii (Düben & Koren)
  Solaster papposus (Linnaeus)
  Henricia sanguinolenta (Müller)

Asterias rubens Linnaeus
Class Ophiuroidea
 Ophiothrix fragilis (Abildgaard)
 Ophiocomina nigra (Abildgaard)
 Amphiura chiajei Forbes
 Amphiura filiformis (Müller)
 Ophiura texturata Lamarck
 Ophiura albida Forbes
 Ophiura affinis Lütken
Class Echinoidea
 Cidaris cidaris (Linnaeus)
 Psammechinus miliaris (Gmelin)
 Echinus esculentis Linnaeus
 Echinus acutus Lamarck
 Echinus elegans Düben & Koren
 Echinocyamus pusillus (Müller)
 Spatangus purpureus Müller
 Spatangus raschii Lovén
 Echinocardium cordatum (Pennant)
 Echinocardium flavescens (Müller)
 Brissopsis lyrifera (Forbes)
Class Holothurioidea
 Cucumaria elongata Düben & Koren (*Cucumaria pentactes*)
Subphylum Cephalochordata
 Amphioxus lanceolatus (Pallas)
Phylum Chordata
 Class Pisces
  Merlangius merlangus (Linnaeus)
  Gadus morhua Linnaeus
  Melanogrammus aeglefinus (Linnaeus)
  Archosargus probatocephalus (Walbaum)
  Gymnammodytes semisquamatus (Jourdain)
  Ammodytes marinus Raitt
  Hyperoplus lanceolatus (Lesauvage) (*Ammodytes lanceolatus*)
  Callionymus lyra Linnaeus
  Pomatoschistus minutus (Pallas)
  Lepidorhombus whiffiagonis (Walbaum)
  Limanda limanda (Linnaeus)
  Pleuronectes platessa Linnaeus
  Microstomus kitt (Walbaum)
  Hippoglossoides platessoides (Fabricius)
  Solea solea (Linnaeus)
  Buglossidium luteum (Risso)

Chapter 7

Ancient offshore tidal deposits

7.1 Introduction

Powerful tidal currents, winnowing shelly gravels and transporting sands and other grades along well defined paths, are a major feature of many partially enclosed continental shelf seas bordering large oceans. Clearly similar environments must have existed in the geological past, but the geologist must not be tempted merely to identify tidal deposits; there is other vital information to be gained. Studies of modern continental shelf tidal currents and sediments have yielded data concerning the lithology and geometry of deposits and the directions and rates of sediment transport, as discussed in earlier chapters. There are, however, two aspects about which relatively little is known, namely, the process of bedform preservation and the nature of the sedimentary sequences eventually produced by tidal currents operating in different continental shelf settings. The recognition of ancient offshore tidal current deposits is important because such discoveries offer the opportunity to seek information on these aspects.

7.2 Recognition of ancient offshore tidal current activity

In recent years investigators of ancient nearshore and shoreline sediments have used the association of various diagnostic characteristics to identify the influence of tidal activity. Such characteristics include 'herringbone' cross-stratification, reactivation surfaces, interference ripples, double-crested ripples, wavy bedding, mud cracks, channels, burrows and fining-upwards sequences (numerous authors in Ginsburg, 1975; Klein, 1970a; 1970b; 1971; 1977; de Raaf and Boersma, 1971). No single characteristic is an unequivocal criterion, but 'herringbone' cross-stratification has gained a significant reputation in this respect; in effect the fluttering flag for deposits of tidal origin. However, in the offshore realm the problems of recognizing the influence of tidal activity are compounded. First, less is known about the internal structure of modern offshore deposits. Secondly, some of the features that are so characteristic of tidal flats will not occur offshore. Examples include mud cracks, interference ripples and double-crested ripples. Thirdly, there are non-tidal agents such as waves, wind-driven and thermohaline flows which can produce deposits similar to those produced by tidal currents.

Studies of modern offshore tidal sediments have contributed information on the geometry of bedforms (Sections 3.3 and 3.4), and the directions and magnitudes of net sand transport (Chapter 4). However, students of ancient shelf sediments commonly experience difficulty in determining the geometry of the original bedform and the pattern of sediment movement. Preserved sets of cross-stratification may not represent the predominant regional direction of net sand transport (Section 7.3.2). Conversely, sedimentary structures and grain-size distributions are more fully documented from ancient shelf sediments than their modern equivalents because accurately located seismic reflection profiles and associated cores taken from open shelf seas are still relatively scarce (but see Chapter 5; Houbolt, 1968; Gadow and Reineck, 1969; Swift, Freeland and Young, 1979). The difference in the nature of information derived from the Recent and the geological record is contributing to the basic problem of recognizing ancient offshore tidal current deposits.

It has therefore become clear that the identification of ancient offshore tidal deposits must include use of other criteria; indeed, some ancient continental shelf sediments have been considered to be offshore tidal in origin despite the fact that there is little unequivocal evidence for the reversal of flow, the most celebrated property of tidal currents (e.g. Narayan, 1971; Anderton, 1976; Levell, 1980). In these cases the tidal interpretation has resulted from two general considerations. First, few other offshore marine agents are considered to rework sand so consistently and effectively, producing compositionally mature sediment with a tabular bed geometry, many large-scale sets of cross-stratification and conspicuous horizons of pebbles. Secondly, the geological context is known to be compatible with a tidal interpretation. More specifically, knowledge of the geological context requires consideration of the proximity of an ocean basin in which the tidal wave would have originated, the width of the continental shelf (Section 2.3), the configuration of the shoreline and the existence of contemporaneous tidal flat or inlet associated deposits. As the ecology of the modern temperate-latitude sea floor becomes better understood (Chapter 6), investigators will also be able to make more use of a classical palaeobiological tool.

7.3 Structures preserved in ancient offshore tidal current deposits

Sand waves, longitudinal sand banks, sand and mud sheets and scoured surfaces which have been tentatively interpreted as tidal in origin are located in rocks ranging back in age to 1500 million years (Appendix 7.1). The following sections are concerned with the structures which are preserved in these ancient offshore deposits.

7.3.1 SAND WAVES

The remains of ancient sand waves are identified by the presence of large-scale sets of cross-strata dipping at angles of 4° to 35° (cf. Section 5.2.4). Attention has been focused on the information which internal structures relate about the growth and demise of these bedforms on the floors of ancient shelf seas, but as yet there has been little comparison possible with modern sand waves because their internal structure is poorly known.

First, the cause of 'herringbone' cross-stratification requires consideration. In the open sea as well as in estuaries, tidal basins and lagoons, an asymmetry in tidal current velocity is widespread, with some areas dominated by flood-oriented flows while other areas are dominated by ebb-oriented flows (Section 4.4). Often the weaker flow is only recorded by the truncation of foresets and the subsequent reactivation of the sand wave. Studies conducted in the strongly tidal Minas Basin, in the Bay of Fundy suggest that 'herringbone' cross-stratification is uncommon (Klein, 1970a). It tends to form only during the period of transition from spring to neap tides, as the depth of reworking with each successive (ebb or flood) tide becomes less. Thus, sets of flood-oriented cross-strata may develop on an ebb dominated part of a sand bar during a spring tidal cycle and become preserved as the subsequent tidal cycles become progressively more neap (Klein, 1970a, p. 1121). In the long term, however, the potential for preservation of these flood-oriented sets of cross-stratification is probably low, because as the bar gradually accretes, each successive phase of deep level reworking accompanying the spring tidal cycles will tend to destroy the flood-oriented set formed during the previous spring tidal cycle.

Nearshore sand banks and channels are constantly shifting (Hayes and Kana, 1976), hence at any fixed point the net migration direction of sand waves may reverse after periods of weeks, months or years

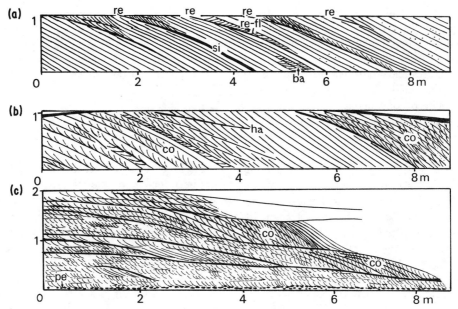

Fig. 7.1 The internal structures of some fossil offshore tidal sand waves. Compound sets of cross-stratification showing reactivation surfaces (re) some with silt drapes (si), back-flow ripples (ba), reverse-flow ripples (re-fl), hanging set boundaries (ha), convex-up boundaries (co) and pebble horizon (pe). Source – the Lower Sandfjord Formation, Precambrian, Northern Norway (after Levell, 1980).

depending on the occurrence of storms which modify the geometry of the banks and thereby influence tidal paths. However, in the offshore realm large tidally generated sand waves do not reverse with any detectable frequency. In the field of large sand waves situated off the coast of the Netherlands for example, the direction of large sand wave migration appears to have been uniform over a 50 year period of observation (A. H. Stride, personal communication). Small sand waves which ascend the slopes of large sand waves probably have a moderately high potential for preservation (Section 5.2.4). Further, small sand waves on the lee side of longitudinal sand banks have a good chance of preservation (Section 5.3.3), and it is possible that 'herringbone' cross-stratification in this environment forms in a manner analogous to the examples documented on intertidal sand bars by Klein (1970a).

Experiments with unidirectional flow demonstrate that the back-flow component of the separation eddy, developed in the lee of sand waves can generate ripples which migrate towards the actively encroaching slipface (Jopling, 1961; Guy, Simons and Richardson, 1966). For a given depth of flow, the nature of these regressive, back-flow ripples and their reverse cross-stratification depends upon the current velocity. If the latter is not much above the threshold for sand wave migration the ripples are confined to the region of the trough and are smothered as a single form set by the avalanche foresets. If the flow velocity is substantially higher the ripples ascend the lower part of the lee slope, producing a coset of climbing cross-lamination (Allen, 1968, Fig. 16.11.12). However, in a tidal environment characterized by reversing flows of unequal competence, regressive, reverse-flow ripples and small sand waves may be generated by the weaker flow. Regressive climbing ripples probably of tidal origin are recorded from the Ordovician Peninsula Formation of Table Mountain, South Africa (Hobday and Tankard, 1978), and the Cretaceous Lower Greensand of England where they are associated with sand waves featuring reactivation surfaces, avalanche foresets of variable dip and silt drapes (Allen and Narayan, 1964; Narayan, 1971).

Similar features were recently described by Levell

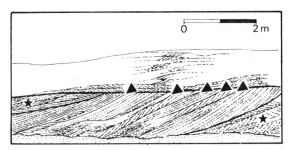

Fig. 7.2 Tidal sand wave with concave foresets and regressive climbing ripples. Arrow heads indicate reactivation surfaces. The sand wave crest appears to have migrated forward about 1.0 m between successive phases of erosion. The sand wave may have advanced under conditions of tidal flow augmented by wind induced flow. The sands with bipolar cross-stratification (*) appear to document fairweather accretion by flood-and ebb-flows of almost equal competence (after Raaf and Boersma, 1971).

(1980) who came to the conclusion that the Lower Sandfjord Formation in North Norway was deposited on a tidal continental shelf. Levell noted, that while simple sets of planar cross-stratification were predominant, many sandstones exhibited compound sets of cross-stratification. Some displayed reactivation surfaces, silt drapes, reverse-flow ripples and back-flow ripples (Fig. 7.1(a)) while other sets featured hanging set boundaries and convex-up set boundaries (Fig. 7.1(b), (c)). The convex-up set boundaries were attributed to the erosional effects of eddies in the lee of small sand waves migrating down the flanks of larger sand waves.

The internal structure of one small Cretaceous sand wave has been documented in detail (de Raaf and Boersma, 1971, Fig. 5). Whilst making a *net* advance of 2.0 m this sand wave suffered five periods of crestal erosion and reactivation, and two periods when back-flow associated with the separation eddy was sufficiently powerful to generate cosets of climbing ripples (Fig. 7.2). Each erosional phase caused the crest to recede 0.5 to 0.8 m, the result possibly of the weaker opposing tidal current. Between these erosional phases the sand wave advanced about 1.0 m, a figure not commensurate with fair weather migration. Modern tidal sand waves on Warts Bank, located to the south of the Isle of Man, migrate on average 0.05 to 0.1 m/day although an advance of 0.74 m was recorded during one flood tide (Jones, Kain and Stride, 1965). Similar mean values of around 0.04 m/day are suggested for sand waves in the Southern Bight of the North Sea (McCave, 1971b). Thus the Cretaceous sand wave may have been advancing in storm influenced conditions with waves and/or wind-driven currents enhancing sand erosion and transport.

The internal structure of a late Precambrian Dalradian tidal sand wave was traced 20 m in the direction of migration (Anderton, 1976). The angle of dip of the foresets progressively decreased from 15° to 5°, presumably indicating a strengthening current.

The Dalradian offshore tidal sequence in northwest Scotland also exhibits climbing cross-stratification (Anderton, 1976). The cosets are 1 to 2 m thick and are composed of sets of thin sigmoidal foresets 0.5 to 0.6 m high. Well preserved sets display upstream and downstream wedging, hanging set terminations and reactivation surfaces interpreted by Anderton as reflecting random periodic growth and decay of the sand waves. The formation of climbing cross-stratification has been inferred from the study of slowly migrating flood tidal sand waves in St Andrew Bay, Florida (Salsman, Tolbert and Villars, 1966). But Anderton suggested that the partial preservation of stoss laminae and the consistent facing direction of the large-scale foresets indicated deposition from a powerful but slowly waning sand laden current. He invoked an exceptional storm induced current reinforcing an ebb tidal current.

The genesis and environmental context of three ancient shallow marine sand formations were reviewed by Nio (1976). From his analysis of the early Tertiary Roda Sandstone of Spain, he concluded that it represented a fossil sand wave complex probably generated by tidal currents. Growth of the sand wave field commenced with the formation of small sand waves producing sets of cross-stratification between 0.3 and 1.5 m thick. These represent his initial sand wave facies. As growth continued the initial sets of cross-strata passed downstream into larger sets up to 10 m thick with foresets dipping at 25° to 28°. These large sets constitute his main sand wave facies. The increasing occurrence of reactivation surfaces as the sets are traced further 'downstream' suggests that the growth of the fully developed sand waves became strongly episodic. Eventually however, the sand waves passed into a state of decline. Smaller sets of cross-stratification developed on both the stoss and lee sides of the former sand wave, gradually reducing the angle of slope of the now complex bedform. The comparison between the Roda sand waves and their supposed modern analogues is interesting because of the relatively massive scale of the ancient foresets. The geographical context of the Roda Sandstone sand waves is not entirely clear. An alternative ebb-dominated tidal delta origin could be entertained, but would only be convincing if an associated barrier, inlet and, perhaps, a corresponding flood-dominated tidal delta were also recognized.

7.3.2 SAND BANKS

Modern examples of longitudinal sand banks are generally asymmetrical in transverse profile with gentle stoss slopes of about 0.5° and steeper lee slopes of up to 6° (Section 5.3.2). The modern Norfolk sand banks show reflectors parallel to the lee slopes indicating large-scale lateral bank migration (Fig. 5.17). Cores, however, display cross-stratification developed by large and small sand waves migrating obliquely along the steeper flank of the bank (Fig. 5.19). Early Holocene sand banks in the Celtic Sea have been preserved in almost their original form, although the slopes have been reduced by degradation (Section 5.3.1). Further, Gullentops (1957) has surmised that an extensive group of sand bodies of Diestian age in Belgium is analogous to groups of present-day sand banks located in the neighbouring North Sea.

In recent years numerous authors have described the internal structures, geometry and context of Mesozoic sandstone sequences in the western interior of North America, and concluded that some represent the remains of longitudinal sand banks (Berven, 1966; Campbell, 1971; Sabins, 1972; Brenner and Davies, 1973, 1974; Berg, 1975; Cotter, 1975; Brenner, 1978; Seeling, 1978). The sequences commonly exhibit a coarsening-upwards from muds to coarse sands, within thicknesses ranging from 2 to 15 m. Where the geometry of the sandstone bodies can be ascertained they are clearly elongate and some sand banks extended for 20 to 50 km (Fig. 7.3(a), (b)) (e.g. Berven, 1966; Campbell, 1971; Seeling, 1978). The lower, muddy parts of sequences are intensely bioturbated, but in higher parts the intensity of burrowing declines and cross-stratification, which is generally unimodal, assumes importance. In some sequences there are additional units. Brenner and Davies (1973, 1974) recorded thick, sharply based, shell-packed sandstones in the Upper Jurassic sequences in Wyoming (Fig. 7.3(c)). All the sequences under consideration have one further characteristic – they are geographically remote from deposits indicating the contemporary shoreline.

The origin of these ancient offshore longitudinal sand banks is far from clear. Brenner and Davies (1973) proposed that the Upper Jurassic sand banks in Wyoming were formed and shaped predominantly by tidal currents, and that storm events were responsible for the accumulation of the shell-packed sandstones. A year later Brenner and Davies (1974) were less sure about the role of tidal currents; the bars were considered to have resulted from the interplay of storms, tidal currents and regional circulation currents. During the early 1970s the storm-dominated regime of the continental shelf on the east side of North America was being considered as a fine model for comparison, in terms of process and scale, with the great Mesozoic interior seaways. Cotter (1975) drew a comparison between the sandstones of the Ferron Sandstone Member and the modern deposits off the coast of Georgia. There was evidence of tidal activity in the vicinity of the contemporary shoreline, but no clear indication of tidal activity in the sediments deposited offshore. Brenner (1980) in his review of the depositional systems in ancient epicontinental seaways developed the comparison between the Mesozoic seaways and the modern continental shelf off eastern USA. There is indeed a strong probability that a storm flow field was at least in part responsible for the formation of offshore sand banks along the western shoreline of the Cretaceous epicontinental seaway as argued by those authors, but attention can also be drawn to the geographical context of the seaway which is compatible with the view that offshore tidal currents were also effective in shifting sediment. The context of the seaways is

Ancient offshore tidal deposits 177

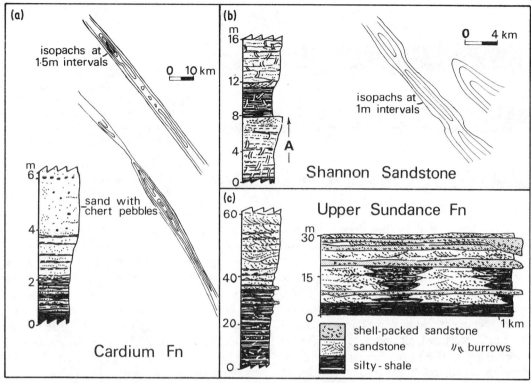

Fig. 7.3 Geometry and internal structures of fossil longitudinal sand banks possibly of tidal origin. (a) Lower Cardium Sandstone of the Crossfield–Garrington area, Alberta: Turonian (after Berven, 1966). Note the extremely elongate form of the banks and the coarsening-upwards sequence. (b) Shannon Sandstone, Heldt Draw and Hartzog Draw fields, Wyoming: Lower Campanian (after Seeling, 1978). Isopachs relate to lower part of sequence (A). Note the coarsening-upwards sequences. (c) Upper Sundance Formation, Wyoming: Oxfordian (after Brenner and Davies, 1973, 1974). Note the coarsening-upwards sequence and the way in which the banks appear to have accreted vertically rather than migrated laterally (compare with Figs 7.4 and 7.5).

discussed in Section 7.6.2, but it is worth pointing out below the resemblances that exist between the Mesozoic sand banks and those of the modern North Sea.

In his description of some Cretaceous sand banks Campbell (1971) noted bioturbated sets of cross-stratification with the facing direction of the foresets consistently directed towards the south-east, parallel to the former bank crests. The analogy with the Norfolk sand banks, located off the eastern coast of England, is close. These north-west to south-east oriented sand banks have sand waves moving north-westwards on the stoss slopes and predominantly south-eastwards on the preservable seaward facing lee slope, although it should be noted that the migration direction of sand waves on the upper parts of the stoss and lee slopes veers towards the crestline (Fig. 4.25). Each of these modern sand banks has some resemblance to an elongate clockwise circulation cell of sand (Section 3.4.5), although not in the simple form envisaged by some workers (Swift, 1975). The Norfolk sand banks are growing north-westwards and migrating north-eastwards, thus it is the south-eastward directed cross-strata which have a high potential for preservation (Section 5.3.3) although a 180° spread of dip directions is evident (Fig. 5.19). If a similar sand deposition pattern operated during the formation of the Cretaceous sand banks it would be erroneous to use the facing direction of the foresets as an indication of the dominant regional net sand transport direction by tidal currents.

Modern longitudinal sand banks in temperate

178 Offshore Tidal Sands

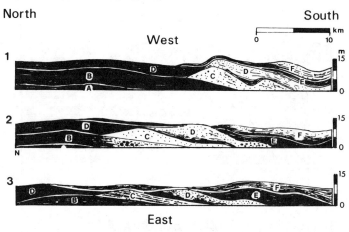

Fig. 7.4 Evidence of the growth and mode of preservation of longitudinal sand banks in the Cretaceous Viking Formation of Saskatchewan. Sediments divided into six lithostratigraphic units (A–F): the sandstones in units C and D appear to represent former longitudinal sand banks, orientated in an east-west sense, which gradually migrated southwards. Distance from Section 1 to Section 3 is 100 km (after Evans, 1970).

latitudes are bioturbated by the deeply infaunal deposit feeder *Echinocardium* and the activity of sand eels (Section 6.13). The effect of these extant organisms is to cause substantial biogenic disturbance to sediments down to depths of 55 to 60 cm below the sediment–water interface. Similar burrowing is a fairly common feature of the Cretaceous sand banks. The bioturbated siltstones of the Ferron Sandstone (Cotter, 1975) probably represent inter-bank deposits comparable with the sands adjacent to the Norfolk sand banks which are colonized by polychaetes, bivalves, gastropods, brittle-stars and echinoids (Section 6.13).

The manner in which longitudinal sand banks may develop in relation to one another, and their mode of preservation has been demonstrated by a study which Evans (1970) conducted into the Albian sediments of Saskatchewan. Evans showed how elongate bodies of sand oriented east–west migrated southward with time (Fig. 7.4). As one bank (C) ceased active growth and became smothered with mud, another bank (D) was initiated just to the south of the site of major development of the first.

In his reinterpretation of some offshore marine sand banks which have been well preserved in the late Precambrian Tanafjord Group of North Norway, Johnson (1977) emphasized the problem of pinpointing the agents of sediment transport. The contemporary shoreline configuration was uncertain and so Johnson sought distinguishing criteria within the sand banks. The bank sandstones exhibited abundant cross-stratification; and at one horizon there were a number of deeply dissected channels infilled by coarse large-scale cross-stratified sandstones. The channels, Johnson discovered, each displayed quite distinct foreset dip directions. A system of interconnected channels in the west showed easterly dipping foresets, while a second system in the east only exhibited westerly dipping foresets. From this pattern the sand banks were tentatively regarded as tidal in origin. In an elegant model (Fig. 7.5) Johnson invoked storm enhanced tidal currents to generate the abundant cross-stratification of the bank sandstones, while separate flood-dominant and ebb-dominant tidal flow paths were considered responsible for the respective systems of cross-connected channels.

7.3.3 SAND AND MUD SHEETS

Extensive offshore sand sheets grading into or intercalated with muds occur at the distal ends or sides of net sand transport paths on the continental shelf, and on the seaward margin of estuaries and tidal inlets (cf. Fig. 5.3 with Fig. 4.10). The fine grained facies of the Precambrian Jura Quartzite (Anderton, 1976) and the Lower Cambrian Duolbasgaissa Formation (Banks, 1973) have been interpreted as offshore tidal muds with interbedded sheets of sand. Modern sand sheets of weak current areas of the continental shelf are populated by a diagnostic assemblage of organisms including the polychaetes *Ditrupa* and *Hyalinoecia*, the scaphopod *Dentalium*, the coral *Caryophyllia*, barnacles and numerous molluscs (Section 6.12.2), but of these only non-skeletal polychaetes comparable with *Hyalinoecia*, and a few

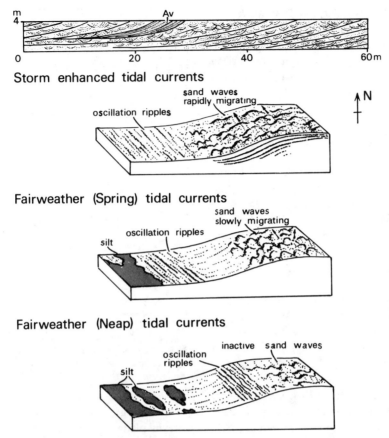

Fig. 7.5 The lateral migration of a fossil longitudinal sand bank, probably of tidal origin, preserved in the late Precambrian Dakkovarre Formation of Northern Norway. Section through sequence shows a succession of gently inclined surfaces facing towards the west. Between each surface cross-stratified sands indicate sediment transport northwards, along the strike of the contemporary slope. Avalanche foresets (Av) indicate the presence of currents, presumed to be tidal, cutting across the bank(s). The model shows northward migration of small sand waves during storms when tidal currents were enhanced by waves and wind driven flows: movement of sand waves slowed during spring tidal cycles and ceased during neap tidal cycles. Sands on the top of the banks were worked into oscillation ripples by the fairweather wave regime (after H. D. Johnson, 1977).

molluscs, were probably extant in late Precambrian and Lower Cambrian times (Brasier, 1979).

By Jurassic times the faunas were much more diverse. The mudstones and lenticular sandstones of the Neill Klinter Formation have been interpreted as offshore estuarine in origin (Sykes, 1974). These sediments display deposit feeding trails including *Planolites* and *Gyrochorte* on the tops of rippled sandstones and compare with the thin bioturbated sands which occur in the Southern North Sea off the coast of Germany (Reineck, 1967; Gadow and Reineck, 1969). The organisms which colonize the modern offshore estuarine sands are mainly infaunal and include *Cerianthus*, echiurids, crustaceans and echinoids (Fig. 6.1).

The late Lower Cretaceous sediments of the Lower Greensand in south-east England appear to have formed in a tidal gulf (Section 7.7). To judge from the palaeogeographical context, the horizontally laminated and rippled sands of Lower Aptian age which are located at Petworth were probably formed in an outer zone of a sand sheet. In Saskatchewan, Canada, the interlaminated muds and sands of the Viking Formation, which is Upper

180 Offshore Tidal Sands

Albian in age, have been interpreted as intercalated sand ribbons and muds (Evans, 1970, p. 484), but since sand ribbons are generally associated with zones of scouring, there may be a closer comparison with longitudinal sand patches alternating with mud layers (Sections 3.4.4 and 7.6.2).

7.3.4 SCOURED HORIZONS AND BED-LOAD PARTINGS

In modern seas vigorous tidal erosion is concentrated at bed-load partings (Fig. 4.10), in narrows (Fig. 2.6), off headlands, in estuaries and more locally in narrow zones between sand ribbons and between sand banks. The tidal currents are not only able to scour sand and gravel, but they are also capable of eroding consolidated and lithified substrates (Sections 3.4.1 and 3.4.2).

The Plio-Pleistocene Coralline and Red Crags of East Anglia document the transgression and still-stand of the pre-Holocene North Sea. Large NNE to SSW depressions up to 20 km long, 6 km wide and 45 m deep characterize the chalk and clay surface on which rest the cross-stratified shell-rich, quartz sands of the Crags. Post-Lower Pleistocene tectonic movements have been invoked (West, 1968, p. 245), but it is possible that the depressions were, in part, scoured by offshore tidal currents. The depressions compare in magnitude with submarine trenches of Quaternary age eroded into the Jurassic limestones and clays by tidal currents (Donovan and Stride, 1961; Section 3.4.1).

In his study of the tidal current formed Jura Quartzite, Anderton (1976) recognized three types of major intraformational erosion surface. There are large channel scours, laterally persistent and irregularly undulating erosion surfaces, and laterally persistent pebble strewn surfaces which are almost plane. The channels are up to 12 m wide and 2 m deep, scoured into coarse cross-stratified sands and infilled by sand waves which migrated along the channel floor. Some of the undulating erosion surfaces may represent the longitudinal sections of these channels. In magnitude these channels compare favourably with the groups of tidally scoured subparallel furrows reported from the English Channel Bed-Load Parting (Section 3.4.2) which measure up

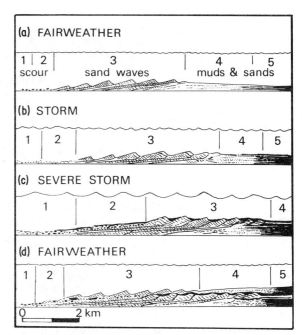

Fig. 7.6 Model illustrating how major sedimentary facies, in a tidal shelf sea, may intercalate with one another as a result of an exceptional storm event (after Anderton, 1976).

to 8 km long, 14 m wide and 1 m deep. The pebble strewn planar surfaces probably signify long term erosion. Comparison of the relative concentration of pebbles on an erosion surface and pebbles scattered in the underlying sediment gives an estimate of the thickness of sediments eroded (Anderton, 1976). Similarly, a distinctive pebble horizon in the Cretaceous Viking Formation (Section 7.3.3) also appears to record the effects of tidal scour (Evans, 1970).

Fossil bed-load partings have not yet been identified, but they should be sought because they would provide valuable information about tidal regimes in the past.

7.4 Tidal currents aided by storm processes

It is clear from studies of modern seas that sediment transport will be greatest on the geologically frequent occasions when the competence of the tidal currents is enhanced by storm waves and wind induced currents (Sections 4.5.2 and 4.5.3). Indeed, Anderton (1976) has suggested that exceptionally severe storms occurring with a frequency of between 10^3 to 10^4 years were responsible for major discontinuities and the thick cross-stratified sand units located within the mudstone facies of the Precambrian Jura Quartzite. His intuitive model for these rocks (Fig. 7.6), is based on bedform zones in the tidally active seas of north-west Europe (e.g. Fig. 3.1), and has given a useful hint of what might be expected for other shallow marine gulfs. Anderton invoked a temporary downstream shift of bedform zones in response to the combined effect of tidal currents and storm processes, resulting in the advance of the sand wave zone into the zones of rippled sand and mud. As conditions returned to the fairweather state, the zones of rippled sand and mud smothered the distal sand waves generated during the storm. This model is important because it supports the earlier ideas of Johnson and Belderson (1969) and because it demonstrates that with the passage of time, a tidal sea may deposit a sequence of intercalated facies without long term changes in sea-level or shoreline configuration. However, the severity of the changes in the location of bedform zones is more extreme than is known from modern tidal seas.

7.5 Factors controlling the structure and composition of offshore tidal sediments through geological time

The internal structure of ancient offshore tidal deposits is not only a function of the water movements. The rate of sand influx, the rate of sea-level change, climate and substrate are all important factors which influence the type and grade of sediment, the nature of the fauna and the preservation and distribution of bedforms.

One further and important factor is time. Major faunal and floral innovations have appeared since early Phanerozoic times. Mudstones of late Precambrian to early Cambrian formations display few definite animal traces (Anderton, 1976; Banks, 1973; Johnson, 1977), but their modern counterparts bear a rich assemblage of shells and burrows (Fig. 6.1). The frequently shifting surface of tidal-current swept sand waves has proved to be a relatively hostile substrate for organisms. Large sand waves of modern seas are colonized by a sparse fauna of polychaetes and burrowing bivalves, but the faunas of modern small sand waves are much richer, including molluscs and echinoderms (Sections 6.11.1 and 6.11.2). Nevertheless, the deposits of small intertidal, near-shore and offshore Cambrian sand waves, which display the deep penetrating vertical burrows *Skolithos* and *Monocraterion*, demonstrate that even 580 million years ago this type of environment was not totally inimical to life (Swett, Klein and Smit, 1971; Swett and Smit, 1972; Goodwin and Anderson, 1974). The evolution of the semi-infaunal echinoderms and the rapid-burrowing bivalves during the Mesozoic later added to the diversity of organisms which could survive in spite of the mobile substrate. Consequently, as successive sediments formed in tidally active regions through the geological record, there is an increasing propensity for them to show the effects of biogenic reworking. In short, the physical conditions which would have given rise to cross-stratified fine sands in the Lower Cambrian are probably now recorded by bioturbated sands.

7.6 Some possible palaeotidal regimes

Further progress in the study of ancient tidal-current deposited sediments is strongly dependent upon an improvement in the understanding of palaeoenvironmental settings. If, for an ancient continental shelf there is knowledge of its proximity to an ocean then an attempt can be made to reconstruct some features of the tidal regime that operated. While necessarily speculative this exercise is worthwhile because the regime invoked must not only be compatible with the overall context, it must also accord with the geometry, the grain-size and directional structures of the sedimentary bodies. The two examples which follow illustrate the considerations involved.

7.6.1 UPPER JURASSIC GULF OF WESTERN NORTH AMERICA

The Oxfordian seas of western North America formed a major epicontinental gulf about 4000 km long and between 500 and 1000 km wide, which opened northwards to the Arctic Ocean (Brenner and

Davies, 1973). It is assumed that the tidal oscillation in the Arctic Ocean caused twice-daily and daily tidal waves to enter the gulf from the north. The Coriolis effect would have given a tendency for tidal wave amplitude and current velocities to decrease from west to east across the gulf, in a manner similar to theoretical models of progressive tidal waves moving in water of uniform depth. As the tidal waves progressed into the narrowing gulf, there would have been a tendency for them to increase in amplitude, although frictional effects would have partially countered this amplification.

First, neglecting friction and assuming that the water depth (h) was uniform, these tidal waves would have progressed at a wave speed $(gh)^{\frac{1}{2}}$ dependent only on h, where g is the gravitational acceleration. For water depths of 100 m or 150 m and the present-day values of g in middle latitudes, the tidal waves would have taken 1.48 and 1.21 days respectively to pass from the northern entrance to the southern margin. Bed friction must have increased these periods, but not by more than a few percent on usual linear-friction models.

The water depth was probably lower towards the margin of the gulf than near the centre. One model representing such a depth distribution by a parabolic formula predicts the reduction of progressive-wave speed from the value it would have had, for a uniform depth equal to the mean actual depths (Defant, 1961, p. 208). For the twice-daily tidal wave the model indicates an increase in the travel times by about 18% for a gulf about 500 km wide having a maximum depth of 150 m but a mean depth of 100 m. The travel time is thus likely to have been between 1.25 and 2 days, a period appreciably greater than either the twice-daily or the daily periods which in Jurassic times were probably a little less than they are today.

The incident twice-daily and daily tidal waves would probably have produced reflected waves travelling north along the eastern side of the gulf. As in modern tidal gulfs and basins (Defant, 1961) the variation in water-surface level experienced by this Jurassic gulf may have been in the form of a series of amphidromic systems with central points of zero tidal range (Fig. 7.7). Since friction would have caused the reflected waves to be weaker than the

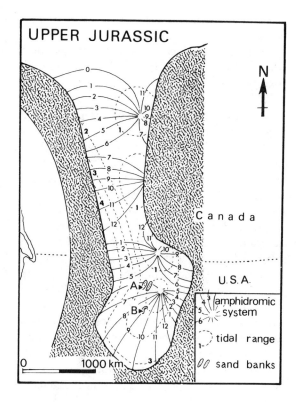

Fig. 7.7 Speculative reconstruction of the palaeotidal regime which operated in the epicontinental seaway over North America in Upper Jurassic times, with tidal range in arbitrary units. Shoreline configuration after Brenner and Davies (1973, Figs 5(a) and (b)). Longitudinal sand banks at A – Oxfordian Upper Sundance Formation Member (Brenner and Davies, 1973, 1974) (See also Fig. 7.3): at B – Callovian–Oxfordian Hulett Member (Stone and Vondra, 1972). (Full explanation in text.)

incident waves the amphidromic points would have been displaced towards the eastern side of the gulf.

If the sand ridges described by Brenner and Davies (1973) are correctly interpreted as offshore longitudinal tidal sand banks, the orientation of their crests indicates the approximate flow direction of the strongest local tidal current (Section 3.4.5). However, reliable deductions cannot be made about the exact positions of the amphidromic points, nor the cotidal lines which radiate from them because neither have a unique relationship to the distributions of current speed and direction.

Ancient offshore tidal deposits 183

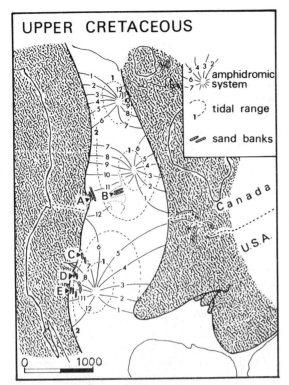

Fig. 7.8 Speculative reconstruction of the palaeotidal regime which operated in the epicontinental seaway over North America in Upper Cretaceous times, with tidal range in arbitrary units. Shoreline configuration from Cotter (1975, 1976), Kauffman (1974, Fig. 12-1A), and Pryor (1960, Fig. 17). Longitudinal sand banks at A – Turonian, Cardium Sandstone (Berven, 1966, Fig. 13) (see also Fig. 7.3(a)): at B – Albian, Viking Formation (Evans, 1970, Figs 1 and 5: see also Fig. 7.4): at C – Lower Campanian, Sussex and Shannon Sandstones (Berg, 1975; Seeling, 1978) (see also Fig. 7.3b): at D – Coniacian, Gallup Sandstone (Campbell, 1971, Figs 1 and 2): at E – Woodside Unit in Ferron Sandstone (Cotter, 1975). Note: The sand banks of Albian age (B) are shown in association with an inferred estuarine shoreline (dotted lines) which later retreated westwards to the position shown in the reconstruction (solid line).

7.6.2 UPPER CRETACEOUS EPICONTINENTAL SEAWAY OF WESTERN NORTH AMERICA

During the Albian to Maestrichtian ages a major epicontinental seaway, 4800 km long and up to 600 km wide, extended northwards from the present site of the Gulf of Mexico to the Arctic Ocean (Kauffman, 1974, Fig. 12-1A). Appreciable twice-daily and daily tidal waves may have entered from the Arctic Ocean in the north, and the growing Atlantic in the south. The reconstruction (Fig. 7.8) pertains to the transgressive phases in the Upper Albian and the Coniacian. Some estimate of the contemporary water depth is desirable. The ratio of planktonic to benthic foraminifera and foraminiferal taxonomic analogues suggest that the Coniacian water depth in the south central part of the seaway was about 200 m (Kent, 1968). If, as for the Jurassic gulf there was a series of amphidromic systems developed in the seaway, then strong reversing tidal currents associated with high tidal ranges would have flowed close and parallel to the western coast of the seaway, in the vicinity of the Gallup longitudinal sand banks (Fig. 7.8D). If these offshore sand banks were indeed tidal then by analogy with their modern analogues (Section 3.4.5) the dominant tidal current velocities at the surface would have attained values of about 50–200 cm/s during mean spring tides.

Recent investigations of the growth patterns recorded by Maestrichtian bivalves living close to the western shore of the seaway constitute an independent check on these suggestions. Pannella (1976) has identified the influence of tidal rhythms on the shell growth of *Limopsis* collected from the Fox Hills Formation (Maestrichtian) of South Dakota. The structures recorded compare closely with those exhibited by bivalves dwelling in the modern Caribbean Sea which has a mixed tide. More recently, Klein and Ryer (1978) have pointed out that there are many Upper Cretaceous units which provide evidence of tidal current activity in the vicinity of the shoreline. Thus, there is growing evidence for tidal activity in the Cretaceous seaway of western North America.

7.7 Sedimentology of a tidal sea: the Lower Greensand of southern England

The glauconitic quartz sands of the Lower Greensand record the spread of a shallow sea over southern England in late Lower Cretaceous times. Sorby (1858) was the first to suggest that the cross-stratification in the sands indicated the activity of tidal currents. Later, Narayan (1971) attempted a

Table 7.1 Late Lower Cretaceous zones.

Stage		Zone	Approximate stratigraphic level of formations
ALBIAN	LOWER	*Douvilleiceras mammillatum* *Leymeriella tardefarcata*	Folkestone Beds
APTIAN	UPPER	*Hypacanthoplites jacobi* *Parahoplites nutfieldensis* *Cheloniceras martinioides*	Sandgate Beds and Faringdon Sponge Gravels
	LOWER	*Tropaeum bowerbanki* *Deshayesites deshayesi* *Deshayesites forbesi* *Prodeshayesites fissicostatus*	Hythe Beds Atherfield Clay

comparison between the deposits of the Lower Greensand and the tidally influenced sediments on the sea floors around Britain. A tidal interpretation is indeed attractive. Even in Lower Cretaceous times the Atlantic basin (Hallam and Sellwood, 1976) was probably sufficiently large to generate daily and twice-daily tidal waves; and it is doubtful if any other marine agent would have so consistently and effectively reworked the sands, thereby preventing widespread colonization by infauna and destroying the burrows of the more successful animal groups. Stratigraphic studies have pinpointed transgressive phases in the Lower Aptian, the Upper Aptian and the Lower Albian (Table 7.1). Nearshore sediments have been recognized in the southern part of the Isle of Wight (Dike, 1972), but as yet tidal flat sediments have not been identified within the Lower Greensand. Sediments of the shoreline environments were probably substantially eroded when offshore conditions became established, but it is possible that small pockets of shoreline sediments remain undetected.

7.7.1 LOWER APTIAN PHASE

During the Lower Aptian the sea is thought to have transgressed along the line of the present English Channel (Middlemiss, 1962; Kirkaldy, 1963) and flooded the Wealden alluvial mudplain (P. Allen, 1975). The basal horizon which records this event, the Perna Bed, consists of pebbles, granules, sand and fragments of the derived ammonite *Pavlovia* mixed with numerous Aptian shells (stratigraphy, Fig. 7.9).

The latter include byssally attached bivalves, cemented oysters, shallow burrowing bivalves, brachiopods, hydrozoans and corals (Casey, 1961, p. 505). Thus, there was a rich epifauna, well adapted for a suspension feeding mode of life (see reconstruction by Kennedy, 1978, Fig. 93). The shell concentrate compares with the tidally winnowed basal shelly gravels that underlie the finer sediments forming in response to present tidal regimes (Section 6.12). During this initial rise of Cretaceous sea-level, silt and clay were trapped in estuaries along the western shoreline (Casey, 1961, p. 515). However, after the transgressive phase had ceased a mud blanket of variable thickness, the Atherfield Clay, extended eastwards.

It is suggested here that the Lower Aptian sea penetrated northwards and westwards as far as Faringdon. Although the earliest deposits at Faringdon are Upper Aptian (*nutfieldensis* zone), a pronounced WNW to ESE aligned depression (Krantz, 1972, Fig. 3) provides evidence of an earlier period of marine erosion. The depression is significant because it may represent the 'upstream' end (zone 1 of Belderson and Stride, 1966) of an ebb current dominated bed transport path operating in a narrow gulf which opened to the south and east (Fig. 7.10). To the north of Guildford, in the area where sand ribbons (zone 2) would be predicted, boreholes at numerous localities demonstrate that there was no long term deposition of sands until Lower Albian times (Middlemiss, 1975). Near Farnham 80 km to the south-east of Faringdon, the Lower Aptian

Ancient offshore tidal deposits 185

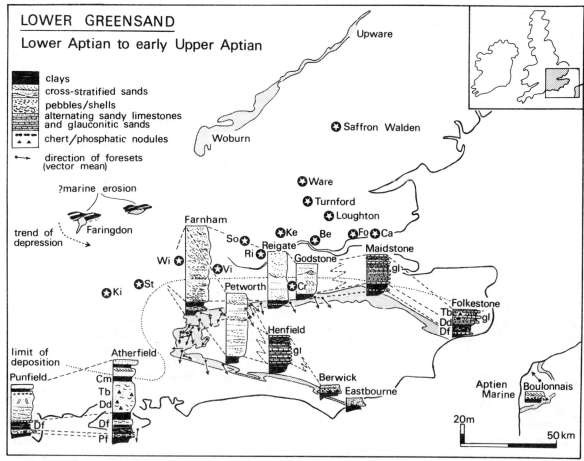

Fig. 7.9 Correlation of the lithofacies in the Lower Greensand for the Lower Aptian – early Upper Aptian period. Stratigraphic information from Casey (1961). Symbols on left margin of logs are late Lower Cretaceous zones (see Table 7.1). gl – glauconite. Azimuths shown are mean values obtained by Narayan (1971) from cross-stratified sands. (*) – Borehole sites with information relating to the Lower Greensand (Middlemiss, 1975): Be, Beckton; Ca, Canvey Island; Cr, Croydon; Fo, Fobbing; Ke, Kentish Town; Ki, Kingsclere; Ri, Richmond; So, Southall; St, Stratfield Mortimer; Vi, Virginia Water; Wi, Winkfield.

Hythe Beds display large-scale foresets attributable to large and small sand waves, indicating a net transport direction of sand towards the south and south-east. These sands represent zone 3. Near Petworth, 105 km to the south-east of Faringdon, the equivalent sands exhibit horizontal lamination and small-scale rippling (zone 4), and are suggestive of weaker tidal currents (Narayan, 1971). Further to the south-east near Henfield, the lithology changes to interbedded quartz calcarenites and glauconitic sands while around Eastbourne 160 km south-east of Faringdon, the basal gravels were only lightly coated with silts and argillaceous sands (zone 5) during Lower Aptian times.

This supposed Lower Aptian gulf was therefore comparable in length with the present Bristol Channel, most of which is subjected to an ebb-dominated net sand transport path approximately 180 km long (Fig. 4.10). The extent of the Lower Aptian cross-stratified sands transverse to the inferred net sand transport path suggests that the zone of sand waves was about 60 km wide, again comparable with the sand wave zone in the outer part of the modern Bristol Channel. Shallowing and

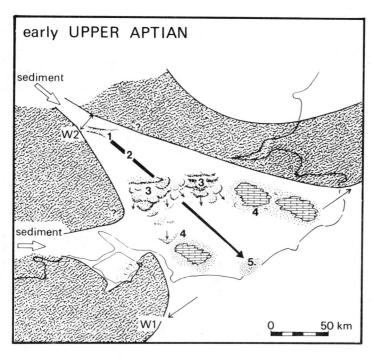

Fig. 7.10 Palaeogeographic reconstruction of south-east England in early Upper Aptian times. Erosional and depositional zones: 1 – tidal scour; 2 – tidal scour and development of sand ribbons (hypothetical); 3 – sand waves and sand ripples; 4 – interbedded calcareous and glauconitic–argillaceous sands; 5 – thin silts and argillaceous sands; small arrows – local sediment transport; large arrow – inferred regional sediment transport path. W1 and W2 are widths of the gulf at entrance and neck respectively used in calculating tidal wave amplification (Appendix 7.2). Shoreline configuration in vicinity of Isle of Wight based on the work of Dike (1972).

convergence of the coastline in the up-estuary direction probably caused a significant twofold or threefold amplification of the twice-daily tidal wave (Appendix 7.2). At present, just 5000 years after the general cessation of the Flandrian transgression the thickness of Holocene sands around the British Isles is less than 12 m (Table 5.1). The Lower Aptian sand wave zone accreted up to 100 m of cross-stratified sands in a period of about 3.3×10^6 years (assuming the Aptian stage represents 10^7 years), but the thickness of the deposits decreased substantially to the south-east. Although the overall rate of accretion in the zone of sand waves during the Aptian was quite low, the Holocene record suggests that the sediment might have accreted quite rapidly after minor transgressive phases. The respective zones of sediment deposition shifted with time. At Petworth, for example, cross-stratified sands alternate with horizontally bedded sands, but at Henfield further to the south-east cross-stratification is uncommon. The substantial thickness of sands in the zone of sand waves suggests that subsidence, sediment supply and sea-level were maintained in a delicate balance for a considerable period

7.7.2 UPPER APTIAN AND LOWER ALBIAN PHASES

The second transgressive phase commenced in middle Upper Aptian times and united the northern and southern seas. The sediments which record this event are rich in carbonate detritus (Fig. 7.11). This suggests that the alluvial sand which had formerly veneered the lithified substrate had been substantially incorporated into the shelf sea bedforms during the first transgressive event. Thus, following the second transgressive phase there was a paucity of sand with a consequent increase in the relative abundance of skeletal carbonate. Near Faringdon large-scale cross-stratified shelly gravels, packed with sponges, bryozoa and brachiopods infilled the WNW to ESE aligned depressions (Krantz, 1972). To the south-east near Farnham, the equivalent sediments known as the Bargate Beds consist of cross-stratified, pebbly sandstones and very coarse skeletal grainstones. Middlemiss (1962, Fig. 1) noted that comminution of shelly material increased towards the south and east. Further south-east near Eastbourne, the carbonate sediments gave way to silts which rest on the Lower Aptian.

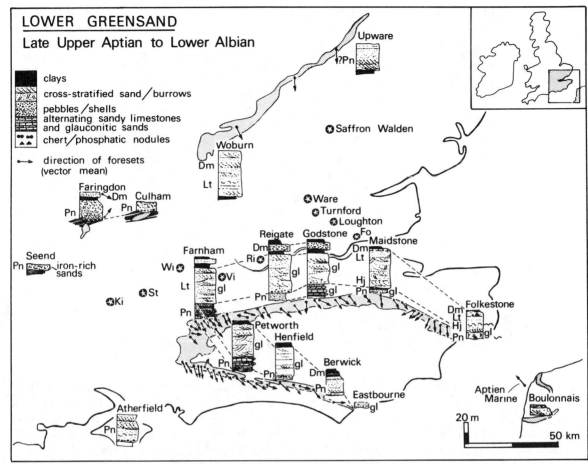

Fig. 7.11 Correlation of the lithofacies in the Lower Greensand for the late Upper Aptian and Lower Albian period. Stratigraphic information from Casey (1961). Symbols adjacent to logs are late lower Cretaceous zones (see Table 7.1). gl – glauconite. Azimuths shown are mean values obtained by Narayan (1971) from cross-stratified sands. (*) – boreholes sites (see legend to Fig. 7.9).

The third transgressive phase is marked by a renewed influx of quartz sand and pebbles. Thick sequences of cross-stratified sands, the Folkestone Beds, indicate that large and small sand waves migrating generally towards the south and south-east, were more widespread than in Lower Aptian times (Fig. 7.12). The palaeocurrent evidence suggests that the sand transport path extended from Upware southwards across the London platform to the Weald, a distance of about 200 km. Thus, there was a consistent transport of sediment southwards across the Weald in Lower Greensand times. The predominance of north-westerly dipping cross-stratification in the Boulonnais area raises the possibility of a contemporary bed-load convergence in the vicinity of the present Dover Strait.

7.8 Tidal currents through geological time: implications for future studies

The recognition of the influence of tidal activity in the formation of ancient nearshore and shoreline sediments is relatively simple compared with the task of identifying ancient tidal deposits of offshore origin. Essentially, this is because there is a well known modern suite of diagnostic sedimentary

188   Offshore Tidal Sands

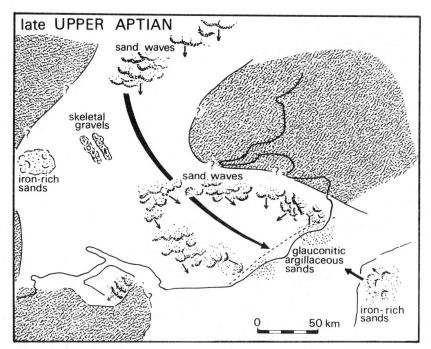

Fig. 7.12 Palaeogeographic reconstruction of depositional zones in south-east England during late Upper Aptian times.

features which characterize intertidal and nearshore subtidal sediments. But, as we have seen, the recognition of offshore tidal current deposits depends on larger scale considerations such as facies distributions and the palaeogeographical context. It is therefore understandable that attempts to trace the earliest tidal sediments have largely used deposits considered to be intertidal in origin.

The Transvaal Dolomite which is 2200 million years old, bears flat, domical and columnar stromatolites comparable with the intertidal stromatolites of Hamelin Pool, in Shark Bay, Australia (Eriksson and Truswell, 1973). But the Pongola Supergroup and the Moodies Group which are 3000 and 3300 million years old respectively, record tidal activity in mid-Archaean times (von Brunn and Hobday, 1976; Eriksson, 1977). The sediments of the Moodies Group are remarkably well preserved and display many of the classic features which are indicative of intertidal flats bordering an epicontinental sea. These important finds of shoreline tidal deposits indicate that elsewhere their offshore equivalents may exist, but they have not been located.

If these deposits are correctly identified as tidal, then the earliest of them are unlikely to be remote in time from the period when the Earth–Moon system originated. Since the angular momentum of the Earth–Moon–Sun system is considered to be constant, the reduction in the rate of the Earth's rotation caused by tidal friction is balanced by an increase in the Earth–Moon distance and an increase in the orbital angular momentum of the Moon (Lamar, McGann-Lamar and Merifield, 1970; Sündermann and Brosche, 1978). This has led investigators to suspect that the amplitude of tidal waves and related tidal current velocities were significantly greater in the past (Merifield and Lamar, 1970). As yet this has not been demonstrated. Application of Klein's (1971) model for the determination of the contemporary palaeotidal range in clastic rocks did not reveal tidal ranges in excess of 13 m for formations of late Precambrian age. Even for the tidal sediments of the Archaean the data are inconclusive. The intertidal sequences of the Pongola Supergroup range up to between 12 and 25 m (von Brunn and Mason, 1977), but those of the older Moodies Group only range up to about 5 m (Eriksson, 1977). Clearly, the tidal range at a particular site is much more a function of geographical context, than the proximity of the tide-raising body.

Even if the tidal ranges indicated by intertidal sequences which fine upwards are not very high, it is well established that tidal currents were a very important agent of sediment transport on the continental shelves of late Precambrian and Cambrian times. Tidally active seas probably covered both the north-western and south-eastern continental shelves of the Iapetus Ocean (Anderton, 1976; Banks, 1973; Johnson, 1977; Levell, 1980; Swett and Smit, 1972).

7.9 Main conclusions

In this study a number of ancient marine formations have been interpreted as offshore tidal current in origin. At the present time these ancient deposits are providing information on the nature of the internal structure of offshore tidal current deposits, but the arguments must not become circular. The basis for the identification of offshore tidal current deposits must be on facies characteristics and distributions, and geographical context; and as more is learnt about the internal structure and faunas of modern offshore tidal bedforms then these lines of evidence may also be used.

The chief importance of ancient offshore tidal current deposits is the information they can impart concerning the hydraulic regimes which operated in ancient shelf seas. This has been attempted in a crude fashion in this review, but in future it should be possible to refine the data and theoretical arguments and discover how ancient continental shelf seas evolved with time.

Appendix 7.1 Possible ancient offshore tidal current deposits.

Age	Formation, authors	Context; nature of sediments; tidal current paths
Plio-Pleistocene	Coralline and Red Crags of East Anglia, UK; Dixon, 1979*.	Westward extension of shallow North Sea into eastern England; substrate composed of chalk and clays; NNE to SSW orientated depressions of 50 m relief; cross-stratified shell-rich sands and silts; interpreted as small sand waves migrating in 15–25 m of water.
Tertiary	Roda Sandstone; Lower Eocene of southern Pyrenean basin, Spain; Nio, 1976*.	Shallow transgressive sea; shorelines uncertain; large-scale cross-stratified sandstone resting on bioturbated and fossiliferous marls formed in a restricted marine lagoon or bay; moderate to narrow range of palaeocurrent vectors; sequence interpreted as the deposits of a complex of large sand waves.
Cretaceous	Cardium Sandstones; Turonian of Alberta, Canada; Berven, 1966† (Fig. 7.3a).	Close to western shoreline of Cretaceous seaway (Fig. 7.8A); shales pass up into sands capped by horizon of chert pebbles; sands formed three NNW to SSE tidal banks; powerful reversing longshore tidal currents invoked.
	Viking Formation; Albian of Saskatchewan, Canada; Evans, 1970* (Fig. 7.4).	Close to western shoreline of major Cretaceous seaway (Fig. 7.8B); formation rests with erosional discordance on marine shales and bentonitic clays; muddy sands pass up into sands with shale partings; the sands formed two large ENE to WSW longitudinal banks which overlapped southwards; dominant tidal currents swept eastwards almost perpendicular to the general shoreline trend.

Appendix 7.1—cont.

Age	Formation, authors	Context; nature of sediments; tidal current paths
	Sussex and Shannon Sandstones; Lower Campanian of Wyoming, USA; Berg, 1975†; Seeling, 1978† (Fig. 7.3b).	Close to western shoreline of Cretaceous seaway (Fig. 7.8C); cross-stratified and rippled sandstone formations, each exhibiting coarsening-upwards, intercalated with shale formations; sandstones, interpreted by authors as offshore longitudinal sand banks orientated NNE to SSW, migrating towards the SE; tidal origin a possibility but no close analogy with modern tidal sand banks established.
	Gallup Sandstones; Bisti and Tocito Sands, Coniacian of New Mexico, USA; Campbell, 1971†; Sabins, 1972†.	Close to western shoreline of Cretaceous seaway (Fig. 7.8D); offshore sediments rest on regressive beach complex; massive lenses of cross-stratified sandstones in shale interpreted as longitudinal tidal banks orientated parallel to shoreline.
	Woodside Unit, Ferron Sandstone; Upper Cretaceous of Utah, USA; Cotter, 1975†.	Close to western shoreline of Cretaceous seaway (Fig. 7.8E); coarsening-upwards sequences separated by bioturbated siltstone; bipolar cross-stratification; interpreted as longitudinal sand banks.
	Lower Greensand; Aptian – Albian of S. England and N. France; Narayan, 1971*; Raaf and Boersma, 1971*; Dike, 1972*; Middlemiss, 1975; Nio, 1976*.	Shallow marine gulf opening to the south-east (Figs 7.10, 7.12); formation rests with erosional discordance on clays; sediments comprise glauconitic quartz sands and muds; sand waves present; general bed transport path directed towards SE; in Isle of Wight, S. England, the upper part of the glauconitic sands feature large-scale cross-bedding which is interpreted by Dike in Nio (1976) as a sand wave complex resting on sediments of estuarine or tidal flat origin; shoreline sediments beneath and above the sand wave deposits make the inferred offshore origin for the complex speculative.
Jurassic	Hulett Member; Callovian – Oxfordian of Wyoming, USA; Stone and Vondra, 1972*.	Southern part of the extensive shallow marine Jurassic gulf (Fig. 7.7); member rests on intertidal fine sands and muds; oolitic grainstone; interpreted as asymmetrical sand waves, some composite; one crescentic grainstone bank has crestline orientated transverse to local bimodal cross-strata dip vectors.
	Upper Sundance Formation; Oxfordian of Wyoming, USA; Brenner and Davies, 1973*, 1974* (Fig. 7.3c).	Tidal sediments rest on calcareous glauconitic shale; coarsening-upwards sequences interpreted as offshore tidal sand banks situated in southern part of Jurassic gulf (Fig. 7.7).
	Neill Klinter Formation; L. Jurassic of Greenland; Sykes, 1974*.	Offshore estuarine context; cross-stratified sandstones and interbedded lenticular sandstones and mudstones; coarsening-upwards sequences mark periodic seaward migration of nearshore facies.
Permian	White Rim Sandstones; L. Permian of SE Utah, USA; Conybeare, 1976, pp. 269–270*.	Cross-stratified sands deposited in a shallow sea; sands appear to have formed a bank approximately 2 km wide and 20 m high; comparison with longitudinal sand banks of North Sea is speculative.

Appendix 7.1—cont.

Age	Formation, authors	Context; nature of sediments; tidal current paths
Devonian	Cooksburg Member; Givetian of New York State, USA; McCave, 1973*.	Shallow sea encroaching alluvial plains fringing the rising Acadian mountains to the east; fine to medium grained sandstones, pebble beds and sand waves are interpreted as tidal; bimodal current vectors parallel to embayed shoreline.
Ordovician	Peninsula Formation of Cape Town, South Africa; Hobday and Tankard, 1978*.	A NE to SW aligned barrier-shoreline retreated towards the NW as a shallow sea transgressed; lenticular sandstone units 2 to 20 m thick show lateral accretion surfaces, low-angle cross-stratification with intrasets and high-angle cross-stratification with regressive ripples; the units are interpreted as deposits of offshore tidal sand banks orientated parallel to the coast; the banks migrated towards the SW (longshore) and SE (offshore), and were cut by large channels 35 m deep and 200 m wide; the channel sandstones display some bipolar cross-stratification.
Cambrian	Ironton and Galesville Sandstones of Illinois, USA; Emrich, 1966†.	Extensive shallow sea with local shoreline to north; increase in sandstone thickness and dolomite content southwards; decrease in maximum grain diameter southwards; overall north to south transport of sand but much local variation in tidal current vectors.
	L. Cambrian Eriboll Formation of Scotland; Bradore Formation of Newfoundland; Kloftlev Formation of Greenland; Tapeats Sandstone of Grand Canyon; Swett and Smit, 1972*; Merifield and Lamar, 1970*; Hereford, 1977*.	Sands deposited on broad shelf bordering the north-western margin of the Iapetus Ocean; sandstones generally rest on irregular hard substrate; cross-strata in Greenland, Scotland and Newfoundland indicate overall sediment transport towards the south and east (offshore).
	Duolbasgaissa Formation; L. Cambrian of Norway; Banks, 1973*.	Shallow sea, shorelines uncertain; coarse to fine sandstones and mudstones; small and large-scale sand ripples; east to west bed transport paths; repeated coarsening-upwards sequences resulting from downcurrent migration of facies zones.
Precambrian	Glenelg and Reynold Point Formations; Hadrynian of Victoria Island, Canada; Young, 1973*.	Three northward facing shallow marine embayments – each separated by a ridge; sediments rest on quartzite and igneous basement; sediments comprise cross-stratified sandstones, mudstones and carbonates including stromatolites; sand ripples and sand waves; current vectors variable.
	Jura Quartzite; Dalradian of Jura and Islay of NW Scotland; Anderton, 1976*; Klein, 1970b*.	Shallow sea flanked to the NW by land of moderate relief; shoreline-parallel ridge inferred to SE; coarse to fine sandstones and shales; sand ripples, sand waves and climbing dunes; bed transport path directed NE parallel to shoreline.

Appendix 7.1—*cont.*

Age	Formation, authors	Context; nature of sediments; tidal current paths
	Dakkovarre Formation of Northern Norway; Hobday and Reading, 1972; Johnson, 1977* (Fig. 7.5).	Shallow sea bordered by a shoreline to the NE; open marine sands rest on sediments of deltaic and lagoonal origin; four coarsening-upwards sequences display rippled and cross-stratified sandstones; sequences interpreted as offshore tidal sand banks orientated parallel to the shoreline; banks migrated westwards during gales when tidal currents were enhanced by storm generated flows; the tidal sand banks were cut by a series of large tidal channels.
	Lower Sandfjord Formation of Northern Norway; Levell, 1980* (Fig. 7.1).	Shallow sea bordered by coastal plain; braided rivers supplied coarse-grained sediment to a tidally influenced delta; sands transported offshore by tidal currents and worked into large sand waves with small sand waves superimposed.

* Tidal origin proposed by author cited.   † Tidal origin inferred by present writer from description of sediments by cited author.

Appendix 7.2 Estimate of the amplification of the twice-daily tidal wave in the Lower Aptian gulf of south-east England.

The calculation below follows the method outlined by Silvester (1974).

Symbols and assumed values:
- d_1 initial depth at offshore entrance to gulf (90 m) (estimate based on present-day Celtic Sea)
- d_2 final depth at the head of gulf (30 m) (estimate based on present-day Bristol Channel)
- d mean depth (60 m)
- f_w resistance coefficient
- g gravitational acceleration (9.81 m/s²)
- H_1 height of tidal wave at entrance to gulf (3 m) (nominal figure)
- H_2 height of tidal wave at head of gulf (friction not considered)
- H height of tidal wave at head of gulf (friction considered)
- H_0 mean height of tidal wave over gulf
- k mean bedform roughness height (1 m)
- L_1 initial tidal wavelength
- L final tidal wavelength – resulting from effects of friction
- X length of tidal wave path in gulf (180 km)
- S_V slope of sea floor along tidal path $(d_1 - d_2)/X$
- W_1 width of gulf at entrance (200 km) (see Fig. 7.10)
- W_2 width of gulf at head (20 km) (see Fig. 7.10)
- S_H coastline convergence $(W_1 - W_2)/2X$
- T tidal period (44700 s or 12.4 h)
- Z_A amplification factor due to shallowing effect
- Z_B amplification factor due to funnelling effect

1. Amplification factor due to shallowing (Z_A)

$$Z_A = \frac{4\pi d_1^{\frac{1}{4}}}{g^{\frac{1}{4}} T S_V} = 2.55; \; d_1/d_2 = 3$$

From Silvester, Fig. 4.2 (graph) amplification $H_2/H_1 = 1.35$

2. Amplification factor due to funnelling effect (Z_B)

$$Z_B = \frac{\pi W_1}{g^{\frac{1}{4}} d_1^{\frac{1}{4}} T S_H} = 0.94625; \; (W_1/W_2)^2 = 100$$

From Silvester, Fig. 4.2 (graph) amplification $H_2/H_1 = 1.95$

3. Thus mean height of tidal wave (H_0), friction not yet considered, is:

$$H_0 = \frac{(1 + 1.35 \times 1.95)H_1}{2} = 6.87 \text{ m}$$

4. Attenuation of tidal wave due to friction:
   (a) $H_0^2 T/d = 35\,162$
   (b) $\dfrac{H_0 T g^{\frac{1}{2}}}{4\pi k d^{\frac{1}{2}}} = 9880$ From Silvester, Fig. 4.4 (graph) resistance coefficient $(f_w) = 0.005$
   (c) attenuation factor $\dfrac{(f_w) H_0 T \sqrt{(gd)}}{10^3 d^2} = 0.01035$

   From Silvester, Fig. 4.3 (graph), tidal wavelength reduction $(L/L_1) = 0.995$

   (d) Ratio of tidal wave path (X) to initial tidal wavelength (L_1)

   $$\frac{X}{T\sqrt{(gd)}} = 0.166$$

   From Silvester, Fig. 4.5 (graph) attenuation in tidal wave height (H_2) due to friction is 10%, thus $H = 0.9 H_2$ (where $H_2 = 1.35 \times 1.95 \times H_1$)

   (e) $H = 0.9 \times 1.35 \times 1.95 \times H_1 = 7.1$ m

5. Amplification of tidal wave $= (7.1/3.0) = 2.37$

If reflection of the tidal wave occurred then the tidal range at the head of the Lower Aptian gulf may have approached 14 m. Thus, with the assumptions outlined above, the amplification of the twice-daily tidal wave in the Lower Aptian gulf was at least 2.37.

References

Accad, Y. and Pekeris, C. L. (1978), Tides in the world oceans, *Philosophical Transactions Royal Society London*, **290** A, 235–266.

Acton, J. R. and Dyer, C. M. (1975), Mapping of tidal currents near the Skerries Bank, *Journal Geological Society*, **131**, 63–67.

Adey, W. H. and Adey, P. J. (1973), Studies on the systematics and ecology of the epilithic crustose Corallinaceae of the British Isles, *British Phycological Journal*, **8**, 343–407.

Alexandersson, E. T. (1976), Actual and anticipated petrographic effects of carbonate undersaturation in shallow seawater, *Nature, London*, **262**, 653–657.

Alexandersson, E. T. (1979), Marine maceration of skeletal carbonates in the Skagerrak, North Sea, *Sedimentology*, **26**, 845–852.

Allen, E. J. (1899), On the fauna and bottom-deposits near the thirty-fathom line from the Eddystone Grounds to Start Point, *Journal Marine Biological Association U.K.*, **5**, 365–542.

Allen, J. R. L. (1968), *Current Ripples – their Relation to Patterns of Water and Sediment Motion*, North Holland Publishing Company, Amsterdam.

Allen, J. R. L. (1970), *Physical Processes of Sedimentation*, Allen and Unwin, London.

Allen, J. R. L. (1973), Phase differences between bed configuration and flow in natural environments, and their geological relevance, *Sedimentology*, **20**, 323–329.

Allen, J. R. L. (1979), Initiation of transverse bedforms in oscillatory bottom boundary layers, *Sedimentology*, **26**, 863–865.

Allen, J. R. L. (1980a), Sand waves: a model of origin and internal structure, *Sedimentary Geology*, **26**, 281–328.

Allen, J. R. L. (1980b), Large transverse bedforms and the character of boundary-layers in shallow-water environments, *Sedimentology*, **27**, 317–323.

Allen, J. R. L. and Collinson, J. D. (1974), The superimposition and classification of dunes formed by unidirectional aqueous flows, *Sedimentary Geology*, **12**, 169–178.

Allen, J. R. L. and Narayan, J. (1964), Cross-stratified units, some with siltbands in the Folkestone Beds (Lower Greensand) of SE England, *Geologie en Mijnbouw*, **43**, 451–461.

Allen, N. H., Fannin, N. G. T. and Farrow, G. E. (1979), Resin peels of vibrocores used in the study of some shelly sediments on the Scottish shelf, *Marine Geology*, **33**, M57–M65.

Allen, P. (1975), Wealden of the Weald: a new model, *Proceedings Geologists' Association*, **86**, 389–437.

Amos, C. L. (1978), The post-glacial evolution of the Minas Basin, NS. A sedimentological interpretation, *Journal Sedimentary Petrology*, **48**, 965–982.

Anderton, R. (1976), Tidal-shelf sedimentation: an example from the Scottish Dalradian, *Sedimentology*, **23**, 429–458.

Andrews, P. B. (1973), Late Quaternary continental shelf sediments off Otago Peninsula, New Zealand, *New Zealand Journal Geology and Geophysics*, **16**, 793–830.

d'Anglejan, B. F. (1971), Submarine sand dunes in the St Lawrence Estuary, *Canadian Journal of Earth Sciences*, **5**, 1480–1486.

Anguenot, F., Gourlez, P. and Migniot, C. (1972), Déplacement des ridens au large du Havre. Etude de leur dynamique par traceurs radioactifs, *Rapport*

SAR-72-08, *Commissariat à l'Energie Atomique*, Centre d'Etudes Nucléaires de Saclay, France.
Annambhotla, V. S. S. (1969), *Statistical properties of bed forms in alluvial channels in relation to flow resistance*, Ph.D. Thesis, University of Iowa.
Ansell, A. D. and Lagardère, F. (1980), Observations on the biology of *Donax trunculus* and *D. vittatus* at Ile d'Oléron (French Atlantic Coast), *Marine Biology*, 57, 287-300.
Anwar, H. O. and Atkins, R. (1980), Turbulence measurements in simulated tidal flow, *Journal Hydraulics Division, American Society of Civil Engineers*, 106, 1273-1289.
Arntz, W. E., Brunswig, D. and Sarnthein, M. (1976), Zonierung von Mollusken und Schill im Rinnensystem der Kieler Bucht (Westliche Ostsee), *Senckenbergiana Maritima*, 8, 189-269.
ASCE (1966), American Society of Civil Engineers, Task Force on bed forms in alluvial channels. Nomenclature for bed forms in alluvial channels. *Journal, Hydraulics Division, American Society of Civil Engineers*, 92, 51-64.
Auffret, G. A., Berthois, L., Cabioch, L. and Douvillé, J. L. (1972), Contribution à l'étude et à la cartographie des fonds sédimentaires au large de Roscoff, *Mémoire Bureau Recherches Géologiques et Minières*, 79, 293-302.
Auffret, J. P., Alduc, D., Larsonneur, C. and Smith, A. J. (1980), Cartographie du réseau des paléovallées et de l'épaisseur des formations superficielles meubles de la Manche orientale, *Annales de l'Institut Océanographique*, 56 (S), 21-35.
Baak, J. A. (1936), *Regional Petrology of the Southern North Sea*, Veenman and Zonen, Wageningen.
Bagnold, R. A. (1941), *The Physics of Blown Sand and Desert Dunes*, Methuen, London.
Bagnold, R. A. (1956), The flow of cohesionless grains in fluids, *Philosophical Transactions Royal Society London*, 249 A, 235-297.
Bagnold, R. A. (1963), Mechanics of marine sedimentation, in *The Sea, Ideas and Observations* (ed. M. N. Hill), Wiley-Interscience, New York, pp. 507-523.
Bagnold, R. A. (1966), An approach to the sediment transport problem from general physics, *U.S. Geological Survey, Professional Paper* no. 422-I.
Ballade, P. (1953), Etudes des fonds sableux en Loire maritime, nature et évolution des ridens, *Comité Central d' Océanographie et d'Etude des Côtes, Bulletin d'Information*, 5, 163-176.
Banks, N. L. (1973), Tide-dominated offshore sedimentation, Lower Cambrian, North Norway, *Sedimentology*, 20, 213-228.
Barton, J. R. and Lin, P. N. (1955), *A study of the sediment transport in alluvial channels*, Civil Engineering Department, Colorado A & M College, Fort Collins, Report 55 JRB2.

Bathurst, R. G. (1971), *Carbonate Sediments and their Diagenesis*, Elsevier, Amsterdam.
Beche, H. T. de la. (1851), *The Geological Observer*, Longman, Brown, Green & Longmans.
Belderson, R. H. (1964), Holocene sedimentation in the western half of the Irish Sea, *Marine Geology*, 2, 147-163.
Belderson, R. H., Johnson, M. A. and Stride, A. H. (1978), Bed-load partings and convergences at the entrance to the White Sea, USSR, and between Cape Cod and Georges Bank, USA, *Marine Geology*, 28, 65-75.
Belderson, R. H. and Kenyon, N. H. (1969), Direct illustration of one way sand transport by tidal currents, *Journal Sedimentary Petrology*, 39, 1249-1250.
Belderson, R. H., Kenyon, N. H. and Stride, A. H. (1970), Holocene sediments on the continental shelf west of the British Isles, Institute of Geological Sciences Report, No. 70/14, 157-170.
Belderson, R. H., Kenyon, N. H., Stride, A. H. and Stubbs, A. R. (1972), *Sonographs of the Sea Floor*, Elsevier, Amsterdam.
Belderson, R. H., Kenyon, N. H. and Wilson, J. B. (1973), Iceberg plough marks in the north-east Atlantic, *Palaeogeography, Palaeoclimatology, Palaeoecology*, 13, 215-224.
Belderson, R. H. and Stride, A. H. (1966), Tidal current fashioning of a basal bed, *Marine Geology*, 4, 237-257.
Belderson, R. H. and Stride, A. H. (1969), Tidal currents and sand wave profiles in the north-eastern Irish Sea, *Nature, London*, 222, 74-75.
Belderson, R. H. and Wilson, J. B. (1973), Iceberg plough marks in the vicinity of the Norwegian Trough, *Norsk Geologisk Tidsskrift*, 53, 323-328.
Berg, J. H. van de (1979), Rhythmic seasonal layering in a mesotidal channel abandonment facies, Oosterschelde Mouth, SW Netherlands, *Abstracts, International meeting on Holocene marine sedimentation in the North Sea basin*, International Association of Sedimentologists, Texel, The Netherlands.
Berg, R. R. (1975), Depositional environment of Upper Cretaceous Sussex Sandstone, House Creek Field, Wyoming, *Bulletin American Association Petroleum Geologists*, 59, 2099-2110.
Berven, R. J. (1966), Cardium sandstone bodies, Crossfield-Garrington area, Alberta, *Bulletin of Canadian Petroleum Geology*, 14, 208-240.
Birch, G. F. (1977), Surficial sediments on the continental margin off the west coast of South Africa, *Marine Geology*, 23, 305-337.
Birkett, L. (1954), Standing crop, mortality and yield of a *Mactra* patch, *International Council for the Exploration of the Sea*, Contribution to Statutory Meeting North Sea Committee CM1954, No. 72.
Bjørlykke, K., Bue, B. and Elverhøi, A. (1978),

Quaternary sediments in the northwestern part of the Barents Sea and their relation to the underlying Mesozoic bedrock, *Sedimentology*, **25**, 227–246.

Blunden, G., Farnham, W. F., Jephson, N., Fenn, R. H. and Plunkett, B. A. (1977), The composition of maërl from the Glenan Islands of Southern Brittany, *Botanica Marina*, **20**, 121–125.

Boekschoten, G. J. (1966), Shell borings of sessile epibiontic organisms as palaeoecological guides (with examples from the Dutch coast), *Palaeogeography, Palaeoclimatology, Palaeoecology*, **2**, 333–379.

Bohlen, W. F. (1976), Shear stress and sediment transport in unsteady turbulent flows, in *Estuarine Processes*, **2** (ed. M. Wiley), pp. 109–123.

Boillot, G. (1964), Géologie de la Manche occidentale: fonds rocheux, dépôts quaternaires, sédiments actuels, *Annales de l'Institut Océanographique*, **42**, 1–220.

Boillot, G. (1965), Organogenic gradients in the study of neritic deposits of biological origin: the example of the western English Channel, *Marine Geology*, **3**, 359–367.

Boillot, G., Bouysse, P. and Lamboy, M. (1971), Morphology, sediments and Quaternary history of the continental shelf between the Straits of Dover and Cape Finisterre, in *Geology of the east Atlantic continental margin*, **3** (ed. F. M. Delany), Institute of Geological Sciences Report No. 70/15, pp. 79–90.

Boothroyd, J. C. and Hubbard, D. K. (1975), Genesis of bedforms in mesotidal estuaries, in *Estuarine Research*, 2, *Geology and Engineering* (ed. L. E. Cronan), Academic Press, pp. 217–234.

Borley, J. O. (1923), The marine deposits of the Southern North Sea, *Fishery Investigations*, Ser. II, **4**, 62.

Bosence, D. W. J. (1976), Ecological studies on two unattached coralline algae from Western Ireland, *Palaeontology*, **19**, 365–395.

Bosence, D. W. J. (1978), Recent carbonate sedimentation in Connemara, Western Eire: a comment, *Estuarine and Coastal Marine Science*, **7**, 303–306.

Bosence, D. W. J. (1979), Live and dead faunas from coralline algal gravels, Co. Galway, *Palaeontology*, **22**, 449–478.

Bosence, D. W. J. (1980), Sedimentary facies, production rates and facies models for Recent coralline algal gravels, Co. Galway, Ireland, *Geological Journal*, **15**, 91–111.

Bouma, A. H., Hampton, M. A. and Orlando, R. C. (1978), Sand waves and other bedforms in Lower Cook Inlet, Alaska, *Marine Geotechnology*, **2**, 291–308.

Bouma, A. H., Hampton, M. A., Rappeport, M. L., Whitney, J. W., Teleki, P. G., Orlando, R. C. and Torresan, M. E. (1978), Movement of sand waves in Lower Cook Inlet, Alaska, *10th Annual Offshore Technology Conference*, Paper No. 3311, Houston, Texas, 8–11 May 1978.

Bouysse, P., Horn, R., Lapierre, F. and Le Lann, F. (1976), Etude des grands bancs de sable du sud-est de la mer Celtique, *Marine Geology*, **20**, 251–275.

Bouysse, P., Le Lann, F. and Scolari, G. (1979), Les sédiments superficiels des Approches Occidentales de la Manche, *Marine Geology*, **29**, 107–135.

Bowden, K. F. (1978), Physical problems of the benthic boundary layer, *Geophysical Surveys*, **3**, 255–296.

Bowden, K. F. and Fairbairn, L. A. (1952), A determination of the frictional forces in a tidal current, *Proceedings Royal Society, London*, **214** A, 371–392.

Bowden, K. F. and Ferguson, S. R. (1980), Variations with height of the turbulence in a tidally-induced bottom boundary layer, in *Marine Turbulence*, Proceedings of 11th International Liège Colloquium on Ocean Hydrodynamics, Elsevier, Amsterdam, pp. 259–286.

Brasier, M. D. (1979), The Cambrian radiation event, in *The Origin of Major Invertebrate Groups* (ed. M. R. House), *Systematics Association Special Volume*, No. 12, Academic Press, London, pp. 103–159.

Brenner, R. L. (1978), Sussex Sandstone of Wyoming – example of Cretaceous offshore sedimentation, *Bulletin American Association Petroleum Geologists*, **62**, 181–200.

Brenner, R. L. (1980), Construction of process-response models for ancient epicontinental seaway depositional systems using partial analogs, *Bulletin American Association Petroleum Geologists*, **64**, 1223–1244.

Brenner, R. L. and Davies, D. K. (1973), Storm-generated coquinoid sandstone: genesis of high-energy marine sediments from the Upper Jurassic of Wyoming and Montana, *Bulletin American Association Petroleum Geologists*, **57**, 1685–1698.

Brenner, R. L. and Davies, D. K. (1974), Oxfordian sedimentation in western interior United States, *Bulletin American Association Petroleum Geologists*, **58**, 407–428.

Bromley, R. G. (1975), Comparative analysis of fossil and Recent echinoid bioerosion, *Palaeontology*, **18**, 725–739.

Bromley, R. G. (1978), Bioerosion of Bermuda reefs, *Palaeogeography, Palaeoclimatology, Palaeoecology*, **23**, 169–197.

Bromley, R. G. and Asgaard, V. (1975), Sediment structures produced by a spatangoid echinoid: a problem of preservation, *Bulletin Geological Society Denmark*, **24**, 261–281.

Bromley, R. G. and Tendal, O. S. (1973), Example of substrate competition and phobotropism between two clionid sponges, *Journal Zoological Society London*, **169**, 151–155.

Brunn, V. von and Hobday, D. K. (1976), Early

Precambrian tidal sedimentation in the Pongola Supergroup of South Africa, *Journal Sedimentary Petrology*, **46**, 670–679.

Brunn, V. von and Mason, T. R. (1977), Siliciclastic-carbonate tidal deposits from the 3000 M yr. Pongola Supergroup, South Africa, *Sedimentary Geology*, **18**, 245–255.

Buchanan, J. B. (1963), The bottom fauna communities and their sediment relationships off the coast of Northumberland, *Oikos*, **14**, 154–175.

Buchanan, J. B. (1966), The biology of *Echinocardium cordatum* (Echinodermata: Spatangoidea), *Journal Marine Biological Association U.K.*, **46**, 97–114.

Buller, A.T. and McManus, J. (1975), Sediments of the Tay Estuary, I. Bottom sediments of the upper and upper-middle reaches, *Proceedings Royal Society Edinburgh*, Series B, **75**, 41–64.

Cabioch, J. (1970), Le maërl des côtes de Bretagne et le problème de sa survie, *Penn ar Bed*, **7**, 421–429.

Cabioch, L. (1968), Contribution à la connaissance des peuplements benthiques de la Manche Occidentale, *Cahiers de Biologie Marine*, **9**, 493–720.

Cabioch, L., Gentil, F., Glacon, R. and Retière, C. (1975), *Pagurus pubescens* Kröyer: Presence dans la region de Roscoff et distribution dans La Manche, *Travaux Station Biologique Roscoff*, **22**, 17–19.

Cabioch, L., Gentil, F., Glacon, R. and Retière, C. (1977), La macrobenthos des fonds meubles de La Manche: Distribution générale et ecologie, in *Biology of Benthic Organisms* (eds B. F. Keegan, P. O. Ceidigh and P. J. S. Boaden), Pergamon Press, Oxford, pp. 115–128.

Cadée, G. C. (1968), *Molluscan Biocoenoses and Thanatocoenoses in the Ria de Arosa, Galicia, Spain*, E. J. Brill, Leiden.

Cadée, G. C. (1976), Sediment reworking by *Arenicola marina* on tidal flats in the Dutch Wadden Sea, *Netherlands Journal Sea Research*, **10**, 440–460.

Cadée, G. C. (1979), Sediment reworking by the polychaete *Heteromastus filiformis* on a tidal flat in the Dutch Wadden Sea, *Netherlands Journal Sea Research*, **13**, 441–456.

Campbell, C. A. and Valentine, J. W. (1977), Comparability of modern and ancient marine faunal provinces, *Palaeobiology*, **3**, 49–57.

Campbell, C. V. (1971), Depositional model – Upper Cretaceous Gallup beach shoreline, Ship Rock area, northwestern New Mexico, *Journal Sedimentary Petrology*, **41**, 395–409.

Carriker, M. R. (1978), Ultrastructural analysis of dissolution of shell bivalve *Mytilus edulis* by the accessory boring organ of the gastropod *Urosalpinx cinerea*, *Marine Biology*, **48**, 105–134.

Carriker, M. R., Scott, D. B. and Martin, G. N. (1963), Demineralization mechanism of boring gastropods, in *Mechanisms of Hard Tissue Destruction* (ed. R. F. Sognnaes), American Association for the Advancement of Science, Publication No. 75, pp. 55–89.

Carriker, M. R., Smith, E. H. and Wilce, R. T. (eds) (1969), Penetration of calcium carbonate substrates by Lower Plants and Invertebrates, Symposium, American Association for the Advancement of Science, *American Zoologist*, **9**, 629–1020.

Carriker, M. R. and Yochelson, E. L. (1968), Recent gastropod boreholes and Ordovician cylindrical borings, *U.S. Geological Survey Professional Paper* 593-B.

Carter, L. (1975), Sedimentation on the continental terrace around New Zealand: a review, *Marine Geology*, **19**, 209–237.

Carter, R. M. (1968), On the biology and palaeontology of some predators of bivalved Mollusca, *Palaeogeography, Palaeoclimatology, Palaeoecology*, **4**, 29–65.

Cartwright, D. E. (1959), On submarine sand-waves and tidal lee waves, *Proceedings Royal Society, London*, **253** A, 218–241.

Cartwright, D. E. (1961), A study of currents in the Strait of Dover, *Journal Institute of Navigation*, **19**, 130–151.

Cartwright, D. E. (1969), Extraordinary tidal currents near St Kilda, *Nature, London*, **223**, 928–932.

Cartwright, D. E. (1974), Years of peak astronomical tides, *Nature, London*, **248**, 656–657.

Cartwright, D. E. (1978), Oceanic tides, *International Hydrographic Review*, **55**, 35–84.

Cartwright, D. E., Huthnance, J. M., Spencer, R. and Vassie, J. M. (1980), On the St Kilda shelf tidal regime, *Deep-Sea Research*, **27**, 61–70.

Casey, H. J. (1935), Uber Geschiebewegung, *Mitteilungen der Preussischen Versuchsanstalt für Wasserbau und Schiffbau*, Berlin, **19**.

Casey, R. (1961), The stratigraphical palaeontology of the Lower Greensand, *Palaeontology*, **3**, 487–621.

Caston, G. F. (1975), Igneous dykes and associated scour hollows of the North Channel, Irish Sea, *Marine Geology*, **18**, M77–M85.

Caston, G. F. (1976), The floor of the North Channel, Irish Sea: a side-scan sonar survey, *Institute of Geological Sciences Report* No. 76/7.

Caston, G. F. (1979), Wreck marks: indicators of net sand transport, *Marine Geology*, **33**, 193–204.

Caston, G. F. (1981), Potential gain and loss of sand by some sand banks in the Southern Bight of the North Sea, *Marine Geology*, **41**, 239–250.

Caston, V. N. D. (1972), Linear sand banks in the Southern North Sea, *Sedimentology*, **18**, 63–78.

Caston, V. N. D. and Stride, A. H. (1970), Tidal sand movement between some linear sand banks in the North Sea off north-east Norfolk, *Marine Geology*, **9**, M38–M42.

Caston, V. N. D. and Stride, A. H. (1973), Influence of older relief on the location of sand waves in a part of the Southern North Sea, *Estuarine and Coastal Marine Science*, **1**, 379–386.

Cavalière, A. R. and Alberte, R. S. (1970), Fungi in animal shell fragments, *Journal Elisha Mitchell Scientific Society*, **86**, 203–206.

Chabert, J. and Chauvin, J.-L. (1963), Formation des dunes et des rides dans les modèles fluviaux, *Bulletin Centre de Recherches et d'Essais, Chatou*, **4**, 31–52.

Channon, R. D. (1971), *Storm and current sorted sediments of the continental shelf southwest of England*, Ph.D. Thesis, University of Bristol.

Channon, R. D. and Hamilton, D. (1976), Wave and tidal current sorting of shelf sediments southwest of England, *Sedimentology*, **23**, 17–42.

Chardy, P., Guennegan, Y. and Branellec, J. (1980), Photographie sous-marine et analyse des peuplements benthiques, *Rapports scientifiques et techniques, Centre National pour l'Exploitation des Océans*, No. 41-1980.

Chaudry, H. M., Smith, K. V. H. and Vigil, H. (1970), Computation of sediment transport in irrigation canals, *Proceedings Institution of Civil Engineers, London*, **45**, 79–101.

Chave, K. E. (1962), Processes of carbonate sedimentation, in *The environmental chemistry of marine sediments* (ed. N. Marshall), Narragansett Marine Laboratory Occasional Publication, No. 1, pp. 77–85.

Chave, K. E. (1967), Recent carbonate sediments – an unconventional view, *Journal of Geological Education*, **15**, 200–204.

Chesterman, W. D., Clynick, P. R. and Stride, A. H. (1958), An acoustic aid to sea bed survey, *Acustica*, **8**, 285–290.

Chiu, T. Y. (1972), Sand transport by water or air, *Coastal and Oceanographic Engineering Laboratory*, University of Florida, *Report UFL/COEL/TR-040*.

Chowdhuri, K. R. and Reineck, H.-E. (1978), Primary sedimentary structures and their sequence in the shorefaces of barrier island Wangerooge (North Sea), *Senckenbergiana Maritima*, **10**, 15–29.

Clarke, R. H. (1970), Quaternary sediments off southeast Devon, *Quarterly Journal Geological Society*, **125**, 277–318.

Cloet, R. L. (1980), Comparison of two close line surveys of the South Sandettie sandwave field, *Hydrographic Journal*, **17**, 19–30.

Clokie, J. J. P., Scoffin, T. P. and Boney, A. D. (1981), Depth maxima of *Conchocelis* and *Phymatolithon rugulosum* on the NW shelf and Rockall Plateau, *Marine Ecology – Progress Series*, **4**, 131–133.

Coastal Research Group, Massachusetts University (1969), *Coastal environments: N.E. Massachusetts and New Hampshire*, Publication 1-CRG, University of Massachusetts, Department of Geology.

Cobb, W. R. (1969), Penetration of calcium carbonate substrates by the boring sponge *Cliona*, *American Zoologist*, **9**, 783–790.

Colby, B. R. (1964), Discharge of sands and mean-velocity relationships in sand-bed streams, *United States Geological Survey Professional Paper* 462-A.

Coleman, J. M. (1969), Brahmaputra River: channel processes and sedimentation, *Sedimentary Geology*, **3**, 131–239.

Comely, C. A. (1978), *Modiolus modiolus* (L) from the Scottish west coast, 1. Biology, *Ophelia*, **17**, 167–193.

Conolly, J. R. and Borch, C. C. von der (1967), Sedimentation and physiography of the sea floor south of Australia, *Sedimentary Geology*, **1**, 181–220.

Conybeare, C. E. B. (1976), Geomorphology of oil and gas fields in sandstone bodies, *Developments in Petroleum Science*, **4**, Elsevier, Amsterdam.

Cook, P. J. and Mayo, W. (1977), Sedimentology and Holocene history of a tropical estuary (Broad Sound, Queensland), *Bulletion 170, Bureau of Mineral Resources*, Geology and Geophysics, Canberra.

Cooke, R. U. and Warren, A. (1973), *Geomorphology in Deserts*, Batsford, London.

Cornish, V. (1914), *Waves of Sand and Snow, and the Eddies which make them*, T. Fisher Unwin, London.

Costello, W. R. (1974), *Development of bed configurations in coarse sands*, Report R-74-2, Department of Earth Sciences, Massachusetts Institute of Technology.

Cotter, E. (1975), Late Cretaceous sedimentation in a low-energy coastal zone: the Ferron Sandstone of Utah, *Journal Sedimentary Petrology*, **45**, 669–685.

Cotter, E. (1976), The role of deltas in the evolution of the Ferron Sandstone and its coal, *Brigham Young University Geology Studies*, **22**, 15–41.

Craig, G. Y. (1966), Concepts in palaeoecology, *Earth Science Reviews*, **2**, 127–155.

Craig, G. Y. and Jones, N. S. (1966), Marine benthos, substrate and palaeoecology, *Palaeoecology*, **9**, 30–38.

Craig, G. Y. and Oertel, G. (1966), Deterministic models of living and fossil populations of animals, *Quarterly Journal Geological Society London*, **122**, 315–355.

Cram, J. M. (1979), The influence of continental shelf width on tidal range: palaeoceanographic implications, *Journal of Geology*, **87**, 441–447.

Crickmore, J. J. (1967), Measurement of sand transport in rivers with special reference to tracer methods, *Sedimentology*, **8**, 175–228.

Crimes, T. P. and Harper, J. C. (eds) (1970), *Trace Fossils*, Proceedings of an international conference Liverpool University, 6–8 January, 1970, Seel House Press, Liverpool.

Crimes, T. P. and Harper, J. C. (eds) (1977), *Trace*

Fossils 2, Proceedings of an international symposium, Sydney, Australia, 23–24 August 1976, Seel House Press, Liverpool.

Cronan, D. S. (1969), Recent sedimentation in the central north-eastern Irish Sea, *Institute of Geological Sciences Report* 69/8.

Cronan, D. S. (1970), Geochemistry of Recent sediments from the central north-eastern Irish Sea, *Institute of Geological Sciences Report* 70/17.

Cronan, D. S. (1972), Skewness and kurtosis in polymodal sediments from the Irish Sea, *Journal Sedimentary Petrology*, **42**, 102–106.

Culbertson, J. K. and Scott, C. H. (1970), Sandbar development and movement in an alluvial channel, Rio Grande near Bernardo, New Mexico, in *United States Geological Survey Professional Paper* 700-B, pp. B237–B241.

Cullen, D. J. (1962), The influence of bottom sediments upon the distribution of oysters in Foveaux Strait, New Zealand, *New Zealand Journal of Geology and Geophysics*, **5**, 271–275.

Cullen, D. J. (1967), The submarine geology of Foveaux Strait, *New Zealand Oceanographic Institute Memoir* No. 33.

Cullen, D. J. (1970), Radiocarbon analyses of individual molluscan species in relation to post-glacial eustatic changes, *Palaeogeography, Palaeoclimatology, Palaeoecology*, **7**, 13–20.

Dalrymple, R. W., Knight, R. J. and Lambiase, J. J. (1978), Bedforms and their hydraulic stability relationships in a tidal environment, Bay of Fundy, Canada, *Nature, London*, **275**, 100–104.

Dangeard, L. (1925), Observations de géologie sous-marine et d'océanographie relatives a la Manche, *Annales de l'Institut Océanographique Paris*, **6**, 1–295.

Darwin, G. H. (1898), *The Tides and Kindred Phenomena in the Solar System*, Murray, London.

Davies, A. M. (1976), Application of a fine mesh numerical model of the North Sea to the calculation of storm surge elevations and currents, *Institute of Oceanographic Sciences, Report* 28.

Davies, A. M. and Furnes, G. K. (1980), Observed and computed M_2 tidal currents in the North Sea, *Journal of Physical Oceanography*, **10**, 237–257.

Davis, F. M. (1923), Quantitative studies on the fauna of the sea bottom, 1. – Preliminary investigation of the Dogger Bank, Ministry of Agriculture and Fisheries, *Fishery Investigations* Series II, **4** (2), HMSO, London.

Davis, F. M. (1925), Quantitative studies on the fauna of the sea bottom, 2. – Results of the investigations in the Southern North Sea, 1921–24. Ministry of Agriculture and Fisheries, *Fishery Investigations*, Series II, **8** (4), HMSO, London.

Deegan, C. E., Kirby, R., Rae, I. and Floyd, P. (1973), The superficial deposits of the Firth of Clyde and its sea lochs, *Institute Geological Sciences, Report* 73/9.

Defant, A. (1961), *Physical Oceanography*, **2**, Pergamon Press, Oxford.

Delanoë, Y. and Pinot, J.-P. (1980), Aperçus sur la dynamique sédimentaire du précontinent Atlantique breton: 2. la mobilité des sables, son influence sur la morphologie actuelle et les structures sédimentaires, *Annales de l'Institut Océanographique, Paris*, **56**, 61–72.

Derna, F. (1974), *Marine geology and sediment transport patterns in Nymphe Bank area of the Celtic Sea*, Ph.D. Thesis, Department of Geology, University College of Wales, Aberystwyth.

Deutsches Hydrographisches Institut (1958), Strombeobachtungen in der südwestlichen Nordsee in den Jahren 1951–55, *Meereskundliche Beobachtungen und Ergebnisse*, **9**.

Dike, E. F. (1972), *Sedimentology of the Lower Greensand of the Isle of Wight*, Ph.D. Thesis, University of Oxford.

Dingle, R. V. (1965), Sand waves in the North Sea mapped by continuous reflection profiling, *Marine Geology*, **3**, 391–400.

Dixon, R. G. (1979), Sedimentary facies in the Red Crag (Lower Pleistocene, East Anglia), *Proceedings Geologists' Association*, **90**, 117–132.

Dobson, M. R., Evans, W. E. and James, K. H. (1971), The sediment on the floor of the southern Irish Sea, *Marine Geology*, **11**, 27–69.

Dominik, J., Forstner, U., Mangini, A. and Reineck, H. E. (1978), ^{210}Pb and ^{137}Cs chronology of heavy metal pollution in a sediment core from the German Bight (North Sea), *Senckenbergiana*, **20**, 213–227.

Donovan, D. T. (1973), The geology and origin of the Silver Pit and other closed basins in the North Sea, *Proceedings Yorkshire Geological Society*, **39**, 267–293.

Donovan, D. T. and Stride, A. H. (1961a), An acoustic survey of the sea floor south of Dorset and its geological interpretations, *Philosophical Transactions Royal Society*, B **244**, 299–330.

Donovan, D. T. and Stride, A. H. (1961b), Erosion of a rock floor by tidal sand streams, *Geological Magazine*, **98**, 393–398.

Donovan, D. T. and Stride, A. H. (1975), Three drowned coast lines of probable late Tertiary age around Devon and Cornwall, *Marine Geology*, **19**, M35–M40.

Doodson, A. T. and Warburg, H. D. (1941), *Admiralty Manual of Tides*, HMSO, London.

Dooley, H. D. and McKay, D. W. (1975), Herring larvae and currents west of the Orkneys, *International Council for the Exploration of the Sea, Pelagic Fish (Northern) Committee*, Paper CM 1975/H:43.

Draper, L. (1967), Wave activity at the sea bed around north-western Europe, *Marine Geology*, **5**, 133–140.

Draper, L. (1973), Extreme wave conditions in British and adjacent waters, *Proceedings, 13th Coastal Engineering Conference*, Vancouver, Canada, 157–165.

Dyer, K. R. (1970a), Linear erosion furrows in Southampton Water, *Nature, London*, **225**, 56–58.

Dyer, K. R. (1970b), Current velocity profiles in a tidal channel, *Geophysical Journal, Royal Astronomical Society*, **22**, 153–161.

Dyer, K. R. (1970c), Sediment distribution in Christchurch Bay, S England, *Journal, Marine Biological Association U.K.*, **50**, 673–682.

Dyer, K. R. (1971), The distribution and movement of sediment in the Solent, southern England, *Marine Geology*, **11**, 175–187.

Dyer, K. R. (1980), Velocity profiles over a rippled bed and the threshold of movement of sand, *Estuarine and Coastal Marine Science*, **10**, 181–199.

Eagle, R. A., Hardiman, P. A., Norton, M. G., Nunny, R. S. and Rolfe, M. S. (1979), The field assessment of effects of dumping wastes off the north-east coast of England, *Fisheries Research Technical Report* No. 51, 1–34, HMSO, London.

Einarsson, H. (1941), Survey of the benthonic animal communities of Faxa Bay (Iceland), *Meddelelser fra Kommissionen for Danmarks Fiskeriog Havundersøgelser, Serie: Fiskeri*, **11** (1), 1–46.

Eisma, D. (1966), The distribution of benthic marine molluscs off the main Dutch coast, *Netherlands Journal Sea Research*, **3**, 107–163.

Eisma, D. (1968), Composition, origin and distribution of Dutch coastal sands between Hoek van Holland and the island of Vlieland, *Netherlands Journal Sea Research*, **4**, 123–267.

Eisma, D., Jansen, J. H. F. and van Weering, T. C. E. (1979), Sea-floor morphology and recent sediment movement in the North Sea, in *The Quaternary history of the North Sea* (eds E. Oek, R. T. E. Schüttenhelm and A. J. Wiggens), Acta Universitatis Uppsaliensis, Uppsala, Sweden, pp. 217–231.

Elliott, T. (1978), Clastic shorelines, in *Sedimentary Environments and Facies* (ed. H. G. Reading), Blackwell Scientific Publications, Oxford, pp. 143–177.

Emery, K. O. (1968), Relict sediments on the continental shelves of the world, *American Association Petroleum Geologists Bulletin*, **52**, 445–464.

Emery, K. O., Merrill, A. S. and Trumbull, J. V. A. (1965), Geology and biology of the sea floor as deduced from simultaneous photographs and samples, *Limnology and Oceanography*, **10**, 1–21.

Emrich, G. H. (1966), Ironton and Galesville (Cambrian) Sandstones in Illinois and adjacent areas, *Illinois State Geological Survey Circular*, 403, 1–55.

Engelund, F. (1970), Instability of erodible beds, *Journal of Fluid Mechanics*, **42**, 225–244.

Engelund, F. and Fredsøe, J. (1971), Three-dimensional stability analysis of open channel flow over an erodible bed, *Nordic Hydrology*, **2**, 93–108.

Engelund, F. and Fredsøe, J. (1974), Transition from dunes to plane bed in alluvial channels, Institute of Hydrodynamics and Hydraulic Engineering, Technical University of Denmark, Series Paper 4.

Eriksson, K. A. (1977), Tidal deposits from the Archaean Moodies Group, Barberton Mountain Land, South Africa, *Sedimentary Geology*, **18**, 257–281.

Eriksson, K. A. and Truswell, J. F. (1973), Did lunar capture predate 2.2 aeons? *South African Journal of Science*, **69**, 150–152.

Evans, G. (1965), Intertidal flat sediments and their environments of deposition in the Wash, *Quarterly Journal Geological Society London*, **121**, 209–245.

Evans, W. E. (1970), Imbricate linear sandstone bodies of Viking Formation in Dodsland–Hoosier area of southwestern Saskatchewan, Canada, *American Association Petroleum Geologists Bulletin*, **54**, 469–486.

Ewing, J. A. (1973), Wave-induced bottom currents on the outer shelf, *Marine Geology*, **15**, M31–M35.

Farrow, G. E., Cucci, M. and Scoffin, T. P. (1978), Calcareous sediments on the nearshore continental shelf of western Scotland, *Proceedings Royal Society Edinburgh*, **76** B, 55–76.

Farrow, G., Scoffin, T., Brown, B. and Cucci, M. (1979), An underwater television survey of facies variation on the inner Scottish shelf between Colonsay, Islay and Jura, *Scottish Journal of Geology*, **15**, 13–29.

Fjeldstad, J. E. (1964), Internal waves of tidal origin, Part I, Theory and analysis of observations, *Geofysiske Publikasjoner*, **25** (5), 1–73.

Flather, R. A. (1976), A tidal model of the north-west European continental shelf, *Mémoires Société Royale des Sciences de Liège*, **10**, 141–164.

Flather, R. A. and Davies, A. M. (1978), On the specification of meteorological forcing in numerical models for North Sea storm surge prediction, with application to the surge of 2 to 4 January 1976, *Deutsche Hydrographische Zeitschrift, Erganzungsheft*, Reihe A, Nr. 15.

Fleming, R. H. (1938), Tides and tidal currents in the Gulf of Panama, *Journal of Marine Research*, **1**, 192–206.

Flemming, B. W. (1976), Side-scan sonar: a practical guide, *International Hydrographic Review*, **53**, 65–92.

Flemming, B. W. (1978), Sand transport patterns in the Agulhas current (south-east African continental margin), in *Marine Geoscience Group, University of Cape Town, Technical Report* No. 10, pp. 57–60.

Flemming, B. W. (1980), Sand transport and bedform patterns on the continental shelf between Durban and Port Elizabeth (south-east African continental margin), *Sedimentary Geology*, **26**, 179–205.

Flemming, N. C. and Stride, A. H. (1967), Basal sand

and gravel patches with separate indications of tidal current and storm-wave paths, near Plymouth, *Journal Marine Biological Association U.K.*, **47**, 433–444.

Flood, R. D. (1981), Distribution, morphology and origin of sedimentary furrows in cohesive sediments, Southampton Water, *Sedimentology*, **28**, 511–529.

Ford E. (1923), Animal communities of the level sea-bottom in waters adjacent to Plymouth, *Journal Marine Biological Association U.K.*, **13**, 164–224.

Ford, E. (1925), On the growth of some lamellibranchs in relation to the food-supply of fishes, *Journal Marine Biological Association U.K.*, **13**, 531–559.

Franco, J. J. (1968), Effects of water temperature on bed-load movement, *Journal Waterways and Harbors Division, American Society of Civil Engineers*, **94**, 343–352.

Fredsøe, J. and Engelund, F. (1975), Bed configurations in open and closed alluvial channels, *Institute of Hydrodynamics and Hydraulic Engineering, Technical University of Denmark, Series Paper 8.*

Freeland, G. L. and Swift, D. J. P. (1978), Surficial sediments, MESA New York Bight Atlas, Monograph 10, New York Sea Grant Institute, Albany, New York.

Frey, R. W. (ed.) (1975), *The Study of Trace Fossils*, Springer-Verlag, Berlin.

Frey, R. W. and Seilacher, A. (1980), Uniformity in marine invertebrate ichnology, *Lethaia*, **13**, 183–278.

Fürsich, F. T. (1978), The influence of faunal condensation and mixing on the preservation of fossil benthic communities, *Lethaia*, **11**, 243–250.

Gadd, P. E., Lavelle, J. W. and Swift, D. J. P. (1978), Estimates of sand transport on the New York shelf using near-bottom current meter observations, *Journal Sedimentary Petrology*, **48**, 239–252.

Gadow, S. and Reineck, H. E. (1969), Ablandiger Sandtransport bei Sturmfluten, *Senckenbergiana Maritima*, **1**, 63–78.

Gaemers, P. A. M. (1978), Late Quaternary and Recent otoliths from the seas around southern Norway, *Mededelingen Werkgroep Tertiare en Kwartaire Geologie*, **15**, 101–117.

Gammelsrød, T. (1975), Instability of Couette flow in a rotating fluid and origin of Langmuir circulations, *Journal Geophysical Research*, **80**, 5069–5075.

Garrett, C. (1972), Tidal resonance in the Bay of Fundy and Gulf of Maine, *Nature, London*, **238**, 441–3.

George, M. and Murray, J. (1977), Glauconite in Celtic Sea sediments, *Proceedings Ussher Society*, **4**, 94–101.

Gerstenkorn, H. (1955), Uber gezeitenreibung beim zweikorperproblem, *Astrophysical Journal*, **36**, 245–274.

Gienapp, H. (1973), Strömungen während der Sturmflut vom 2 November 1965 in den Deutschen Bucht und ihre Bedeutung für den Sedimenttransport, *Senckenbergiana Maritima*, **5**, 135–151.

Ginsburg, R. N. (ed.) (1975), *Tidal deposits: A casebook of Recent examples and fossil counterparts*, Springer-Verlag, Berlin.

Ginsburg, R. N. and James, N. P. (1974), Holocene carbonate sediments of continental shelves, in *The geology of continental margins* (ed. C. A. Burk and C. C. Drake), Springer-Verlag, New York, pp. 137–155.

Glémarec, M. (1969a), Le plateau continental Nord-Gascogne et la Grande Vasière étude bionomique, *Revue Travaux Institut Pêches Maritimes*, **33**, 301–310.

Glémarec, M. (1969b), Les peuplements benthiques du plateau continental Nord-Gascogne, Thèse de Doctorat d'Etat et Sciences Naturelles, Centre Nationale du Recherches Scientifiques, AO 3422.

Glémarec, M. (1973), The benthic communities of the European North Atlantic continental shelf, *Oceanography and Marine Biology Annual Review* (ed. H. Barnes), **11**, 263–289.

Goldring, R. (1964), Trace fossils and the sedimentary surfaces in shallow-water marine sediments, in *Deltaic and shallow marine deposits* (ed. L. M. J. U. van Straaten), Elsevier, Amsterdam, pp. 136–143.

Golubic, S. (1969), Distribution, taxonomy and boring patterns of marine endolithic algae, *American Zoologist*, **9**, 747–751.

Goodwin, P. W. and Anderson, E. J. (1974), Associated physical and biogenic structures in environmental subdivision of a Cambrian tidal sand body, *Journal of Geology*, **82**, 779–794.

Gordon, C. M. (1974), Intermittent momentum transport in a geophysical boundary layer, *Nature, London*, **248**, 392–394.

Goreau, T. F. and Hartman, W. D. (1963), Boring sponges as controlling factors in the formation and maintenance of coral reefs, in *Mechanisms of Hard Tissue Destruction* (ed. R. F. Sognnaes), American Association for the Advancement of Science Publication No. 75, pp. 25–54.

Grace, R. A. (1973), Real and theoretical water motion near the sea floor under long ocean swell, in *Engineering Dynamics of the Coastal Zone*, National Conference Publication 73/1, Institution of Australia, pp. 1–7.

Green, C. D. (1975), Sediments of the Tay Estuary III. Sedimentological and faunal relationships at the entrance to the Tay, *Proceedings Royal Society Edinburgh*, **75 B**, 91–112.

Greensmith, J. T. and Tucker, E. V. (1969), The origin of Holocene shell deposits in the chenier plain facies of Essex (Great Britain), *Marine Geology*, **7**, 403–425.

Greenspan, H. P. (1968), *The Theory of Rotating Fluids*, Cambridge University Press, Cambridge.

Gullentops, F. (1957), L'origine des collines du Hageland, *Bulletin Société belge Géologie*, **66**, 81–85.

Gunatilaka, A. (1977), Recent carbonate sedimentation in Connemara, Western Ireland, *Estuarine and Coastal Marine Science*, **5**, 609–629.

Guy, H. P., Simons, D. B. and Richardson, E. V. (1966), Summary of alluvial channel data from flume experiments 1956–61, *United States Geological Survey Professional Paper* 462-I, 1–96.

Hadley, M. L. (1964), Wave-induced bottom currents in the Celtic Sea, *Marine Geology*, **2**, 164–167.

Haight, F. J. (1942), Currents in Narragansett Bay, Buzzards Bay and Nantucket and Vineyard Sounds, *U.S. Coast and Geodetic Survey Special Publication* No. 208.

Hails, J. R. (1975), Sediment distribution and Quaternary history in Submarine geology, sediment distribution and Quaternary history of Start Bay, Devon, *Journal Geological Society*, **131**, 19–35.

Hallam, A. and Sellwood, B. W. (1976), Middle Mesozoic sedimentation in relation to tectonics in the British area, *Journal of Geology*, **84**, 301–321.

Hamilton, D. and Smith, A. J. (1972), The origin and sedimentary history of the Hurd Deep, English Channel, with additional notes on other deeps in the Western English Channel, *Mémoires Bureau Recherches Géologiques et Minières*, **79**, 59–78.

Hamilton, D., Somerville, J. H. and Stanford, P. N. (1980), Bottom currents and shelf sediments, southwest of Britain, *Sedimentary Geology*, **26**, 115–138.

Hammond, T. M. and Collins, M. B. (1979), On the threshold of transport of sand-sized sediment under the combined influence of unidirectional and oscillatory flow, *Sedimentology*, **26**, 795–812.

Hanna, S. R. (1969), The formation of longitudinal sand dunes by large helical eddies in the atmosphere, *Journal of Applied Meteorology*, **8**, 874–883.

Häntzschel, W. (1975), *Treatise on Invertebrate Palaeontology, Part W Miscellanea, Supplement 1, Trace Fossils and Problematica*, Geological Society of America and University of Kansas, Boulder, Colorado and Lauwrence, Kansas.

Harden-Jones, R. and Mitson, R. B. (In Press), The movement of noisy sand waves in the Straits of Dover, *Journal du Conseil*.

Harms, J. C. (1969), Hydraulic significance of some sand ripples, *Bulletin, Geological Society America*, **80**, 363–396.

Harris, P. M. (1979), Facies anatomy and diagenesis of a Bahamian ooid shoal, *Sedimenta*, **7** (ed. R. N. Ginsburg), University of Miami.

Harrison, W., Byrne, R. J., Boon, J. D. and Moncure, R. W. (1970), Field study of a tidal inlet, Bimini, Bahamas, *Proceedings, 12th Coastal Engineering Conference*, Washington, D.C., 1201–1222.

Harvey, J. G. (1966), Large sand waves in the Irish Sea, *Marine Geology*, **4**, 49–55.

Hayes, M. O. (1976), Morphology of sand accumulation in estuaries: an introduction to the symposium, in *Estuarine Research*, **2** (ed. L. E. Gronin), Academic Press, London, pp. 3–22.

Hayes, M. O. and Kana, T. W. (1976), Terrigenous clastic depositional environments – some modern examples, *Coastal Research Division, University of South Carolina, Technical Report* 11-CRD, I-131; II-184.

Hayes, S. P. and Halpern, D. (1976), Observations of internal waves and coastal upwelling off the Oregon coast, *Journal of Marine Research*, **34**, 247–267.

Heath, R. A. (1980), Phase relations between the over- and fundamental-tides. *Deutsche Hydrographische Zeitschrift*, **33**, 177–191.

Heathershaw, A. D. (1976), Measurements of turbulence in the Irish Sea benthic boundary layer, in *The Benthic Boundary Layer* (ed. I. N. McCave), Plenum Press, New York, pp. 11–31.

Heathershaw, A. D. and Carr, A. P. (1977), Measurements of sediment transport rates using radioactive tracers, in *Coastal Sediments '77*, American Society Civil Engineers, pp. 399–416.

Heathershaw, A. D. and Hammond, F. D. C. (1980), Transport and deposition of non-cohesive sediments in Swansea Bay, in *Industrial Embayments and their Environmental Problems: a case study of Swansea Bay* (ed. M. B. Collins *et al.*), Pergamon Press, Oxford, pp. 215–248.

Heathershaw, A. D. and Simpson, J. H. (1978), The sampling variability of the Reynolds stress and its relation to boundary shear stress and drag coefficient measurements, *Estuarine and Coastal Marine Science*, **6**, 263–274.

Heckel, P. H. (1972), Recognition of ancient shallow marine environments, in *Recognition of Ancient Sedimentary Environments* (ed. J. K. Rigby and W. K. Hamblin), SEPM Special Publication No. 16, pp. 226–286.

Hecker, R. F. (1965), *Introduction to Paleoecology*, American Elsevier Publishing Co., New York.

Herdman, W. A. (1920), The Marine Biological Station at Port Erin (Isle of Man), *34th Annual Report, Liverpool Marine Biological Committee*.

Hereford, R. (1977), Deposition of the Tapeats Sandstone (Cambrian) in central Arizona, *Bulletin Geological Society America*, **88**, 199–211.

Hertweck, G. (1972), Distribution and environmental significance of *lebenspuren* and *in-situ* skeletal remains, *Senckenbergiana Maritima*, **4**, 125–167.

Hider, A. (1882), Report of assistant engineer A. Hider upon observations at Lake Providence, Nov. 1879–Nov. 1880, *Progress Report of Mississippi River Commission for 1882*.

Hill, H. M., Robinson, A. J. and Srinivasan, V. S. (1971), On the occurrence of bed forms in alluvial channels, *International Association for Hydraulic Research, Proceedings 14th Congress,* **3,** 91–100.

Hinschberger, F. (1963), Les hauts-fonds sableux de l'Iroise et leurs rapports avec les courants de marée. *Bulletin Section Géographie,* **75,** 53–80.

Hobday, D. K. and Reading, H. G. (1972), Fairweather versus storm processes in shallow marine sand bar sequences in the late Precambrian of Finnmark, North Norway, *Journal Sedimentary Petrology,* **42,** 318–324.

Hobday, D. K. and Tankard, A. J. (1978), Transgressive-barrier and shallow shelf interpretation of the Lower Paleozoic Peninsula Formation, South Africa, *Geological Society America Bulletin,* **89,** 1733–1744.

Holme, N. A. (1949), The fauna of the sand and mud banks near the mouth of the Exe Estuary, *Journal Marine Biological Association U.K.,* **28,** 189–237.

Holme, N. A. (1954), The ecology of British species of *Ensis, Journal Marine Biological Association U.K.,* **33,** 145–172.

Holme, N. A. (1959), The British species of *Lutraria* (Lamellibranchia), with a description of *L. angustior* Philippi, *Journal Marine Biological Association U.K.,* **38,** 557–568.

Holme, N. A. (1961), The bottom fauna of the English Channel, *Journal Marine Biological Association U.K.,* **41,** 397–461.

Holme, N. A. (1966), The bottom fauna of the English Channel, Part II, *Journal Marine Biological Association U.K.,* **46,** 401–493.

Hoskin, C. M. (1980), Flux of barnacle plate fragments and fecal pellets measured by sediment traps, *Journal Sedimentary Petrology,* **50,** 1213–1218.

Hoskin, C. M. and Nelson, R. V. (1969), Modern marine carbonate sediment, Alexander Archipelago, Alaska, *Journal Sedimentary Petrology,* **39,** 581–590.

Houbolt, J. J. H. C. (1968), Recent sediments in the Southern Bight of the North Sea, *Geologie en Mijnbouw,* **47,** 245–273.

Howard, J. D. and Reineck, H.-E. (1972), Physical and biogenic sedimentary structures of the nearshore shelf, *Senckenbergiana Maritima,* **4,** 81–123.

Howard, J. D. (1975), The sedimentological significance of trace fossils, in *The Study of Trace Fossils* (ed. R. W. Frey), Springer-Verlag, New York, pp. 131–146.

Huthnance, J. M. (1973), Tidal current asymmetries over the Norfolk sandbanks, *Estuarine and Coastal Marine Science,* **1,** 89–99.

Huthnance, J. M. (1975), On trapped waves over a continental shelf, *Journal of Fluid Mechanics,* **69,** 689–704.

Huthnance, J. M. (1982), On one mechanism forming linear sand banks, *Estuarine and Coastal Marine Science,* **14,** 79–99.

Hydrographic Department, Admiralty (1945), Torres Strait and Approaches, *Atlas of Tidal Streams,* HD 419, HMSO, London.

Hydrographic Department, Admiralty (1948), *Tidal streams of the waters surrounding the British Islands and off the west and north coasts of Europe, Gibraltar to Yugorski Strait,* Second Edition, Part 2.

Iannello, J. P. (1979), Tidally induced residual currents in estuaries of variable breadth and depth, *Journal of Physical Oceanography,* **9,** 962–974.

Imbrie, J. and Buchanan, H. (1965), Sedimentary structures in modern carbonate sands of the Bahamas, in *Primary Sedimentary Structures and their Hydrodynamic Interpretation* (ed. G. W. Middleton), Special Publication, Society of Economic Palaeontologists and Mineralogists, No. 12, pp. 149–172.

Imbrie, J. and Newell, N. D. (1964), *Approaches to Palaeoecology,* John Wiley, New York.

Inman, D. L. (1957), Wave-generated ripples in nearshore sands, *United States Army Corps of Engineers Beach Erosion Board, Technical Memorandum* 100.

Inman, D. L. and Bowen, A. J. (1963), Flume experiments on sand transport by waves and currents, *Proceedings 8th Coastal Engineering Conference,* 137–150.

Irwin, M. L. (1965), General theory of epeiric clear water sedimentation, *Bulletin American Association Petroleum Geologists,* **49,** 445–459.

Jackson, R. G. (1976), Large scale ripples of the Lower Wabash River, *Sedimentology,* **23,** 593–623.

Jansen, J. H. F. (1976), Late Pleistocene and Holocene history of the Northern North Sea, based on acoustic reflection records, *Netherlands Journal of Sea Research,* **10,** 1–43.

Jarke, J. (1956), Der Boden der südlichen Nordsee, *Deutsche Hydrographische Zeitschrift,* **9,** 1–9.

Jelgersma, S. (1979), Sea-level changes in the North Sea basin, in *The Quaternary History of the North Sea* (ed. E. Oele et al.), Acta Universitatis Upsaliensis, Uppsala, Sweden, pp. 233–248.

Johns, B. (1967), Tidal flow and mass transport in a slowly converging estuary, *Geophysical Journal Royal Astronomical Society,* **13,** 377–386.

Johnson, D. W. (1919), *Shore Processes and Shoreline Development,* John Wiley, New York.

Johnson, H. D. (1977), Shallow marine sand bar sequences: an example from the late Precambrian of North Norway, *Sedimentology,* **24,** 245–270.

Johnson, H. D. (1978), Shallow siliciclastic seas, in *Sedimentary Environments and Facies* (ed. H. G. Reading), Blackwell Scientific Publications, Oxford, pp. 209–258.

Johnson, M. A. and Belderson, R. H. (1969), The tidal origin of some vertical sedimentary changes in epicontinental seas, *Journal of Geology,* **77,** 353–357.

Johnson, M. A. and Stride, A. H. (1969), Geological

significance of North Sea sand transport rates, *Nature, London*, **224**, 1016–1017.

Johnson, M. A., Stride, A. H., Belderson, R. H. and Kenyon, N. H. (1981), Predicted sand wave formation and decay on a large offshore tidal current sand sheet, *Special Publication* **5**, *International Association Sedimentologists*, pp. 247–256.

Johnson, R. G. (1964), The community approach to paleoecology, in *Approaches to Paleoecology* (ed. J. Imbrie and N. D. Newell), John Wiley, New York, pp. 107–134.

Jolliffe, I. P. (1963), A study of sand movements on the Lowestoft sandbank using fluorescent tracers, *Geographical Journal*, **129**, 480–493.

Jones, N. S. (1950), Marine bottom communities, *Biological Reviews*, **25**, 283–313.

Jones, N. S., Kain, J. M. and Stride, A. H. (1965), The measurement of sand waves on Warts Bank, Isle of Man, *Marine Geology*, **3**, 329–336.

Jones, R. (1954), The food of the Whiting and a comparison with that of the Haddock, *Scottish Home Department, Marine Research 1954*, No. 2.

Jong, J. D. de (1977), Dutch tidal flats, *Sedimentary Geology*, **18**, 13–23.

Jopling, A. V. (1961), Origin of regressive ripples explained in terms of fluid mechanics processes, *United States Geological Survey Professional Paper* 424-D, 15–17.

Jordan, G. F. (1962), Large submarine sand waves, *Science*, **136**, 839–848.

Jordan, P. R. (1965), Fluvial sediment of the Mississippi River at St Louis, Missouri, *United States Geological Survey, Water Supply Paper* 1802.

Joseph, J. (1955), Extinction measurements to indicate distribution and transport of water masses, *Proceedings UNESCO Symposium on Physical Oceanography*, Tokyo, 59–75.

Joubin, L. (1922), Les coraux de mer profonde nuisibles aux chalutiers, *Notes et mémoires, Office scientifique et technique des pêches maritimes*, No. 18.

Kachel, N. B. and Sternberg, R. W. (1971), Transport of bedload as ripples during an ebb current, *Marine Geology*, **19**, 229–244.

Kahn, P. G. K. and Pompea, S. M. (1978), Nautiloid growth rhythms and dynamical evolution of the Earth–Moon system, *Nature, London*, **275**, 606–610.

Karcz, I. (1967), Harrow marks, current-aligned sedimentary structures, *Journal of Geology*, **75**, 113–121.

Kauffman, E. G. (1974), Cretaceous assemblages, communities and associations: Western interior United States and Caribbean Islands, in *Principles of Benthic Community Analysis* (ed. A. M. Zeigler *et al.*), University of Miami, pp. 12.1–12.25.

Kauffman, E. G. and Scott, R. W. (1976), Basic concepts of community ecology and paleoecology, in *Structures and Classification of Paleocommunities* (eds R. W. Scott and R. W. West), Dowden, Hutchinson and Ross, Stroudsburg, Pennsylvania, USA, pp. 1–28.

Keary, R. (1967), Biogenic carbonate in beach sediments of the west coast of Ireland, *Scientific Proceedings Royal Dublin Society*, A, **7**, 75–85.

Keary, R. and Keegan, B. F. (1975), Stratification by in-fauna debris: a structure, a mechanism and a comment, *Journal Sedimentary Petrology*, **45**, 128–131.

Keegan, B. F. (1974), The macrofauna of maerl substrates on the west coast of Ireland, *Cahiers de Biologie Marine*, **15**, 513–530.

Kelland, N. and Bailey, A. (1975), An underwater study of sand wave mobility in Start Bay, *Report of Underwater Association*, **1** (New Series), 74–80.

Kellogg, C. W. (1976), Gastropod shells: A potentially limiting resource for hermit crabs, *Journal of Experimental Marine Biology and Ecology*, **22**, 101–111.

Kennedy, J. F. and Locher, F. A. (1972), Sediment suspension by water waves, in *Waves on Beaches* (ed. R. E. Meyer), Academic Press, New York, pp. 249–295.

Kennedy, W. J. (1978), Cretaceous, in *The Ecology of Fossils* (ed. W. S. McKerrow), Duckworth, pp. 280–322.

Kent, H. C. (1968), Biostratigraphy of Niobrara-equivalent part of Mancos Shale (Cretaceous) in north-western Colorado, *American Association Petroleum Geologists' Bulletin*, **52**, 2098–2115.

Kenyon, N. H. (1970), Sand ribbons of European tidal seas, *Marine Geology*, **9**, 25–39.

Kenyon, N. H. (1980), Bedforms of shelf seas viewed with SEASAT synthetic aperture radar, in *Proceedings of seminar, Advances in hydrographic surveying*, Society for Underwater Technology, London, pp. 67–74.

Kenyon, N. H. and Belderson, R. H. (1973), Bedforms of the Mediterranean undercurrent observed with side-scan sonar, *Sedimentary Geology*, **9**, 77–99.

Kenyon, N. H., Belderson, R. H., Stride, A. H. and Johnson, M. A. (1981), Offshore tidal sand banks as indicators of net sand transport and as potential deposits, *Special Publication* **5**, *International Association of Sedimentologists*, pp. 257–268.

Kenyon, N. H. and Pelton, C. D. (1979), Seabed conditions west of the Outer Hebrides, *Institute of Oceanographic Sciences Report* No. 95, Unpublished Manuscript.

Khayrallah, N. and Jones, A. M. (1975), A survey of the benthos of the Tay Estuary, *Proceedings Royal Society Edinburgh*, Series B, **75**, 113–135.

Kirby, R. and Parker, W. R. (1974), Seabed density measurements related to echo sounder records, *Dock and Harbour Authority*, **54**, 423–424.

Kirkaldy, J. F. (1963), The Wealden and marine Lower Cretaceous beds of England, *Proceedings Geologists' Association*, **74**, 127–146.

Kirtley, D. W. and Tanner, W. F. (1968), Sabellarid worms: builders of a major reef type, *Journal Sedimentary Petrology*, **38**, 73–78.

Klein, G. de V. (1970a), Depositional and dispersal dynamics of intertidal sand bars, *Journal Sedimentary Petrology*, **40**, 1095–1127.

Klein, G. de V. (1970b), Tidal origin of a Precambrian quartzite – the lower fine grained quartzite (Middle Dalradian) of Islay, Scotland, *Journal Sedimentary Petrology*, **40**, 973–985.

Klein, G. de V. (1971), A sedimentary model for determining paleotidal range, *Geological Society America Bulletin*, **82**, 2585–2592.

Klein, G. de V. (1977a), Tidal circulation model for deposition of clastic sediment on epeiric and mioclinal shelf seas, *Sedimentary Geology*, **18**, 1–12.

Klein, G. de V. (1977b), *Clastic Tidal Facies*, Continuing Education Publishing Co., Champaign, USA.

Klein, G. de V. and Ryer, T. A. (1978), Tidal circulation patterns in Precambrian, Paleozoic and Cretaceous epeiric and mioclinal shelf seas, *Geological Society America Bulletin*, **89**, 1050–1058.

Kohlmeyer, J. (1969), The role of marine fungi in the penetration of calcareous substances, *American Zoologist*, **9**, 741–746.

Komar, P. D. (1976), *Beach Processes and Sedimentation*, Prentice-Hall, Englewood Cliffs, New Jersey, USA.

Komar, P. D. and Miller, M. C. (1975), Sediment threshold under oscillatory waves, *Proceedings, 14th Conference on Coastal Engineering*, 756–775.

Krantz, R. (1972), Die sponge-gravels von Faringdon (England), *Neues Jahrbuch für Geologie und Paläontologie Abhandlungen*, **140**, 207–231.

Kuenen, P. H. (1950), *Marine Geology*, John Wiley, London.

Kuijpers, A. (1980), *Sediment patterns and bedforms and their relationship to the flow regime in the Balt Sea and Sound*, Dissertation, Kiel University.

Ladd, H. S. (ed.) (1957), *Treatise on Marine Ecology and Paleoecology, Vol. 2 Paleoecology*, Geological Society America Memoir 67.

Lagaaij, R. (1968), Fossil Bryozoa reveal long-distance sand transport along the Dutch coast, *Koninklijk Nederlands Akademie von Wetenschappen Proceedings*, B, **71**, 31–50.

Lamar, D. L., McGann-Lamar, J. V. and Merifield, P. M. (1970), Age and origin of Earth–Moon system, in *Palaeogeophysics* (ed. S. K. Runcorn), Academic Press, London.

Lande, R. (1976), Food and feeding habits of the Long Rough Dab *Hippoglossus platessoides* (Fabricus) (Pisces, Pleuronectidae) in Borgenfjorden, Norway, *Sarsia*, **62**, 19–24.

Lane, E. W. and Eden, E. W. (1940), Sand waves in the lower Mississippi River, *Journal, Western Society of Engineers*, **45** (6), 281–291.

Langeraar, W. (1966), Sandwaves in the North Sea, *Hydrographic Newsletter*, **1**, 243–246.

Langhorne, D. N. (1976), Consideration of meteorological conditions when determining the navigational water depth over a sandwave field, *15th Annual Canadian Hydrographic Conference*, Ottawa.

Langhorne, D. N. (1978), Offshore engineering and navigational problems: the relevance of sandwave research, *Society for Underwater Technology*, London.

Langhorne, D. N. (In Press), A study of the dynamics of a marine sandwave, *Sedimentology*.

Lapierre, F. (1969), *Répartition des sédiments sur le plateau continental du Golfe de Gascogne*, Thèse, Faculté de Science, Bordeaux.

Lapierre, F. (1975), Contribution a l'étude géologique et sédimentologique de la Manche orientale, *Philosophical Transactions Royal Society*, **279** A, 177–187.

Larsonneur, C. (1971), *Manche Centrale et Baie de Seine: géologie du substratum et dépôts meubles*, Thèse de Docteur d'Etat, Université de Caen, Centre National de la Recherche Scientifique, Archives Originales, 5404.

Larsonneur, C. (1972), La modèle sédimentaire de la Baie de Seine à la Manche centrale dans son cadre géographique et historique, Colloque du la géologie de la Manche, Paris, *Mémoires du Bureau de Recherches Géologiques et Minières*, **79**, 241–255.

Lawford, A. L. (1954), Currents in the North Sea during the 1953 gale, *Weather*, **9**, 67–72.

Lee, A. J. and Folkard, A. R. (1969), Factors affecting turbidity in the Southern North Sea, *Journal du Conseil International pour l'Exploration de la Mer*, **32**, 291–302.

Lee, A. J. and Ramster, J. W. (1979), Atlas of the seas around the British Isles, Ministry of Agriculture, Fisheries and Food, *Fisheries Research Technical Report*, **20**, HMSO, London.

Lees, A. (1973), Les dépots cabonatés de plate-forme, *Bulletin Centre Recherches de Pau, Société Nationale des Pétroles d'Aquitaine*, **7**, 177–192.

Lees, A. (1975), Possible influence of salinity and temperature on modern shelf carbonate sedimentation, *Marine Geology*, **19**, 159–198.

Lees, A. and Buller, A. T. (1972), Modern temperate-water and warm-water shelf carbonate sediments contrasted, *Marine Geology*, **13**, M67–M73.

Lees, A., Buller, A. T. and Scott, J. (1969), Marine carbonate sedimentation processes, Connemara, Ireland, *University of Reading, Geology Department Report* No. 2.

Lefort, J.-P. (1970), Etude géologique de la Manche au nord du Trégor, II Le problème des sables calcaires: la sédimentation actuelle, *Bulletin Société Géologique minéralogique Bretagne*, **2**, 11–23.

Lemoine, P. (1923), Répartition des algues calcaires dans La Manche occidentale d'après les dragages du

Pourquoi-Pas? Bulletin Muséum National d'Histoire Naturelle, **29**, 462–469.

Levell, B. K. (1980), A late Precambrian tidal shelf deposit, the Lower Sandfjord Formation, Finnmark, North Norway, *Sedimentology*, **27**, 539–557.

Longuet-Higgins, M. S. (1953), Mass transport in water waves, *Philosophical Transactions Royal Society, London*, **254A**, 535–581.

Longuet-Higgins, M. S. (1968), Double Kelvin waves with continuous depth profiles, *Journal of Fluid Mechanics*, **34**, 49–80.

Lovell, J. P. B. (1979), Composition of Holocene sands of Mull and adjacent offshore areas: a study of provenance, *Institute Geological Sciences Report* 79/2.

Ludwick, J. C. (1971), Migration of tidal sand waves in Chesapeake Bay entrance, *Technical Report 2, Institute of Oceanography, Old Dominion University*, Virginia, USA.

Ludwick, J. C. (1972), Migration of tidal sand waves in Chesapeake Bay entrance, in *Shelf Sediment Transport*, (eds D. J. P. Swift, D. B. Duane and O. H. Pilkey), Dowden, Hutchinson and Ross, Stroudsburg, Pennsylvania, USA, pp. 377–410.

Lyell, C. (1853), *Principles of Geology*, John Murray, London.

McCave, I. N. (1970), Deposition of fine-grained suspended sediment from tidal currents, *Journal of Geophysical Research*, **75**, 4151–4159.

McCave, I. N. (1971a), Wave effectiveness at the sea bed and its relationship to bed-forms and deposition of mud, *Journal Sedimentary Petrology*, **41**, 89–96.

McCave, I. N. (1971b), Sand waves in the North Sea off the coast of Holland, *Marine Geology*, **10**, 199–225.

McCave, I. N. (1973), The sedimentology of a transgression: Portland Point and Cooksburg Members (Middle Devonian), New York State, *Journal Sedimentary Petrology*, **43**, 484–504.

McCave, I. N. (1978), Grain size trends and transport along beaches: an example from eastern England, *Marine Geology*, **28**, M43–M51.

McCave, I. N. (1979), Tidal currents at the North Hinder Lightship, Southern North Sea: flow directions and turbulence in relation to maintenance of sand banks, *Marine Geology*, **31**, 101–114.

MacDonald, G. J. F. (1966), Origin of the Moon: dynamical considerations, in *The Earth–Moon System* (eds B. G. Marsden and A. G. W. Cameron), Plenum Press, New York, pp. 165–209.

Macer, C. T. (1966), Sand eels (*Ammodytidae*) in the south-western North Sea; their biology and fishery, *Ministry of Agriculture, Fisheries and Food, Fishery Investigations, Series 2*, **24** (6), 1–55, HMSO, London.

MacGinitie, G. E. (1935), Ecological aspects of a California marine estuary, *American Midland Naturalist*, **16**, 629–765.

MacGinitie, G. E. (1939), Littoral marine communities, *American Midland Naturalist*, **21**, 28–55.

McIntosh, W. C. (1913), Notes from the Gatty Marine Laboratory, St Andrews, No. 34, *Annals Magazine Natural History*, Series 8, **11**, 83–130.

McIntyre, A. D. (1978), The benthos of the western North Sea, *Rapports et Procès-Verbaux Réunions I.C.E.S.*, **172**, 405–417.

McKerrow, W. S. (ed.) (1978), *The Ecology of Fossils*, Gerald Duckworth and Co, London.

McManus, J., Buller, A. T. and Green, C. D. (1980), Sediments of the Tay Estuary VI, Sediments of the lower and outer reaches, *Proceedings Royal Society Edinburgh*, Series B, **78**, э133–э153.

Macmillan, D. H. (1966), *Tides*, C. R. Books Ltd, London.

Maddock, L. and Pingree, R. D. (1978), Numerical simulation of the Portland tidal eddies, *Estuarine and Coastal Marine Science*, **6**, 353–363.

Marshall, J. F. and Davies, P. J. (1978), Skeletal carbonate variation on the continental shelf of eastern Australia, *BMR Journal Australian Geology and Geophysics*, **3**, 85–92.

Matthews, R. K. (1974), *Dynamic Stratigraphy*, Prentice-Hall, New Jersey.

Maxwell, W. G. (1969), Radiocarbon ages of sediment: Great Barrier Reef, *Sedimentary Geology*, **3**, 331–333.

Mazzullo, S. J. (1980), Stratigraphic traps in carbonate rocks, American Association of Petroleum Geologists, Reprint Series No. 23, Tulsa, Oklahoma, USA.

Meckel, L. D. (1975), Holocene sand bodies in the Colorado Delta area, northern Gulf of California, in *Deltas, Models for Exploration* (ed. M. L. Broussard), Houston Geological Society, Houston, pp. 239–265.

Menzies, R. J. (1957), Marine borers, in *Treatise on Marine Ecology and Paleoecology*, **1** (ed. J. W. Hedgpeth), Geological Society of America Memoir 67, pp. 1029–1034.

Merifield, P. M. and Lamar, D. L. (1970), Palaeotides and the geologic record, in *Palaeogeophysics* (ed. S. K. Runcorn), Academic Press, London, pp. 31–40.

Middlemiss, F. A. (1962), Brachiopod ecology and Lower Greensand palaeontology, *Palaeontology*, **5**, 253–267.

Middlemiss, F. A. (1975), Studies in the sedimentation of the Lower Greensand of the Weald, 1875–1975: a review and commentary, *Proceedings Geologists' Association*, **86**, 457–473.

Middleton, G. V. and Southard, J. B. (1977), *Mechanics of Sediment Movement*, Society of Economic Palaeontologists and Mineralogists, Special Publication No. 3.

Miller, M. C., McCave, I. N. and Komar, P. D. (1977), Threshold of sediment motion under unidirectional currents, *Sedimentology*, **24**, 507–527.

Milliman, J. D. (1974), *Recent Sedimentary Carbonates*, Part I *Marine Carbonates*, Springer-Verlag, Berlin.

Mishra, S. K. (1968), Granulometric studies of Recent sediments in the Firth of Tay region (Scotland), *Sedimentary Geology*, 2, 191–200.

Mogi, A. (1979), *An Atlas of the Sea Floor around Japan: Aspects of Submarine Geomorphology*, University of Tokyo Press.

Mogridge, G. R. and Kamphuis, J. W. (1972), Experiments on bed forms generated by wave action, *Proceedings, 13th Coastal Engineering Conference*, 2, 1123–1142.

Molander, A. R. (1928), Animal communities on soft bottom areas in the Gullmar Fjord, *Kristinebergs Zoologiska Station 1877-1927*, No. 2, 1–90, Uppsala, Sweden.

Molander, A. R. (1962), Studies of the fauna in the fjords of Bohuslän with reference to the distribution of different associations, *Arkiv för Zoologi* Serie 2, 15 (1), 1–64.

Moore, J. R. (1968), Recent sedimentation in northern Cardigan Bay, Wales, *Bulletin British Museum (Natural History), Mineralogy*, 2, 19–131.

Morra, R. H. J., Oudshoorn, H. M., Svasek, J. N. and de Voss, F. J. (1961), De zandbeweging in het getijgebied van Zuidwest Nederland, *Rapport Deltacommissie*, 5, 327–380, s'Gravenhage.

Müller, J. and Milliman, J. D. (1973), Relict carbonate-rich sediments on south-western Grand Bank, Newfoundland, *Canadian Journal Earth Sciences*, 10, 1744–1750.

Müller, P. M. and Stephenson, F. R. (1975), The accelerations of the Earth and Moon from early astronomical observations, in *Growth Rhythms and the History of the Earth's Rotation* (eds D. Rosenberg and S. K. Runcorn), Wiley, New York, pp. 459–534.

Mullins, H. T. and Neumann, A. C. (1979), Deep carbonate bank margin structure and sedimentation in the northern Bahamas, in *Geology of Continental Slopes* (eds L. J. Doyle and O. H. Pilkey), Society of Economic Paleontologists and Mineralogists Special Publication No. 27, pp. 165–192.

Murray, J. W. (1976), A method of determining proximity of marginal seas to an ocean, *Marine Geology*, 22, 103–119.

Narayan, J. (1971), Sedimentary structures in the Lower Greensand of the Weald, England and Bas-Boulonnais, France, *Sedimentary Geology*, 6, 73–109.

Neale, J. W. (1974), Ostracod faunas from the Celtic Sea, *Geoscience and Man*, 6, 81–98.

Negus, M. (1975), An analysis of boreholes drilled by *Natica catena* (da Costa) in the valves of *Donax vittatus* (da Costa), *Proceedings Malacological Society London*, 41, 353–356.

Nelson, C. S. (1978), Temperate shelf carbonate sediments in the Cenozoic of New Zealand, *Sedimentology*, 25, 737–771.

Neumann, A. C. (1966), Observations on coastal erosion in Bermuda and measurements of the boring rate of the sponge *Cliona lampa*, *Limnology and Oceanography*, 11, 92–108.

Neumann, A. C., Kofoed, G. H. and Keller, G. H. (1977), Lithoherms in the Straits of Florida, *Geology*, 5, 4–10.

Neumann, H. and Meier, C. (1964), Die Oberflächenströme in der Deutschen Bucht, *Deutsche Hydrographische Zeitschrift*, 17, 1–40.

Nihoul, J. C. J. (1977), Three-dimensional model of tides and storm surges in a shallow well-mixed continental sea, *Dynamics of Atmospheres and Oceans*, 2, 29–48.

Nio, Swie-Djin, (1976), Marine transgressions as a factor in the formation of sandwave complexes, *Geologie en Mijnbouw*, 55, 18–40.

Nio, Swie-Djin, Berg, J. H. van den and Siegenthaler, C. (1979), *Excursion Guide to the Oosterschelde Basin, S. W. Netherlands*: an example of Holocene tidal sedimentation, International Association of Sedimentologists, Texel, The Netherlands, Sept. 17–23, 1979.

Nixon, M., Maconnachie, E. and Howell, P. G. T. (1980), The effects on shells of drilling by *Octopus*, *Journal Zoological Society London*, 191, 75–88.

Noort, G. J. van, Leeuwen, F. van and Creutzberg, F. (1979a), 'Aurelia' – Cruise Reports on the benthic fauna of the Southern North Sea. Report 1. Trawl survey April–May 1972, *Internal Report. Netherlands Institute for Sea Research*, Texel, 1979-5.

Noort, G. J. van, Leeuwen, F. van and Creutzberg, F. (1979b), 'Aurelia' – Cruise Reports on the benthic fauna of the Southern North Sea, Report 2. Trawl Survey June–July 1972, *Internal Report Netherlands Institute for Sea Research*, Texel, 1979-6.

Noort, G. J. van, Leeuwen, F. van and Creutzberg, F. (1979c), 'Aurelia' – Cruise Reports on the benthic fauna of the Southern North Sea, Report 3, Trawl Survey October–November 1972, *Internal Report Netherlands Institute for Sea Research*, Texel, 1979-7.

Noort, G. J. van, Leeuwen, F. van and Creutzberg, F. (1979d), 'Aurelia' – Cruise Reports on the benthic fauna of the Southern North Sea, Report 4, Trawl Survey January–February 1973, *Internal Report Netherlands Institute for Sea Research*, Texel, 1979-8.

Norris, R. M. (1972), Shell and gravel layers, western continental shelf, New Zealand, *New Zealand Journal Geology and Geophysics*, 15, 572–589.

Odin, G. S. (1973), Répartition, nature minéralogique et genèse des granules verts recueillis dans les sédiments marins actuels, *Sciences de la Terre*, 18, 79–94.

Oele, E. (1969), The Quaternary geology of the Dutch

part of the North Sea, north of the Frisian Isles, *Geologie en Mijnbouw*, **48**, 467–480.

Oele, E. (1971), The Quaternary geology of the southern area of the Dutch part of the North Sea, *Geologie en Mijnbouw*, **50**, 461–474.

Off, T. (1963), Rhythmic linear sand bodies caused by tidal currents, *Bulletin, American Association Petroleum Geologists*, **47**, 324–341.

Oliver, G. and Allen, J. A. (1980a), The functional and adaptive morphology of the deep-sea species of the Arcacea (Mollusca: Bivalvia) from the Atlantic, *Philosophical Transactions Royal Society London*, **291**B, 45–76.

Oliver, G. and Allen, J. A. (1980b), The functional and adaptive morphology of the deep-sea species of the family Limopsidae (Bivalvia: Arcoida) from the Atlantic, *Philosophical Transactions Royal Society London*, **291**B, 77–125.

Olson, W. S. (1970), Tidal amplitudes in geological history, *Transactions, New York Academy of Sciences*, Series 2, **32**, 220–233.

Otto, L. (1971), The probability of high current speeds at the Netherlands lightships, *De Ingenieur*, **83**, W26–W31.

Pannella, G. (1976), Tidal growth patterns in Recent and fossil mollusc bivalve shells: a tool for the reconstruction of paleotides, *Die Naturwissenschaften*, **63**, 539–543.

Pantin, H. M. (1978), Quaternary sediments from the north-east Irish Sea: Isle of Man to Cumbria, *Institute of Geological Sciences, Bulletin* 64.

Pasternak, F. A. (1971), Marine biology of the East Atlantic continental margin, in *Geology of the East Atlantic Continental Margin* (ed. F. M. Delany), Institute of Geological Sciences Report No. 70/13, pp. 67–77.

Pendlebury, D. C. and Dobson, M. R. (1976), Sediment and macrofaunal distributions in the eastern Malin Sea, as determined by side-scan sonar and sampling, *Scottish Journal Geology*, **4**, 315–332.

Perkins, E. J. (1974), *The Biology of Estuaries and Coastal Waters*, Academic Press, London.

Petersen, C. G. J. (1913), Valuation of the sea II, the animal communities of the sea bottom and their importance for marine zoogeography, *Report Danish Biological Station to Board of Agriculture*, **21**.

Petersen, C. G. J. (1918), The sea-bottom and its production of fish food, *Report Danish Biological Station to Board of Agriculture*, **25**.

Petersen, G. H. (1977), The density, biomass and origin of bivalves of the central North Sea, *Meddelelser fra Danmarks Fiskeri-og Havundersøgelser* NS, **7**, 221–273.

Petrie, B. (1975), M_2 surface and internal tides on the Scotian shelf and slope, *Journal Marine Research*, **33**, 303–323.

Pettijohn, F. J., Potter, P. E. and Siever, R. (1972), *Sand and Sandstone*, Springer-Verlag, Berlin.

Pickerill, R. K. and Brenchley, P. J. (1975), The application of the community concept in palaeontology, *Maritime Sediments*, **2**, 5–8.

Piessens, P. and Lees, A. (1977), Sedimentation de carbonates biogeniques Récents dans la Baie de Clifden (Connemara, Irlande), *Mémoire de l'Institut Géologique de l'Université de Louvain*, **29**, 357–367.

Pingree, R. D. (1978), The formation of the Shambles and other banks by tidal stirring of the seas, *Journal Marine Biological Association U.K.*, **58**, 211–226.

Pingree, R. D. and Griffiths, D. K. (1979), Sand transport paths around the British Isles resulting from M_2 and M_4 tidal interactions, *Journal Marine Biological Association U.K.*, **59**, 497–513.

Pingree, R. D. and Griffiths, D. K. (1980), Currents driven by a steady uniform wind stress on the shelf seas around the British Isles, *Oceanologica Acta*, **3**, 227–236.

Pingree, R. D. and Maddock, L. (1977), Tidal residuals in the English Channel, *Journal Marine Biological Association U.K.*, **57**, 339–354.

Pingree, R. D. and Maddock, L. (1979), The tidal physics of headland flow and offshore tidal bank formation, *Marine Geology*, **32**, 269–289.

Piper, J. D. A. (1978), Geological and geophysical evidence relating to continental growth and dynamics and the hydrosphere in Precambrian times: a review and analysis, in *Tidal Friction and the Earth's Rotation* (eds P. Brosche and J. Sündermann), Springer-Verlag, Berlin, pp. 197–241.

Prandle, D. (1978), Residual flows and elevations in the Southern North Sea, *Proceedings Royal Society London*, **359**A, 189–228.

Pratje, O. (1931), Die sedimente der Deutschen Bucht: eine regional-statistische untersuchung, *Wissenschaftliche Meeresuntersuchungen der Kommission zur wissenschaftlichen Untersuchung der deutschen Meere in Kiel und der biologischen Anstalt auf Helgoland*, **18** (6).

Pratje, O. (1950), Die Bodenbedeckung des Englischen Kanals und die maximalen Gezeitenstromgeschwindigkeiten, *Deutsche Hydrographische Zeitschrift*, **3**, 201–205.

Pratt, C. J. (1970), Summary of experimental data for flume tests over 0.49 mm sand, *Report CE/9/70, Department of Civil Engineering, University of Southampton*.

Prentice, J. E., Beg, I. R., Colleypriest, C., Kirby, R., Sutcliffe, P. J. C., Dobson, M. R., D'Olier, B., Elvines, M. F., Kilenyi, T. I., Maddrell, R. J. and Phinn, T. R. (1968), Sediment transport in estuarine areas, *Nature, London*, **218**, 1207–1210.

Proudman, J. (1953), *Dynamical Oceanography*, Methuen, London.

Pruvot, G. (1897), Essai sur les fonds et la faune de la Manche occidentale (côtes de Bretagne) comparés à ceux du Golf du Lion, *Archives de Zoologie Expérimentale et Générale* 3rd Série, **5**, 511–660.

Pryor, W. A. (1960), Cretaceous sedimentation in Upper Mississippi embayment. *Bulletin American Association Petroleum Geologists*, **44**, 1473–1504.

Raaf, J. F. M. de, and Boersma, J. R. (1971), Tidal deposits and their sedimentary structures, *Geologie en Mijnbouw*, **50**, 479–504.

Rae, B. B. (1956), The food and feeding habits of the Lemon Sole, *Scottish Home Department, Marine Research 1956*, No. 3.

Rae, B. B. (1963), The food of the Megrim, *Department of Agriculture and Fisheries for Scotland, Marine Research 1963*, No. 3.

Rae, B. B. (1967), The food of the Cod in the North Sea and on the west of Scotland grounds, *Marine Research 1967*, No. 1.

Reade, T. M. (1888), Tidal action as an agent of geological change, *Philosophical Magazine*, **25**, 338–343.

Redfield, A. C. (1958), The influence of the continental shelf on the tides of the Atlantic coast of the United States, *Journal of Marine Research*, **17**, 432–448.

Reid, W. J. (1958), Coastal experiments with radioactive tracers, recent work on the coast of Norfolk, *Dock and Harbour Authority*, **39**, 84–88.

Reineck, H.-E. (1958), Wühlbau-Gefüge in Abhangigkeit von Sediment-Umlagerungen, *Senckenbergiana Lethaea*, **39**, 1–23, 54–56.

Reineck, H.-E. (1963), Sedimentgefüge im Bereich der südlichen Nordsee, *Abhandlungen der Senckenbergischen Naturforschen den Gesellschaft*, No. 505.

Reineck, H.-E. (1967), Parameter von Schichtung und Bioturbation, *Geologische Rundschau*, **56**, 420–438.

Reineck, H.-E. (1973), Bibliographie geologischer Arbeiten über rezente und fossile Kalk-und Silikwatten, *Courier Forschungsinstitut Senckenberg* No. 6, Frankfurt am Main.

Reineck, H.-E. (1976), Primärgefüge und Makrofauna als Indikatoren des Sandversatzes im Seegebiet von Norderney (Nordsee), I. Zonierung von Primärgefügen und Bioturbation, *Senckenbergiana Maritima*, **8**, 155–169.

Reineck, H.-E., Dörjes, J., Gadow, S. and Hertweck, G. (1968), Sedimentologie, Faunenzonierung und Faziesabfolge vor der Ostküste der inneren Deutschen Bucht, *Senckenbergiana Lethaea*, **49**, 261–309.

Reineck, H.-E., Gutmann, W. F. and Hertweck, G. (1967), Das Schlickgebiet südlich Helgoland als Beispiel rezenter Schelfablagerungen, *Senckenbergiana Lethaea*, **48**, 219–275.

Reineck, H.-E. and Singh, I. B. (1972), Genesis of laminated sand and graded rhythmites in storm-sand layers of shelf mud, *Sedimentology*, **18**, 123–128.

Reineck, H.-E. and Singh, I. B. (1973), *Depositional Sedimentary Environments*, Springer-Verlag, Berlin.

Rhoads, D. C. (1967), Biogenic reworking of intertidal and subtidal sediments in Barnstable Harbor and Buzzards Bay, Massachusetts, *Journal of Geology*, **75**, 461–476.

Rhoads, D. C. (1975), The paleoecological and environmental significance of trace fossils, in *The Study of Trace Fossils* (ed. R. W. Frey), Springer-Verlag, Berlin, pp. 147–160.

Rhoads, D. C. and Stanley, D. J. (1965), Biogenic graded bedding, *Journal Sedimentary Petrology*, **35**, 956–963.

Richter, R. (1920), Ein devonisher Pfeifenquarzit vergleichen mit der heutigen 'Sandkoralle' (Sabellaria, Annelidae), *Senckenbergiana*, **2**, 215–235.

Richter, R. (1927), 'Sandkorallen' riffe in der Nordsee, *Natur und Volk*, **57**, 49–62.

Richter, R. (1928), Aktuo-paläontologie und Paläobiologie, eine Abgrenzung, *Senckenbergiana*, **10**, 285–292.

Rigby, J. K. and Hamblin, W. K. (eds) (1972), *Recognition of Ancient Sedimentary Environments*, Society of Economic Palaeontologists and Mineralogists, Special Publication No. 16.

Ritchie, A. (1937), The food and feeding habits of the haddock (*Gadus aeglefinus*) in Scottish waters, *Scientific Investigations, Fishery Board for Scotland 1937*, No. 2.

Robinson, A. H. W. (1963), Physical factors in the maintenance of port approach channels, *Dock and Harbour Authority*, **43**, 351–357.

Robinson, I. S. (1979), The tidal dynamics of the Irish and Celtic Seas, *Geophysical Journal*, **56**, 159–197.

Sabins, F. F. (1972), Comparison of Bisti and Horseshoe Canyon stratigraphic traps, San Juan Basin, New Mexico, in *Stratigraphic Oil and Gas Fields* (ed. R. E. King), American Association of Petroleum Geologists, Memoir 16, pp. 610–622.

Sager, G. (1963), Die Beziehungen zwischen den Gezeitenströmen und der Meeresbodenbedeckung in der Nordsee, dem Kanal und der Irischen See, *Petermans Geographischen Mitteilungen*, **107** (2), 111–115.

Sager, G. and Sammler, R. (1975), *Atlas de gezeitenströme für die Nordsee, dem Kanal und die Irische See*, 3rd edition, Seehydrographischer Dienst, Deutsche Demokratischen Republik.

Saila, S. B. (1976), Sedimentation and food resources: Animal-sediment relationships, in *Marine Sediment Transport and Environmental Management* (eds D. J. Stanley and D. J. P. Swift), John Wiley and Sons, New York, pp. 479–492.

Salsman, G. G., Tolbert, W. H. and Villars, R. G. (1966), Sand-ridge migration in St Andrews Bay, Florida, *Marine Geology*, **4**, 11–19.

Samu, G. (1968), Ergebnisse der Sandwanderungs-untersuchungen in der südlichen Nordsee, *Mitteilungsblatt der Bundesanstalt für Wasserbau*, No. 26, 13–61.

Schäfer, W. (1962), *Actuo-paläontologie nach Studien in der Nordsee*, Waldemar Kramer, Frankfurt am Main.

Schäfer, W. (1972), *Ecology and Palaeoecology of Marine Environments*, Oliver and Boyd, Edinburgh.

Schlee, J. (1973), Atlantic continental shelf and slope of the United States – sediment texture of the north-eastern part, *U.S. Geological Survey Professional Paper* 529-L, L1–L64.

Schott, F. (1971), On horizontal coherence and internal wave propagation in the North Sea, *Deep-Sea Research*, **18**, 291–307.

Schureman, P. (1940), Manual of harmonic analysis and prediction of tides, *U.S. Coast and Geodetic Survey, Special Publication*, **98**.

Scrutton, C. T. (1978), Periodic growth features in fossil organisms and the length of the day and month, in *Tidal Friction and the Earth's Rotation* (eds P. Brosche and J. Sündermann), Springer-Verlag, Berlin, pp. 154–196.

Seeling, A. (1978), The Shannon Sandstone, a further look at the environment of deposition at Heldt Draw Field, Wyoming, *The Mountain Geologist*, **15**, 133–144.

Seibold, E. (1974), *Der Meeresboden, Ergebnisse und Probleme der Meeresgeologie*, Springer-Verlag, Berlin.

Seilacher, A. (1956), Wirkungen der Benthos-Organismen auf den jungen Schichtverband, *Senckenbergiana Lethaea*, **37**, 183–263.

Selley, R. C. (1970), *Ancient Sedimentary Environments*, Chapman and Hall, London.

Shen, H. W., Mellema, W. J. and Harrison, A. S. (1978), Temperature and Missouri river stages near Omaha, *Journal Hydraulics Division, American Society of Civil Engineers*, **104**, 1–20.

Shepard, F. P. (1932), Sediments on continental shelves, *Bulletin Geological Society America*, **43**, 1017–1034.

Silvester, R. (1974), *Coastal Engineering*, 2, *Sedimentation, Estuaries, Tides, Effluents and Modelling*, Elsevier, Amsterdam.

Simons, D. B. and Richardson, E. V. (1963), Forms of bed roughness in alluvial channels, *Transactions, American Society Civil Engineers*, **128**, Part I, 284–302.

Slatt, R. M. (1973), Continental shelf sediments off eastern Newfoundland: a preliminary investigation, *Canadian Journal Earth Sciences*, **11**, 362–368.

Sly, P. G. (1966), *Marine geological studies in the Eastern Irish Sea and adjacent estuaries, with special reference to sedimentation in Liverpool Bay and the River Mersey*, Ph.D. Thesis, Liverpool University.

Smith, J. D. (1969), Geomorphology of a sand ridge, *Journal of Geology*, **77**, 39–55.

Smith, J. D. (1970), Stability of sand bed subjected to shear flow of low Froude number, *Journal Geophysical Research*, **75**, 5928–5940.

Smith, J. E. (1932), The shell gravel deposits and infauna of the Eddystone Grounds, *Journal Marine Biological Association U.K.*, **18**, 243–278.

Sognnaes, R. F. (ed.) (1963), Mechanisms of hard tissue destruction, *American Association for Advancement of Science Publication* No. 75.

Sorby, H. C. (1858), On the ancient physical geography of the south-east of England, *Edinburgh New Philosophical Journal*, New Series, 1–13.

Southard, J. B. and Boguchwal, L. A. (1973), Flume experiments on the transition from ripples to lower flat bed with increasing sand size, *Journal Sedimentary Petrology*, **43**, 1114–1121.

Spärk, R. (1935), On the importance of quantitative investigations of the bottom fauna in marine biology, *Journal du Conseil*, **10**, 3–19.

Stanley, S. M. (1970), Relation of shell form to life habits of the Bivalvia (Mollusca), *Geological Society America Memoir*, **125**.

Stanton, R. J. (1976), Relationship of fossil communities to original communities of living organisms, in *Structure and Classification of Paleocommunities* (eds R. W. Scott and R. W. West), Dowden, Hutchinson and Ross, Stroudsburg, Pennsylvania, USA, pp. 107–142.

Stanton, R. J. and Dodd, J. R. (1976), The application of trophic structure of fossil communities in paleoenvironmental reconstruction, *Lethaia*, **9**, 327–342.

Stein, R. A. (1965), Laboratory studies of total load and apparent bed load, *Journal Geophysical Research*, **70**, 1831–1842.

Stephen, A. C. (1933), Studies on the Scottish marine fauna: The natural faunistic divisions of the North Sea as shown by the quantitative distribution of the molluscs, *Transactions Royal Society Edinburgh*, **57**, 601–616.

Stephen, A. C. (1934), Studies on the Scottish marine fauna: Quantitative distribution of the echinoderms and the natural faunistic divisions of the North Sea, *Transactions Royal Society Edinburgh*, **57**, 777–787.

Sternberg, R. W. (1968), Friction factors in tidal channels with differing bed roughness, *Marine Geology*, **6**, 243–260.

Stetson, T. R., Squires, D. F. and Pratt, R. M. (1962), Coral banks occurring in deep water on the Blake Plateau, *American Museum Novitates*, No. 2114.

Steven, G. A. (1930), Bottom fauna and the food of fishes, *Journal Marine Biological Association U.K.*, **16**, 677–705.

Stewart, H. B. and Jordan, G. F. (1964), Underwater sand ridges on Georges Shoal, in *Papers in Marine*

Geology (ed. R. L. Miller), Macmillan, New York, pp. 102–114.

Stone, R. and Vondra, C. F. (1972), Sediment dispersal patterns of oolitic calcarenite in the Sundance Formation (Jurassic), Wyoming, *Journal Sedimentary Petrology*, **42**, 227–229.

Straaten, L. M. J. U. van (1952), Biogene textures and the formation of shell beds in the Dutch Wadden Sea, *Proceedings, Koninklijk Nederlandse Akademie van Wetenschappen – Amsterdam*, Series B, **55**, 500–516.

Straaten, L. M. J. U. van (1953), Megaripples in the Dutch Wadden Sea and in the basin of Arcachon (France), *Geologie en Mijnbouw*, **15**, 1–11.

Straaten, L. M. J. U. van (1954), Composition and structure of Recent marine sediments in the Netherlands, *Leidse Geologische Mededelingen*, **19**, 1–110.

Straaten, L. M. J. U. van (1956), Composition of shell beds formed in tidal flat environment in the Netherlands and in the Bay of Arcachon (France), *Geologie en Mijnbouw*, **18**, 209–226.

Straaten, L. M. J. U. van (1961), Sedimentation in tidal flat areas, *Journal Alberta Society Petroleum Geologists*, **9**, 203–226.

Straaten, L. M. J. U. van and Kuenen, P. H. (1957), Accumulation of fine grained sediments in the Dutch Wadden Sea, *Geologie en Mijnbouw*, **19**, 329–354.

Stride, A. H. (1963a), Current-swept sea floors near the southern half of Great Britain, *Quarterly Journal Geological Society*, **119**, 175–199.

Stride, A. H. (1963b), North-east trending ridges of the Celtic Sea, *Proceedings Ussher Society*, **1**, 62–63.

Stride, A. H. (1970), Shape and size trends for sand waves in a depositional zone of the North Sea, *Geological Magazine*, **107**, 469–477.

Stride, A. H., Belderson, R. H. and Kenyon, N. H. (1972), Longitudinal furrows and depositional sand bodies of the English Channel, *Mémoire Bureau Recherches Géologiques et Minières*, **79**, 233–240.

Stride, A. H. and Chesterman, W. D. (1973), Sedimentation by non-tidal currents around northern Denmark, *Marine Geology*, **15**, M53–M58.

Stubbs, A. R., McCartney, B. S. and Legg, J. G. (1974), Telesounding, a method of wide swathe depth measurement, *International Hydrographic Review*, **51**, 23–59.

Summerhayes, C. P. (1969a), Recent sedimentation around northernmost New Zealand, *New Zealand Journal Geology and Geophysics*, **12**, 172–207.

Summerhayes, C. P. (1969b), Marine geology of the New Zealand sub-antarctic sea floor, *Memoir New Zealand Oceanographic Institute*, No. 50.

Sündermann, J. and Brosche, P. (1978), Numerical computation of tidal friction for present and ancient oceans, in *Tidal Friction and the Earth's Rotation* (eds P. Brosche and J. Sündermann), Springer-Verlag, Berlin, pp. 125–144.

Sündermann, J. and Krohn, J. (1977), Numerical simulation of tidal caused sand transport in coastal waters, *Proceedings 17th Congress, International Association for Hydraulic Research*, **1**, 173–181.

Swett, K., Klein, G. de Vries and Smit, D. E. (1971), A Cambrian tidal sand body – the Eriboll Sandstone of north-west Scotland: an ancient–recent analog, *Journal of Geology*, **79**, 400–415.

Swett, K. and Smit, R. E. (1972), Palaeogeography and depositional environments of the Cambro–Ordovician shallow marine facies of the North Atlantic, *Geological Society America Bulletin*, **83**, 3223–3248.

Swift, D. J. P. (1975), Tidal sand ridges and shoal–retreat massifs, *Marine Geology*, **18**, 105–134.

Swift, D. J. P. (1976), Continental shelf sedimentation, in *Marine Sediment Transport and Environmental Management* (eds D. J. Stanley and D. J. P. Swift), John Wiley and Sons, New York, pp. 311–350.

Swift, D. J. P., Freeland, G. L. and Young, R. A. (1979), Time and space distribution of megaripples and associated bedforms, Middle Atlantic Bight, North American Atlantic shelf, *Sedimentology*, **26**, 389–406.

Sykes, R. M. (1974), Sedimentological studies in Southern Jameson Land, East Greenland, II Offshore-estuarine regressive sequences in the Neill Klinter Formation (Pleinsbachian-Toarcian), *Bulletin Geological Society of Denmark*, **23**, 213–224.

Taverner-Smith, R. and Williams, A. (1972), The secretion and structure of the skeleton of living and fossil Bryozoa, *Philosophical Transactions Royal Society London*, **264B**, 97–159.

Taylor, B. D. (1971), Temperature effects in alluvial streams, *W. M. Keck Laboratory, California Institute of Technology*, USA, Report KH-R-27.

Taylor, P. A. and Dyer, K. R. (1977), Theoretical models of flow near the bed and their implications for sediment transport, in *The Sea* (eds E. D. Goldberg, I. N. McCave, J. J. O'Brien and J. H. Steele), **6**, John Wiley, New York, pp. 579–601.

Tee, K. T. (1977), Tide-induced residual current-verification of a numerical model, *Journal Physical Oceanography*, **7**, 396–402.

Teichert, C. (1958), Cold- and deep-water coral banks, *Bulletin American Association Petroleum Geologists*, **42**, 1064–1082.

Terwindt, J. H. J. (1970), Observation on submerged sand ripples with heights ranging from 30 to 200 cm occurring in tidal channels of SW Netherlands, *Geologie en Mijnbouw*, **49**, 489–501.

Terwindt, J. H. J. (1971a), Litho-facies of inshore estuarine and tidal-inlet deposits, *Geologie en Mijnbouw*, **50**, 515–526.

Terwindt, J. H. J. (1971b), Sand waves in the Southern

Bight of the North Sea, *Marine Geology*, **10**, 51–67.

Thamdrup, H. M. (1935), Beiträge zur ökologie der wattenfauna auf experimenteller grundlage, *Meddelelser fra Kommissionen for Danmarks Fiskeriog Havundersøgelser, Serie: Fiskeri*, **10** (2), 1–125.

Thomas, J. M. (1980), Sediments and sediment transport in the Exe Estuary, in *Essays on the Exe Estuary* (ed. G. T. Boalch), Devonshire Association, Special Volume 2, pp. 73–87.

Thorn, M. F. C. (1975), Deep tidal flow over a fine sand bed, *Proceedings 16th Congress, International Association for Hydraulic Research*, **1**, 217–223.

Thors, K. (1978), The sea-bed of the southern part of Faxaflói, Iceland, *Jökull*, **28**, 42–52.

Thorson, G. (1957), Bottom communities (sublittoral or shallow shelf), in *Treatise on Marine Ecology and Palaeoecology*, **1** (ed. J. W. Hedgpeth), Geological Society of America Memoir 67, pp. 461–534.

Tyler, P. A. and Shackley, S. E. (1980), The benthic ecology of linear sandbanks: a modified *Spisula* sub-community, in *Industralised Embayments and their Environmental Problems: a Case Study of Swansea Bay* (eds M. B. Collins *et al.*), Pergamon Press, Oxford, pp. 539–551.

Uchupi, E. (1968), Atlantic continental shelf and slope: physiography, *U.S. Geological Survey Professional Paper*, 529-C.

Ulrich, J. and Pasenau, H. (1973), Morphologische Untersuchungen um Problem der tidebedingten Sandbewegung im Lister Tief, *Die Kuste*, **22**, 95–112.

Ursin, E. (1960), A quantitative investigation of the Echinoderm fauna of the central North Sea, *Meddelelser fra Danmarks Fiskeri-og Havundersøgelser*, Ny serie, **11** (24).

Vanney, J.-R. (1977), *Géomorphologie des plates-formes continentales*, Doin éditeurs, Paris.

Vanoni, V. A. (1944), Transportation of suspended sediment by water, *American Society of Civil Engineers, Transactions*, **111**, 67–133.

Vanoni, V. A. (1975), Closure to discussion of: Factors determining bedforms of alluvial streams, *American Society of Civil Engineers, Journal of the Hydraulics Division*, **101**, 1435–1440.

Vanoni, V. A. and Brooks, N. H. (1957), Laboratory studies of the roughness and suspended load of alluvial streams, *Report E68, Sedimentation Laboratory, California Institute of Technology*.

Vanoni, V. A. and Hwang, L. S. (1967), Relation between bed forms and friction in streams, *American Society of Civil Engineers, Journal of Hydraulics Division*, **93**, 121–144.

Veen, J. van (1935), Sandwaves in the Southern North Sea, *Hydrographic Review*, **12**, 21–29.

Veen, J. van (1936), *Onderzoekingen in de Hoofden*, Algemeene Landsdrukkerij, 's-Gravenhage.

Veen, J. van (1938), Water movements in the Straits of Dover, *Journal du Conseil International Exploration Mer*, **13**, 7–36.

Veenstra, H. J. (1965), Geology of the Dogger Bank area, North Sea, *Marine Geology*, **3**, 245–262.

Vermeij, G. J. (1978), *Biogeography and Adaptation, Patterns of Marine Life*, Harvard University Press, Cambridge, Mass.

Vevers, H. G. (1951), Photography of the sea floor, *Journal Marine Biological Association U.K.*, **30**, 101–111.

Vevers, H. G. (1952), A photographic survey of certain areas of sea floor near Plymouth, *Journal Marine Biological Association U.K.*, **31**, 215–222.

Visser, M. J. (1980), Neap/spring cycles reflected in Holocene subtidal large-scale bedform deposits, a preliminary note, *Geology*, **8**, 543–546.

Völpel, F. (1959), Studie über das Verhalten weitwandernder Flachseesande in der südlichen Nordsee, *Deutsche Hydrographische Zeitschrift*, **12**, 65–76.

Walker, R. G. (1979), Shallow marine sands, in *Facies Models* (ed. R. G. Walker), Geoscience Canada, Reprint Series 1, Geological Association of Canada, pp. 75–89.

Warme, J. E. (1967), Graded bedding in the Recent sediments of Mugu Lagoon, California, *Journal Sedimentary Petrology*, **37**, 540–547.

Warme, J. E. (1975), Borings as trace fossils, and the processes of marine bioerosion, in *The Study of Trace Fossils, A synthesis of principles, problems and procedures in ichnology* (ed. R. W. Frey), Springer-Verlag, Berlin, pp. 181–227.

Warme, J. E., Ekdale, A. A., Ekdale, S. F. and Petersen, C. H. (1976), Raw material of the fossil record, in *Structure and Classification of Palaeocommunities* (eds R. W. Scott and R. W. West), Dowden, Hutchinson and Ross, Stroudsburg, USA, pp. 143–169.

Warner, G. F. (1971), On the ecology of a dense bed of the brittle-star *Ophiothrix fragilis*, *Journal Marine Biological Association U.K.*, **51**, 267–282.

Warwick, R. M. (1980), Population dynamics and secondary production of benthos, in *Marine Benthic Dynamics* (eds K. R. Tenore and B. C. Coull), University of South Carolina Press, pp. 1–24.

Warwick, R. M. and Davies, J. R. (1977), The distribution of sublittoral macrofauna communities in the Bristol Channel in relation to the substrate, *Estuarine and Coastal Marine Science*, **5**, 267–288.

Warwick, R. M. and George, C. L. (1980), Annual macrofauna production in an *Abra* community, in *Industrialised Embayments and their Environmental Problems: a Case Study of Swansea Bay* (eds M. B. Collins *et al.*), Pergamon Press, Oxford, pp. 517–538.

Warwick, R. M., George, C. L. and Davies, J. R. (1978), Annual macrofauna production in a *Venus* community, *Estuarine and Coastal Marine Science*, **7**, 215–241.

Warwick, R. M. and Uncles, R. J. (1980), Distribution of benthic macrofauna associations in the Bristol Channel in relation to tidal stress, *Marine Ecology – Progress Series*, **3**, 97–103.

Wass, R. E., Conolly, J. R. and MacIntyre, R. J. (1970), Bryozoan carbonate sand continuous along southern Australia, *Marine Geology*, **9**, 63–73.

Watkins, R., Berry, W. B. N. and Boucot, A. J. (1973), Why 'Communities'? *Geology*, **1**, 55–58.

Webb, D. J. (1976a), Patterns in the equilibrium tide and the observed tide, *Australian Journal of Marine and Freshwater Research*, **27**, 617–632.

Webb, D. J. (1976b), A model of continental-shelf resonances, *Deep-Sea Research*, **23**, 1–15.

Webb, J. E., Dörjes, D. J., Gray, J. S., Hessler, R. R., van Andel, Tj. H., Werner, F., Wolff, T., Zijlstra, J. J. and Rhoads, D. C. (1976), Organism-sediment relationship, in *Benthic Boundary Layer* (ed. I. N. McCave), Plenum Press, New York, pp. 273–295.

Werner, F., Arntz, W. E. and Tauchgruppe Kiel (1974), Sedimentologie und Ökologie eines ruhenden Riesenrippelfeldes, *Meyniana*, **26**, 39–62.

Werner, F. and Newton, R. S. (1975), The pattern of large scale bed forms in the Langeland Belt (Baltic Sea), *Marine Geology*, **19**, 29–59.

Werner, F., Unsöld, G., Koopmann, B. and Stefanon, A. (1980), Field observations and flume experiments on the nature of comet marks, *Sedimentary Geology*, **26**, 233–262.

West, R. G. (1968), *Pleistocene Geology and Biology*, Longman Group Limited.

Wheeler, A. (1969), *The Fishes of the British Isles and north-west Europe*, Macmillan, London.

White, W. R., Milli, H. and Crabbe, A. D. (1975), Sediment transport theories: a review, *Proceedings Institution Civil Engineers, London*, **59**, 265–292.

Wigley, R. L. (1961), Bottom sediments of Georges Bank, *Journal Sedimentary Petrology*, **31**, 165–188.

Wilkinson, M. and Burrows, E. M. (1972), The distribution of marine shell-boring green algae, *Journal Marine Biological Association U.K.*, **52**, 59–65.

Williams, G. P. (1967), Flume experiments on the transport of a coarse sand, *United States Geological Survey, Professional Paper* 562-B.

Wilson, I. G. (1971), Desert sandflow basins and a model for the development of ergs, *Geographical Journal*, **137**, 180–199.

Wilson, J. B. (1965), *The palaeoecological significance of infaunas and their associated sediments*, Ph.D. Thesis, University of Edinburgh.

Wilson, J. B. (1967), Palaeoecological studies on shell beds and associated sediments in the Solway Firth, *Scottish Journal of Geology*, **3**, 329–371.

Wilson, J. B. (1975), The distribution of the coral *Caryophyllia smithii* S. and B. on the Scottish continental shelf, *Journal Marine Biological Association U.K.*, **55**, 611–625.

Wilson, J. B. (1976), The attachment of the coral *Caryophyllia smithii* S. and B. to tubes of the polychaete *Ditrupa arietina* (O. F. Muller) and other substrates, *Journal Marine Biological Association U.K.*, **56**, 291–303.

Wilson, J. B. (1977), The role of manned submersibles in sedimentological and faunal investigations on the United Kingdom continental shelf, in *Submersibles and their use in Oceanography and Ocean Engineering* (ed. R. A. Geyer), Elsevier, Amsterdam, pp. 151–167.

Wilson, J. B. (1979a), The distribution of the coral *Lophelia pertusa* (L.) (*L. prolifera* (Pallas)) in the north-east Atlantic, *Journal Marine Biological Association U.K.*, **59**, 149–164.

Wilson, J. B. (1979b), 'Patch' development of the deep-water coral *Lophelia pertusa* (L.) on Rockall Bank, *Journal Marine Biological Association U.K.*, **59**, 165–177.

Wilson, J. B. (1979c), Biogenic carbonate sediments on the Scottish continental shelf and on Rockall Bank, *Marine Geology*, **33**, M85–M93.

Wilson, J. B., Holme, N. A. and Barrett, R. L. (1977), Population dispersal in the brittle-star *Ophiocomina nigra* (Abildgaard) (Echinodermata: Ophiuroidea), *Journal Marine Biological Association U.K.*, **57**, 405–439.

Wilson, J. L. (1970), Depositional facies across carbonate shelf margins, *Transactions Gulf Coast Association Geological Sciences*, **20**, 229–233.

Wilson, J. L. (1974), Characteristics of carbonate platform margins, *Bulletin American Association Petroleum Geologists*, **58**, 810–824.

Wilson, J. L. (1975), *Carbonate Facies in Geologic History*, Springer-Verlag, Berlin.

Wodinsky, J. (1969), Penetration of the shell and feeding on gastropods by *Octopus*, *American Zoologist*, **9**, 997–1010.

Wolf, J. (1980), Estimation of shearing stresses in a tidal current with application to the Irish Sea, in *Marine Turbulence* (ed. J. C. J. Nihoul), Elsevier, Amsterdam, pp. 319–344.

Wood, A. (1974), Submerged platform of marine abrasion around the coasts of south-western Britain, *Nature, London*, **252**, 563.

Wunderlich, F. (1978), Deposition of mud in giant ripples of Inner Jade, German Bight, North Sea, *Senckenbergiana Maritima*, **10**, 257–267.

Wunsch, C. (1975), Internal tides in the ocean, *Reviews of Geophysics and Space Physics*, **13**, 167–182.

Yalin, M. S. (1972), *Mechanics of sediment transport*, Pergamon, London.

Yalin, M. S. and Price, W. A. (1976), Time growth of

tidal dunes in a physical model, in *Proceedings Symposium on Modelling Techniques*, San Francisco, September 1975, American Society of Civil Engineers, pp. 936–944.

Yalin, M. S. and Russell, R. C. H. (1966), Shear stress due to long waves, *Journal of Hydraulics Research*, **4**, 55–98.

Young, G. M. (1973), Stratigraphy, paleocurrents and stromatolites of Hadrynian (Upper Precambrian) rocks of Victoria Islands, Arctic archipelago, Canada, *Precambrian Research*, **1**, 13–41.

Zagwijn, W. H. and Veenstra, H. J. (1966), A pollen-analytical study of cores from the Outer Silver Pit, North Sea, *Marine Geology*, **4**, 539–551.

Zimmerman, J. T. F. (1978), Topographic generation of residual circulation by oscillatory (tidal) currents, *Geophysical and Astrophysical Fluid Dynamics*, **11**, 35–48.

Index

Aeolian dunes, 2, 3, 27, 54, 55
Age of temperate water shell gravels, 161, 165, 166
Agulhas Current, 38, 40, 123
Alaska, 80, 135
Alberta, 177, 189
Algae, 137, 155
   boring, 150, 159, *Plate 6.4*
   gravels, 135, 137, 160, *Plate 6.1*
   stromatolites, 188
Allochthonous deposits, 7
Amphidromic points, 14, 83, 84, 182, 183
Ancient net sand transport directions, 177, 189–192
   Lower Greensand seas, 184–187
Ancient sand banks, 176–178, 182, 183, 189–192
Ancient sand sheets, 178–180
   model 180–181
Ancient sand waves, 33, 173–176, 177, 178, 180, 181, 185–187, 189–192
Ancient tidal deposits, 189–192
Ancient tidal ranges, 25, 182, 183, 188, 192
Ancient tidal regimes, 25, 181–183
Ancient tidal shelf seas, 181–187, 189, 190
Antidunes, 34, 35, 43, 84
Arctic Ocean, 181, 182
Argentine shelf (Patagonian), 14, 80
ASCE classification of bedforms, 30, 34
Assemblages, *see* Faunal assemblages
Atlantic basin, 183, 184
Australia
   fauna, 134, 135, 161
   sand banks, 117, *Plate 3.20*
   tidal currents, 69, 86
Autochthonous deposits, 7

Avalanche foresets, 40, 173–176, 177, 178, 181, 185–187, 189–192, *see also* Cross-bedding

Bahama banks, 47, 134
Baie de la Seine, 139, 140
Baltic Sea, 38, 47, 123, 160
Banner banks, *see* Sand banks
Barchan style sand waves, *see* Sand waves
Barents Sea, 80, 116, 135
Barnstable Harbour, Mass., 132
Bars (alternate), 34
   nearshore, 49
Basal gravel, *see* Gravel sheet
Bassur de Baas Bank, 117
Battur Bank, 117
Bay of Biscay, 157
Bay of Fundy, 13, 14, 45, 173, *see also* Minas Basin and Channel
Bay of Mont St Michel, 144
Bedforms
   ASCE classification, 30, 31, 80
   co-existent, 54
   marine, 56, 57
Bedform zones and current strength, 2, 55, 67, 89, 138
   around Scotland, 138
   Bristol Channel, 66, 67, 138, 185
   Celtic Sea, 66
   English Channel, 66, 138
   Gibraltar Strait, 123
   Southern Bight North Sea, 138
Bed-load convergences
   ancient, 187
   modern, 3, 75, 80–83, 102

Index

Bed-load partings (modern)
  British Isles, 3, 15, 80–83, 102, 103, 138, 180
  deserts, 83, 84
  faunas, 138, 139, 155
  Georges Bank, 80, 81
  non-tidal currents, 83, 123
  White Sea entrance, 80
Belgium, 121, 176
Benthic Boundary Working Group, 132, 133
Bioturbation of sediments, 107, 130–132, 153, 166, 178, 179, 181
  biogenic grading, 132, 134
  depth of disturbance, 130
  by echinoids, 131, 146, 153, 181
  trace fossil assemblages, 130, 132–134, 179–181
  types, 132, 133
Bisan Strait, 44
Blake Plateau, 157
Bore, tidal, 75
Boring
  by algae, 150, 159, *Plate 6.4*
  by fungi, 150, 159, *Plate 6.4*
  by molluscs, 159
  by sponges, 156, 159
Boulonnais, 185, 187
Box cores, 106, 153
Brahmaputra River sand waves, 33, 40, 92
Breakdown of shelly material, 150, 159, 165, 186
Bristol Channel
  bedform zones, 66, 67, 185
  comet marks, *Plate 3.13*
  faunas, 129, 138, 141–145, 154, 155
  net sand transport, 74–77, 185
  sand banks, 53, 154
  sand ribbons, *Plate 3.15*
  sand waves, 42
  sediments, 97, 98, 110
  shell gravel, 162
  tidal currents, 18, 77
  tidal scour, 45
Bristol Channel Bed-Load Parting, 76, 102
Brittany shelf, 43, 54, 137, 149, 150, 162, 163
Broken Bank, 116
Brown Ridge, 153
Buiten Ratel Bank, 121

Cable safety, 1
Calcareous algal gravels, *see* Algae
California, 130, 134, 160
Canada, 59, 80, 103, 173, 177, 178, 182, 183, 189, 191
Canyons, 157
  sand transport near heads, 76
Cape Cod, 80, 81, 101
Carbonate content, *see* Sediment
Carbonate producers (major)
  barnacles, 164, 165, *Plates 6.5, 6.11*
  bryozoans, 151, 164, 165
  corals, 149–151, 156, *Plates 6.7, 6.8*
  echinoderms, 151, 163
  molluscs, 151, 164, 165, *Plate 6.10*
  serpulids, 149–151, 164, 165, *Plate 6.2*
Cardigan Bay, 98
Caribbean Sea, 134, 183
Carmarthen Bay, 145
Celtic Sea
  faunas, 126, 149, 156, 163
  net sand transport, 76, 82
  sand banks, 50, 66, 113, 114, 176
  sand patches, 43, 110, *Plates 3.1, 3.6, 3.17*
  sand waves, 66
  sediments, 66, 100, 109, 110
  tidal currents, 18, 66
  tidal lee waves, 56, 110
Chaussée de Seine, 54
Chesapeake and Delaware Canal sand waves, 105
Chloralgal association, 162
Chlorozoan association, 162
Colliery waste dispersion, 74
Comet marks, 47, *Plate 3.13*
Conclusions (main), 26, 55–57, 94, 124–125, 167–168, 189
Continental shelf width and tidal range, 14, 15
Continental shelf west of
  France, 66, 75, 76, 110, 162
  Ireland, 156, 157, *Plate 3.16*
  Scotland, 69, 156, 157, 159, 160, 162–164, 166, 167, *Plates 6.5, 6.6, 6.10, 6.11*
  see also Celtic Sea, Malin Sea
Continental slope, 156, 157
Coral, 150–151, 156, 159, 160, *Plates 6.7, 6.8*
Cork Hole, 108
Cornwall, 48, 162
Corton Road, off Lowestoft, 119
Cross-bedding (ancient)
  herringbone, 173, 174
  reactivation surfaces, 174, 175
  sand banks, 176–178, 189–192
  sand sheets, 180
  sand waves, 173–176, 185–187, 189–192
Cross-bedding (modern)
  destruction, 130, 131
  herringbone, 106
  master-bedding, 105, 111, 119, 120, 124
  reactivation surfaces, 105
  sand banks, 117, 118, 124
  sand sheets, 109, 110, 111, 124
  sand waves, 5, 102–106, 118
Cultivator Shoal, 51
Currents, *see* Tidal currents; Unidirectional currents
Current ripples
  in flumes, 30–33
  in sea, 35, 36, 60–64, 66, 67, 93, 174, 175, 185, 190, 191
  lag effects, 64

Current strength and bedform zones, *see* Bedform zones
Current strength and grain size, *see* Grain size–current strength

Days in lunar month, 11, 97
Death assemblages of shelly faunas, 162–165, 181, *Plates 6.10, 6.11*
  differences from living assemblages, 160, 181
  relative proportions of carbonate producers, 162–165
Denmark, 89, 123, 128, 159
Deposit feeders, 127, 131, 133, 143
Deposit evolution, 4, 122
  carbonates, 166, 167
Deposition rate
  carbonates, 161, 162
  muds, 106, 108
  sands, 98, 99, 108, 112, 130, 186
Deposition surface around the British Isles (Holocene), 7
Depth indicators, 121, *see also* Faunal indicators of environment
Desert bedforms, 27, 44, 54, 55
Dogger Bank, 101, 113
Dover Strait, 74, 75, 81, 89, 141, 187, *Plate 3.10*
Dover Strait Bed-Load Convergence, 75, 102, 187
Draa, 54, 55
Drag
  coefficient, 23, 24, 60, 61
  bedform, 23, 24, 32, 60, 61, 78
  surface, 60, 61
Drowned valleys, 7, 8, 114, 122
Dunes, *see* Aeolian dunes; Sand waves
Durban, 124

Eastbourne, 185, 186
Earth–Moon system
  ancient, 1, 25, 96, 188
  modern, 10, 11, 25
East Anglia, 59, 74, 84, 89, 91, 98, 119
East Bank, 113, 114
Ebb–flood channel systems, 50, 51, 178
Economic importance
  sands, 1
  carbonates, 135
Eddystone, 139
Ekman layer, 23
Electric log, 98
English Channel
  faunas, 129, 136–141, 144, 155, 156, 160, 162, 163, 166, *Plate 6.1*
  Hurd Deep, 44
  longitudinal furrows, 66, 72, 73, *Plates 3.8, 3.9*
  sand banks, 113
  sand patches, 66
  sand ribbons, 66, *Plates 3.14, 3.18*
  sand transport, 2, 74, 75, 77, 80, 87, 155
  sand waves, 42, 66, 77, 80, 81
  sediments, 100, 110
  shell gravel, 135, 162
English Channel Bed-Load Parting, 73, 76, 81, 102, 180
  faunas, 138, 139
Epifauna, 127, 128, 133, 137–139, 141, 145, 151–153, 155, 166, 168, *Plates 6.1, 6.2, 6.3, 6.7*
Epicontinental seas
  Lower Greensand, England, 183–187, 190
  Mesozoic, North America, 181–183, 189–190
Erosional furrows, *see* Longitudinal furrows
Exe Estuary, 135
Extended chlorozoan association, 162
Extinction coefficients, 117

Facies models
  estuarine and embayment sand banks, 120, 121
  offshore sand banks, 119, 120
  sand and gravel sheets, 110–112
Faecal pellets, 133, 134
Faeroe Bank, 156
Fair Isle Channel and Current, 89, 100, 101, 135, 149, 164, *Plates 6.2, 6.3*
Faringdon, 184, 185
Farnham, 184, 186
Faunal assemblages, Holocene dead, 160–165
  comparison with living, 110, 178, 179
Faunal associations, 127–129
  Boreal Deep Coral Association, 129
  Boreal Deep Mud Association, 129
  Boreal Offshore Gravel Association, 129, 139, 141
  Boreal Offshore Mud Association, 129
  Boreal Offshore Muddy Gravel Association (Holme, 1966), 129, 140
  Boreal Offshore Muddy Sand Association, 136, 141
  Boreal Offshore Sand Association, 129, 141
  Boreal Shallow Mud Association, 129, 136
  Boreal Shallow Rock Association, 136
  Boreal Shallow Sand Association, 129, 136
Faunal communities, 128, 129
  Abra community, 144, 145
  Modiolus community, 144, 145
  Reduced Hard Bottom Community, 142–144
  Spisula sub-community, 143, 145
  Tellina sub-community, 143, 145
  Venus community, 145
Faunal (shelly) differences along a sand sheet, 148–149
Faunal diversity, 127, 139, 141, 143, 146, 148, 154, 156
Faunal (shelly) indicators of environment, 155
  edge of continental shelf, 156, 157, *Plate 6.9*
  enclosed sea, 126, 155
  exposed continental shelf, 126, 149, 151, 155, 156
  lower sea level, 165, 166, *Plate 6.11*
Faunal list, 168–171
Faunas (shelly) of ancient offshore tidal current deposits, 167, 181

Faunas (shelly) around the British Isles
  encrusting, 149, 151, 153, 157, *Plates 6.1, 6.2, 6.3*
  gravel, 139, 149, 151–154, 157
  muddy sand, 145
  nearshore, 135
  open oceanic shelf, 126, 149, 155, 156, *Plates 6.5, 6.6*
  rippled sand, 141, 148–151, 155–157, 161, *Plate 6.6*
  rock, 142, 143, 165
  sampling problems, 127, 145, 146
  sand banks, 153, 154
  sand patches, 157–165
  sand ribbons, 139, 140, 141, 144, 145
  sand waves (large), 141, 145, 146, 154, 155
  sand waves (small), 146–148, 161
  tidal flats, 130, 132, 160
Flamborough Head, 49
Flandrian transgression, 114, 186
Florida, 175
Flume bedforms, 27, 29, 43, 45
Foraminifera, 150, 156, 163, 183
Foramol association, 162
Fraser River Delta, 59
Friction factor, *see* Drag coefficient
Froude number, 33–35, 44, 92
Fouveaux Strait, 141
Furrows, *see* Longitudinal furrows

Galway, 137
Geological dogmas, 5
Georges Bank, 51, 80, 81, 101, 102, 116, 135
Georges Shoal, 94
Georgia, 130, 132, 176
German Bight
  sand banks, 89, 107, 108, 122
  sand sheet, 102, 106–108, 137, 154
  sand transport, 71, 75, 77, 89
Gibraltar Strait, 123
Glauconite, 98, 121, 185–187
Grain size–current strength, 74, 100
  gradients, 74, 110, 177
Graded bedding, biogenic, 132
Grand Banks, Newfoundland, 135
Grand Vasière, 110
Gravel ripples, 36, 102, 162, 163, *Plate 3.1*
Gravel sheet, 66, 98, 101–102, 184
Gravel sheet and current strength, 2, 67, 99–101
Gravel sheet fauna
  currents strong, 139, *Plate 6.2*
  currents weak, 151–153
Gravel transport, 62, 65, 67
Gravel waves, 37, 56, 102
Great Bahama Bank, 160
Greenland, 190, 191
Greensand, *see* Lower Greensand
Growth lines in fossils, 183
Guildford, 184

Gulf of California, 116
Gulf of Korea, 114
Gulf of Maine, 14
Gulf of Mexico, 134, 183
Gulf of Panama, 16
Gulf of St Lawrence, 103
Gulf of St Malo, 75
Gulf of Thailand, 12

Hayasui Strait, 44, 45
Heavy minerals, 74, 98
Helgoland, 107
Helical flow, 47, 52, 54, 57
Henfield, 185, 186
Herringbone cross-stratification, ancient, 172–174, *see also* Cross-bedding (modern)
Holistic concept, 129
Holland, 39, 74, 75, 79, 91, 96, 105, 106, 108, 119, 130, 132, 145, 158, 159, 160, 161, 174
Hurd Deep, 44
Hydrographic Department, 49

Iapetus Ocean, 189
Iceberg plough marks, 7, 122
Iceland, 135
Infauna, 127, 129, 130, 133, 144, 154, 166
Inner Silver Pit, 44
Institute of Geological Sciences, 136, 167
Institute of Oceanographic Sciences, 49, 126, 136, 149
Internal tides, 24, 56
Irish Sea
  faunas, 129, 136, 144
  North Channel Bed-Load Parting, 76, 79, 81, 82
  St George's Channel Bed-Load Parting, 76, 79, 81, 82
  sand banks, 53, 113
  sand transport, 71, 74, 77
  scour hollows, 44
  sediments, 100, 108, 109
  tidal currents, 18, 66
Islay, 191
Isle of Man, 53, 154, 175
Isle of Wight, 186, 190

Jade Estuary, 103, 106
Jura, Scotland, 178, 191

Kelvin wave, 14, 16
Kiel Bight, 154
Korea Bay, 80, 114, 122

Labadie Bank, 114
La Chapelle Bank, 40, 41, 110
Lag deposits, 74, 165, 166
Lag effects
  on bedforms, 34, 37
  on sand transport, 62–64
  on sand transport rates, 59, 62–64

Land's End, 110
Langmuir circulations, 47, 52
Lead line samples, 115
Leman Bank, 116, 118
Linear sand banks, *see* Sand banks
Lister Tief, 71
Liverpool Bay, 75
Loire Estuary sand waves, 64, 70
London platform, 187
Longitudinal flume bedforms, 34
Longitudinal furrows, 45, 54, 57, 67, 72, 73, 180, *Plate 3.7*
   faunas, 139
Looe, 139
Lower flow regime bed states, 30–33
Lower Greensand deposits, 183–187, 190
Low sea level deposits
   Celtic Sea, 113–115, 176
   faunal evidence, 161, 165, 166
   North Sea, 113, 114, 115
   West of Scotland, 165, 166
Lundy Island, 53
Lyme Bay, 141

Maërl, 135, 137, *Plate 6.1, see also* Algal gravels
Magnetic fabric of silts, 75, 109
Malin Sea, 129, 163, 165
Malacca Strait, 80
Marine transgressions
   ancient, 183–187
   modern, 95–96, 114
Master-bedding, *see* Cross-bedding
Mediterranean Undercurrent, 123
Megaripples, 36, 37, 103, 161
Mesozoic seaways, *see* Epicontinental seas
Mexico, 160
Minas Basin and Channel, 61, 173
Minerals, 98
Mississippi River sand waves, 37, 40, 64, 92
Missouri River sand waves, 92
Monsoons, *see* North-west monsoon
Moraines, 7
Mud
   clasts, 116
   concentration, 97, 98, 117
   deposition, 66, 72, 96–98, 108, 117, 118
   drapes
      ancient, 178–180, 184
      modern, 116
   fluid, 98
   in sand banks, 116–120, 124, 176, 178
   in sand waves, 72, 119, 124
   thickness, 108
   transport direction, 65, 94
Mugu Lagoon, California, 134, 160

Nantucket Shoals, 81
Newfoundland, 191
Newport Bay, 130
New Guinea sand banks, 86, *Plate 3.20*
New Mexico, 190
New York Bight, 135
New York State, 191
New Zealand, 135, 141, 161, 162, 167
Noise of sand movement, 58, 70, *Plate 4.1*
Non-tidal water movements, *see* Unidirectional currents; Waves
Nordegründe sand banks, 106
Norderney, Isle of, 106
Norfolk Banks, 21, 22, 51, 114, 116, 118, 132, 138, 153, 154, 176, 177, 178
North Africa, 55
North Channel Bed-Load Parting, 76, 81, 82
North Hinder Bank, 73
North Sea
   faunas, 122, 138, 140, 145–149, 155, 160
   non-tidal currents, 88, 89
   obstacle marks, *Plates 3.11, 3.12*
   sand banks, 2, 72, 73, 113, 116–118, 176, 177
   sand sheets, 98, 100, 102
   sand transport, 73–79, 89, 91
   sand waves, 2, 33, 71–73, 89–93, 102, 103, 105, *Plates 3.2 to 3.5*
   Southern Bight Bed-Load Parting, 102, 103, 138
   tidal currents, 18
   see also Dogger Bank; German Bight; Inner and Outer Silver Pits; Southern Bight
North-west monsoon, 86
Norway, 159, 160, 174, 178, 191, 192
   fjord, 25, 157

Obstacle marks, 46, 54, 57, 73, 74, *Plates 3.10–3.13*
Ooliths, 59
Oosterschelde Estuary, 97
Orkney Islands, 81, 89, 101, 135, 137, 149, 164
Otoliths, 159, 160
Outer Gabbard Bank, 98, 153
Outer Hebrides continental shelf, 164, 165
Outer Ruytingen Bank, 73, 118, *Plate 3.19*
Outer Silver Pit, 106
Ower Bank, 118

Palimpsest deposits, 5
Parting-lineation, 34
Patagonian continental shelf, 14, 80
Patchy sands, history of study, 3
Pentland Firth Bed-Load Parting, 76
Persian Gulf, 162
Petworth, 179, 185, 186
Pipelines, 1, 47, *Plates 3.11, 3.12*
Plane bed, 33, 34
Plymouth, 102, 110, 139, 141

Porosity of sands, 4, 98
Portland Bill, 21, 53
Predation role in production of carbonate debris, 157–159
Predators
  asteroids, 157, 158
  crustaceans, 157, 158
  echinoderms, 157, 158, 159
  fish, 157, 158, 159
  molluscs, 146, 148, 157, 158
Progressive wave, 14, 82

Radiocarbon dates, 150, 161, 162, 165, 166
Reactivation surfaces, *see* Cross-bedding
Residual current
  definition, 70
  suspension transport, 65, 94
  tidal, 61, 70, 79, 86
Reworking, *see* Bioturbation of sediments
Relict deposits, critique, 5
Reynold's number, 31, 32
Rhodes Island, 131
Rhodolith, 137, *Plate 6.1*
Ria de Arosa Estuary, 135, 160
Riff, *see* Sabellarids
Rip-currents, 59
Rippled sand zone, 106
  faunas, 141, 148–151, 155–157, *Plate 6.6*
  principal carbonate producers, 149–151
  recognition of edges, 151
River bedforms, 27, 40
Rockall Bank, 98, 126, 150, 161, 162, *Plates 6.7, 6.8*
Roscoff, 137, 140, 141
Ross, *see* Sabellarids
RRS Discovery, 102

Sabellarids, 143, 154
Sahara Desert sand transport, 54, 55, 83, 84
St Andrew Bay, Florida, 175
St George's Channel Bed-Load Parting, 76, 81, 87, 89
St Lawrence Estuary, 103
Sand banks
  axial obliquity to peak flow, 50, 51, 63, 72, 73, 93, 94, 117
  banner banks, 46, 49, 52, 53, 116, 122, *Plate 3.20*
  channels and swatchways, 119, 124
  composition, 49, 115, 116
  dimensions, 49, 115
  estuarine, 49, 50
  facies, 113, 124
    models, 119, 120
  faunas (active banks), 153, 154
  internal structure, 117, 118, 124
  migration, 64, 70, 89, 93
  modern, 49, 50, 57, 113, 115
  moribund, 50, 57, 113–115
  morphological terms, 117
  offshore, 49
  origin, 52, 53, 89
  preservation, 64, 115, 122
  sand waves on, 51, 72, 75, 76, 93, 94, 113, 114, *Plate 3.19*
  separation and water depth, 50, 52, 115
  shape in plan, 49, 50, 70, 72, 93, 116, 117, *Plate 3.19*
  shape in section, 51, 70, 72, 93, 115, 116, 122
  smoothing near crest, 33, 84, 93, 115, 116
  see also Ancient sand banks; Tidal currents
Sand bundles, 96, 97
Sand drifts, 54
Sandettie Bank, 73, 118, *Plate 3.19*
Sand flow divide, 83
Sand hills, 49, 113, 114
Sand loss from land, 59
Sand patches
  composition, 110
  current strength, 43, 48, 67
  dimensions, 43, 48, 110
  faunas, 149–151, *Plates 6.5, 6.6*
  longitudinal, 47, 48, 57, 180, *Plates 3.16–3.18*
  polarity, 70
  preservation, 49
  transverse, 43, 44, 57, 110, *Plate 3.6*
Sand ribbons, 47, 48, 54, 57, 73, *Plates 3.14, 3.15, 3.18*
  current strength, 67
  faunas, 139, 144, 145
  in flumes, 34
  polarity, 73
  preservation, 49
  water depth, 48
Sand ripples, 30, 35, 37, 49, 54, 56, *see also* Current ripples; Wave ripples
Sand shadows, 46, 47, 53, 54
Sand sheet facies, 49, 98, 100, 102, 105, 107, 108
  faunal differences along a sand sheet, 148–149
  model, 110, 124
  see also Ancient sand sheets
Sandstones, 5, 33, 172–192
Sand streams, 49
Sand transport by modern tidal currents
  lag effects, 62–64
  net direction, 3, 21, 59, 67–80, 177, 185
  rates, 60–65, 74, 75, 77–79
  see also Ancient net sand transport directions
Sand transport direction (net) on modern sand banks, 72, 73, 75, 76, 93, 94
  lag effects, 64
  Georges Bank, 80, 81
  White Sea entrance, 80
Sand transport direction (net) on modern sand sheets
  around British Isles, 2–4, 75–79
  Georges Bank, 80, 81
  oblique to peak current, 63, 75, 79, 80, 86, 89, 90

Sand trap, 70, 75
Sand waves (modern)
  barchan type, 41, 42, 47, 48, 54, 55, 80, *Plate 3.5*
  catback type, 103
  composition, 30, 119
  distribution around British Isles, 66, 68
  in estuaries, 64, 70
  faunas, 141, 145, 146, 154, 155
  in flumes, 30, 34
  generation, 89–91
  height, 38, 56, 91, 105
  intertidal, 37
  movement observations, 70–73, 175
  nomenclature, 36, 37
  plan view, 41–43, 71, *Plates 3.2–3.5, 3.19*
  profile, 39–41, 56, 64, 70–73, 76, 77, 80, 81, 87, 93, 102, 103, 105
  in rivers, 37, 40, 64
  on sand banks, 51, 72, 75, 76, 93, 94, 113, 114, *Plate 3.19*
  on sand ribbons, 48
  small on large, 42, *Plate 3.2*
  smoothing, 40, 43, 85, 86, 90–93, 111, 112
  structure (internal), 103, 104, 124
  symmetrical, 39–41
  unidirectional, 37, 38, 40
  very low height sand waves, 31, 56
  water depth, 38
  wavelength, 38, 41, 56, 105, *Plates 3.3, 3.4*
  see also Ancient sand waves
San Francisco Harbour, 44
Sapelo Island, Georgia, 130, 132
Saskatchewan, 178, 179, 189
Scotland, 68, 76, 126, 137, 150, 151, 155–159, 175, 191, *see also* Continental shelf; Fair Isle Channel; Orkney Islands; Outer Hebrides; Shetland Isles; Solway Firth
Scour hollows, 2, 7, 44, 45, 47, 57
Scour lag, *see* Lag effects
Scoured horizons, ancient, 180, 184, *see also* Scour hollows
Seasat side-scan radar, 41, 42, 49, 93, *Plate 3.19*
Sediment
  carbonate content, 135, 150, 153, 159, 160, 163, 164, 166, 167
  modes, transport of, 65
  porosity, 4, 98
  sorting, 61, 65, 74, 102, 116
Sedimentary record, 96, 189–192
Seif dunes, 55
Senckenberg Institute, 137
Seoul Approaches, 80, 114, 122
Settling lag, *see* Lag effects
Severn Estuary, 116
Shambles Bank, 21, 53
Shear stress, *see* Stress (bottom)

Shell gravels, 135, 161–165, 167, *Plates 6.10, 6.11*
  faunal composition
    in strong currents, 162–164
    in weak currents, 165
Shell growth and tidal rhythms, 183
Shells
  breakage, 150, 159, 165
  dissolution, 159, 160
  lenses, 132
  pavements, 139
  transport by hermit crabs, 140, 141
Shelly faunas, *see* Faunas
Shetland Isles, 81, 89, 101, 135, 137, 149, 164
Shields diagram, 31, 32, 60
Side-scan radar, 42, 93, *Plate 3.19*
Side-scan sonar, 2, 27, *see also* Sonographs
  multibeam, *Plate 4.1*
Silt balls, 110
Skagerrak, 159
Skerries Bank, 21, 53
Smiths Knoll Bank, 117
Solway Firth, 75, 116, 135, 161
Sonographs, 27, *Plates 1.1, 1.2, 3.1–3.18, 4.1, 4.2*
Sorting of sands, 65, 74, 102, 116, 166, 167
South Africa, 40, 83, 112, 124, 135, 155, 174, 191
South America, 14, 80, 83, 123
South Equatorial Current, 83, 123
Southern Bight Bed-Load Parting, 39, 73, 74, 76, 81, 89, 102, 138
Southern Bight sand waves, 39–43, 68, 73, 89–93, 103–106, 111, 117–119, *Plates 3.4, 3.19*
South Falls Bank, 73, 118, *Plate 3.19*
Spain, 137, 160, 175, 189
Standing wave, 14, 82, 83
Start Point, 21, 53
Stewart Island shelf, 161
Storm surge effects, 88, 89, 106
Strait of Dogger, 113, 114
Strait of Florida, 157
Stratigraphic traps, 1
Stress (bottom), 23, 31, 32, 44, 47, 60, 61, 78, 79
  and fauna, 127, 142, 143
Suderpiep sand banks, 107
Suspension transport, 59–65, 93, 94
Suspension feeders, 127, 133, 143
Swansea Bay, 74, 143
Swarte Bank, 116
Swatchways, 119
Sylt Island, 71

Tay Estuary Research Group, 135
Thames Estuary, 46, 63, 64, 73, 75, 108, 114, 116, 122, 135
Thermocline, 43
Threshold of grain movement, 32, 33, 59, 61, 62, 65, 68
Tidal amplification, 12

Tidal channels
  off Florida, 70, 175
  off Holland, 96, 106, 119
Tidal current asymmetry and net sand transport, 20, 58, 59, 67, 69, 71–73, 86, 91, 93, 94
Tidal currents
  around British Isles, 16
  near sand banks, 21, 52, 53
  near sea floor, 22
  velocity profiles, 24, 60, 62
Tidal delta, 176
Tidal eddies, 53, 75, 79
Tidal envelope, 19, 51, 59, 63, 64, 79, 90, 94
Tidal flats, 127, 130, 132, 135, 160, 161
Tidal harmonics, 20
  M_4 tidal currents, 20, 61, 67–79, 82, 83
Tidal lee waves, 40, 43, 56, 110
Tidal range
  ancient, 25, 182, 183, 188, 189, 192
  modern, 12, 16, 17, 84
  variation with continental shelf width, 14
Tidal resonance, 13, 67
Tidal sand banks, *see* Sand banks
Tidal (bottom) stress, *see* Stress (bottom)
Tide generating forces, 10, 83, 84
Tides
  daily and twice daily, 10
  internal, 24, 56
  prehistoric, 25
  some peak tides, 12, 84
  spring–neap cycles, 11, 78
  world spring ranges, 16
  18.6 year cycle, 83, 84
Torres Strait
  banner banks, 53, 86, *Plate 3.20*
  currents, 38, 69, 86
Trace fossil assemblages, 179, 181, *see also* Bioturbation
Tracers, 65, 70, 74, 75
Transgressive deposits of tidal seas
  ancient, 183–187
  modern, 95, 96, 113–115, 119–121

Transition state bedforms
  in flumes and rivers, 33, 92
  in sea, 44
Trophic structure, 128
Tunnel valleys, 7
Turbulent bursts, 56

Unidirectional currents, 37, 38, 40, 54, 67, 71, 86–89, 106, 123
United States of America, 14, 37, 40, 44, 51, 61, 64, 69, 80, 81, 92, 94, 101, 102, 105, 116, 130, 132, 134, 135, 157, 160, 175
  rocks, 176, 177, 181–183, 190, 191
Upper flow regime bed states, 33, 43, 84
Upware, 187
Ushant, 163
Utah, 190

Victoria Island, Canada, 191

Wadden Sea, 130, 132, 135
Wangeroog Island, 106
Warts Bank, 53, 154, 175
Wash, The, 75, 135
Water temperature (viscosity) effects, 33, 34, 56, 90, 92
Wave breaking zone, 5, 85
Wave enhanced sand transport, 67, 85, 86, 91–94
Wave heights around British Isles (50-year storm), 85
Wave mass transport, 86, 91
Wave oscillatory speeds, 36, 85, 86
Wave refraction, 43
Wave ripples, sand and gravel, 36, 86, *Plate 3.1*
Well Bank, 98, 116, 117, 118
  internal structures, 117, 118, 153
West Barrow Bank, 117
West Dyck Bank, 73, 118, *Plate 3.19*
White Sea, 80, 81
Wreck marks, 46, 57, *Plate 3.10*, *see also* Obstacle marks

Yellow Sea, 116, 122
Yucatan Peninsula, 160

Zeeland Ridges, 73